Physical-Layer Security for 6G

IEEE Press
445 Hoes Lane
Piscataway, NJ 08854

Physical-Layer Security for 6G

Edited by

Parthajit Mohapatra
Department of Electrical Engineering, Indian Institute of Technology
Tirupati, India

Nikolaos Pappas
Department of Computer and Information Science, Linköping University,
Campus Valla, Sweden

Arsenia Chorti
Information, Communications and Imaging (ICI) Group of the ETIS Lab
UMR8051, CY Cergy Paris Universite, Cergy, France

Stefano Tomasin
Department of Information Engineering, University of Padova, Padova, Italy

IEEE*Press*

WILEY

Published by John Wiley & Sons, Inc., Hoboken, New Jersey.
Published simultaneously in Canada.

Library of Congress Cataloging-in-Publication Data

Names: Mohapatra, Parthajit, editor. | Pappas, Nikolaos, 1982- editor. | Chorti, Arsenia, editor. | Tomasin, Stefano, editor.
Title: Physical-layer security for 6G / edited by Parthajit Mohapatra, Nikolaos Pappas, Arsenia Chorti, Stefano Tomasin.
Description: Hoboken, New Jersey : Wiley, [2024] | Includes index.
Identifiers: LCCN 2024034415 (print) | LCCN 2024034416 (ebook) | ISBN 9781394170913 (hardback) | ISBN 9781394170920 (adobe pdf) | ISBN 9781394170937 (epub)
Subjects: LCSH: 6G mobile communication systems–Security measures.
Classification: LCC TK5103.252 .P48 2024 (print) | LCC TK5103.252 (ebook) | DDC 621.3845/6–dc23/eng/20240819
LC record available at https://lccn.loc.gov/2024034415
LC ebook record available at https://lccn.loc.gov/2024034416

Cover Design: Wiley
Cover Image: © koiguo/Getty Images

Set in 9.5/12.5pt STIXTwoText by Straive, Chennai, India

Contents

About the Editors

Parthajit Mohapatra is an associate professor in the Department of Electrical Engineering, Indian Institute of Technology Tirupati, India. His primary research interests include wireless communication and current research focuses on physical-layer secrecy, short packet communication, and union of physical and network layers. Parthajit received the PhD degree from the Indian Institute of Science, Bangalore, in 2015.

Nikolaos Pappas is an associate professor and a docent in the Department of Computer and Information Science, Linköping University, Linköping, Sweden. His main research interests include the field of wireless communication networks with an emphasis on semantics-aware goal-oriented communications, age of information, energy harvesting networks, and network-level cooperation. Nikolaos received the PhD degree in computer science from the University of Crete, Greece, in 2012.

Arsenia Chorti is a professor (first class) at Ecole Nationale Supérieure de l'Electronique et de ses Applications, Cergy, Ile de France. Dr. Chorti received the Habilitation pour Diriger des Recherches in 2020.

Stefano Tomasin is a full professor in the Department of Information Engineering, University of Padova, Italy. Stefano received the PhD degree in telecommunications engineering from the University of Padova, Italy, in 2003.

List of Contributors

Francesco Ardizzon
Department of Information Engineering
University of Padova
Padova
Italy

Marco Baldi
Department of Information Engineering
Università Politecnica delle Marche
Ancona
Italy

Igor Bjelaković
Technische Universität Berlin
Berlin
Germany

and

Fraunhofer Heinrich-Hertz-Institut
Berlin
Germany

Matthieu Bloch
School of Electrical and Computer Engineering
Georgia Institute of Technology
Atlanta, GA
USA

Arsenia Chorti
Wireless Connectivity Group
Barkhausen Institut
Dresden
Germany

and

ETIS UMR8051, CY University
ENSEA, CNRS
Cergy
France

Laura Crosara
Department of Information Engineering
University of Padova
Padova
Italy

Kemal Davaslioglu
Nexcepta
Gaithersburg, MD
USA

Tugba Erpek
Nexcepta
Gaithersburg, MD
USA

He Fang
School of Electronic and Information Engineering
Soochow University
Sochow
China

Matthias Frey
Department of Electrical and Electronic Engineering
The University of Melbourne
Parkville, Victoria
Australia

Marco Giordani
Department of Information Engineering
University of Padova
Padova
Italy

Shenjie Huang
School of Engineering
The University of Edinburgh
Edinburgh
UK

Eduard Jorswieck
Institute for Communications Technology
TU Braunschweig
Brunswick
Germany

Babak Khalaj
EE Department
Sharif University of Technology
Tehran
Iran

Juliane Krämer
Universität Regensburg
Regensburg, Bayern
Germany

Nicola Laurenti
Department of Information Engineering
University of Padova
Padova
Italy

Mehdi Letafati
EE Department
Sharif University of Technology
Tehran
Iran

Laura Luzzi
ETIS, UMR 8051, CY Cergy Paris Université
ENSEA, CNRS
Cergy-Pontoise
France

Christos Masouros
Department of Electronic and Electrical Engineering
University College London
London
UK

Amitha Mayya
Wireless Connectivity Group
Barkhausen Institut
Dresden
Germany

Miroslav Mitev
Wireless Connectivity Group
Barkhausen Institut
Dresden
Germany

Parthajit Mohapatra
Department of Electrical Engineering
Indian Institute of Technology Tirupati
Tirupati
India

Derrick Wing Kwan Ng
School of Electrical Engineering and Telecommunications
University of New South Wales
Sydney, New South Wales
Australia

Janis Nötzel
Emmy-Noether Group Theoretical Quantum Systems Design
Technische Universität München
München, Bayern
Germany

Nikolaos Pappas
Department of Computer and Information Science
Linköping University
Linköping
Sweden

Majid Safari
School of Engineering
The University of Edinburgh
Edinburgh
UK

Yalin Sagduyu
Nexcepta
Gaithersburg, MD
USA

Abdelhamid Salem
Department of Electronic and Electrical Engineering
University College London
London
UK

and

Department of Electronic and Electrical Engineering
University of Benghazi
Benghazi
Libya

Robert Schober
Institute for Digital Communications
Friedrich-Alexander-University
Erlangen Nürnberg
Erlangen
Germany

Linda Senigagliesi
Department of Information
Engineering
Università Politecnica delle Marche
Ancona
Italy

Mahdi Shakiba Herfeh
ETIS UMR8051, CY University
ENSEA, CNRS
Cergy
France

Sotiris Skaperas
ETIS UMR8051, CY University
ENSEA, CNRS
Cergy
France

Mohammad Dehghani Soltani
School of Engineering
The University of Edinburgh
Edinburgh
UK

Muralikrishnan Srinivasan
Department of Electronics
Engineering
Indian Institute of Technology (BHU)
Varanasi
India

Sławomir Stańczak
Technische Universität Berlin
Berlin
Germany

and

Fraunhofer Heinrich-Hertz-Institut
Berlin
Germany

Stefano Tomasin
Department of Information
Engineering
University of Padova
Padova
Italy

Xianbin Wang
Department of Electrical and
Computer Engineering
Western University
London, Ontario
Canada

Dongfang Xu
Academy of Interdisciplinary Studies
The Hong Kong University of Science
and Technology
Kowloon, Hong Kong
China

Michele Zorzi
Department of Information
Engineering
University of Padova
Padova
Italy

Preface

6G is envisioned to provide hyperconnectivity between users and objects, with interconnected machines being the dominant users. The widespread adoption of wireless devices presents unique challenges for ensuring secure communications. To meet the requirements of various services such as Ultra-Reliable Low Latency Communication (URLLC) and massive Machine Type Communication (mMTC), it is required to redesign existing physical-layer-based techniques to take into account the quality of service requirements such as reliability, latency, and energy consumption. The book aims to provide a comprehensive background on the physical-layer security and advances in the context of 6G networks.

The initial part of the book offers an introduction to physical-layer secrecy and then discusses the coding theory advances in the context of 6G. It provides novel material on the role of physical-layer secrecy in emerging communication scenarios ranging from intelligent reflecting surfaces to non-terrestrial networks. Some of the key topics covered in the book are as follows:

- Application of physical-layer security in emerging communication technologies such as visible light communication (VLC) and intelligent reflecting surfaces (IRS)
- Impact of physical-layer secrecy on latency aspects of communication such as delay and age of information (AoI)
- Physical-layer-based authentication
- Joint secure communication and sensing
- AI/ML-enabled physical-layer secrecy
- Context-aware physical-layer secrecy

The book explores use cases and practical demonstrations to provide insights into how physical-layer security can address the unique challenges of 6G networks. The reader will obtain a view of physical-layer security as a technology ready to be considered for the upcoming 6th generation of cellular networks. The book will benefit a broad audience of engineers and scientists by helping them understand the functioning of a new type of security that will be included in future communication systems.

June 2024

Parthajit Mohapatra
Tirupati, India

Nikolaos Pappas
Linköping, Sweden

Arsenia Chorti
Cergy-Pontoise, France

Stefano Tomasin
Padova, Italy

Part I

Preliminaries

1

Foundations of Physical-Layer Security for 6G*

Matthieu Bloch

School of Electrical and Computer Engineering, Georgia Institute of Technology, Atlanta, GA, USA

Wireless connectivity has become a cornerstone of our modern societies, driving innovation and supporting an ever-growing range of services. With the increasingly sensitive nature of information transmitted over wireless networks, privacy and secrecy mechanisms have naturally become an integral part of new protocols and standards. While identified weaknesses of previous generation wireless protocols are typically addressed with the rollout of the next generation, challenges constantly emerge that must be proactively addressed. For instance, while 5G systems have addressed some of the security weaknesses identified in 4G systems, the attack surface of 5G networks has increased because of the heterogeneity of devices and the larger number of use cases [48], as exemplified by the growth of machine-to-machine communications [19]. Consequently, security has yet again already been identified as one of the main challenges that 6G networks must address [72].

Several security solutions have been considered to provide full-stack security, including lightweight cryptography for Internet of Things (IoT) devices [61], the use of post-quantum cryptography [3, 72], and physical-layer security (PLS) [11, 45], which has again re-emerged as a possible technology [32, 38, 49, 66, 74]. The key concept behind PLS is to exploit the random imperfections inherent to wireless channels and devices (noise, interference) to provide, e.g., secrecy or authentication, using physical-layer signal processing and coding algorithms [11, 56, 68]. While PLS may certainly not solve all 6G security challenges in isolation, its main benefits are (i) to provide a concrete framework in which security can be quantified, e.g., through the notion of secrecy capacity [68]; (ii) to treat security on par with other system-level metrics, such as

*The work of M. Bloch was supported in part by NSF under award 2148400.

Physical-Layer Security for 6G, First Edition. Edited by Parthajit Mohapatra, Nikolaos Pappas, Arsenia Chorti, and Stefano Tomasin.

power consumption, throughput, and latency, at the design stage; (iii) to reduce the attack surface at the physical layer by making eavesdropping extremely costly, if not ineffective; and (iv) to seamlessly integrate with security mechanisms in the upper layers of the protocol stack. In particular, ensuring confidentiality for ultralow-latency communications is a known challenge [1, 54] that PLS could help tackle [15, 58].

PLS was already discussed in the context of 5G networks [35, 66], and one should recognize that, with the exception of niche applications and use cases [73], PLS has not had much impact on deployed systems. This state of affairs can be attributed to a multitude of factors, both technological and conceptual, ranging from scientific challenges related to the foundation of PLS itself (e.g., how do we characterize and learn a passive eavesdropper's channel?) to technological hurdles (e.g., how do we justify integrating new codes at the physical layer of a standard?). Nevertheless, 6G promises new unique features that may finally offer the opportunity to push PLS into widely deployed systems [16, 45]. In particular, the integration of sensing and communication, especially as it relates to enhancing the localization of devices, and the push toward higher frequencies in the mmWave region are offering new avenues to strengthen the case of PLS.

The objective of this chapter is twofold. First, we will review the seminal coding ideas behind PLS, which have been refined over the last two decades to provide a strong basis for discussing secrecy in a principled manner. Second, we will discuss how these principles may be used in the more specific context of 6G systems, with an eye toward engineering channels, developing dedicated hardware, and exploiting channel knowledge for security. Given the breadth of literature on the topic, this chapter does not do justice to many creative ideas, in particular those involving PLS in the context of networks of many devices for which the reader is referred to tutorial articles [46, 51, 66, 71]. The focus of the chapter is on point-to-point links, for they still capture the essence of the challenges that remain to address and the opportunities that have emerged and might represent the realistic use cases for which PLS could be deployed at scale.

1.1 Coding Mechanisms

The appeal of (PLS) can be largely attributed to the early work of Wyner [68], Csiszár and Körner [21], Ahlswede and Csiszár [2], and Maurer [43], that first established and analyzed the notion of *secrecy capacity* and *secret-key capacity*. We defer to Section 1.2 for exact definitions, suffice to say for now that these definitions are the counterparts of the traditional notion of channel capacity and that they quantify the maximum rate of information that can be transmitted or extracted reliably and confidentially over a channel that includes an eavesdropping adversary. While secrecy capacity and secret-key capacity therefore provide system-level metrics that can be optimized as a function of channel parameters to understand how much secrecy can be achieved in a network, the ability to operationalize them is fundamentally tied to the ability to design specific coding schemes to extract or encode information in signal. Said differently, in the same way that the notion of channel capacity is useful because good error-control codes exist, secrecy and secret-key capacity are useful because good secrecy codes exist. The objective of this section is to introduce four coding operations that shall enable PLS by providing operational meaning to what it means to enforce secrecy in Section 1.2.

1.1.1 Channel Coding

The problem of channel coding is illustrated in Figure 1.1. The objective consists in transmitting a uniformly distributed messages $W \in \{1, \dots, M\}$ over n uses of a discrete memoryless channel with known transition probability $P_{Y|X}$ by encoding the message W into a coded sequence X^n. The set of coded sequences $\{X^n\}$ is called the codebook while n is called the blocklength of the code. Upon receiving the corrupted signal Y^n, the receiver attempts construct a correct estimate $\widehat{W}$ of W using its knowledge of the channel and the code. The performance of channel coding may be measured in terms of the rate of transmission $R = \frac{\log_2 M}{n}$ and the probability of decoding error $P_e = P(\widehat{W} \neq W)$.

The seminal result established by Shannon [55] is that, asymptotically, reliable communication is possible as long as the rate does not exceed a channel-dependent quantity called the channel capacity. We state this result more formally as follows.

Theorem 1.1 Given a discrete memoryless channel with known transition probability $P_{Y|X}$, a distribution P_X and any $\epsilon > 0$, there exists a blocklength n and an encoder/decoder pair such that $R \geq I(X;Y) - \epsilon$ and $P_e \leq \epsilon$ where $I(X;Y)$ is the mutual information between the random variables X and Y with joint distribution $P_X P_{Y|X}$. The quantity $\max_{P_X} I(X;Y)$ is called the channel capacity since no higher such constant can be found.

Specific instances of such codes can be designed using low-density parity-check codes [24, 33, 34] or polar codes [4].

Figure 1.1 Channel coding over a discrete memoryless channel.

1.1.2 Soft Covering

The operation of channel coding can be interpreted as introducing *structure* in coded sequences that is resilient to the corruption of the noisy channel. A lesser known coding operation over channels consists in introducing structure in coded sequences that *disappear* when corrupted by noise. Formally, this coding operation called soft covering is illustrated in Figure 1.2. Consider a random variable with distribution P_X transmitted over a discrete memoryless channel with known transition probability $P_{Z|X}$. The output of the channel is a new random variable Z with distribution P_Z obtained by taking the marginal of $P_X P_{Z|X}$. Instead of transmitting the random variable X, one can instead ask whether one can approximately simulate n transmissions of the random variable X using instead a uniformly distributed message $W \in \{1, \dots, M\}$ encoded into sequences X^n of length n. The intuition is that M-coded sequences might be sufficient to approximately cover all possible realizations of n i.i.d. realizations of the random variable Z. The performance of soft covering may be measured in terms of the rate of transmission $R = \frac{\log_2 M}{n}$ and the relative entropy $D = D(\widehat{P}_{Z^n} \| P_Z^{\otimes n})$, where $\widehat{P}_{Z^n}$ is the distribution induced by the random choice of coded sequences X^n while $P_Z^{\otimes n}$ is the n-fold product distribution of P_Z.

The fundamental result of soft covering, first identified by Wyner [67] but studied and refined later on by others [22, 26, 29, 30, 65], is that $\widehat{P}_{Z^n}$ and $P_Z^{\otimes n}$ are virtually indistinguishable as long as the rate does not fall below a quantity called the channel resolvability. We state this result more formally below.

Theorem 1.2 Given a discrete memoryless channel with known transition probability $P_{Z|X}$, a distribution P_X, and any $\epsilon > 0$, there exists a blocklength n and an encoder such that $R \leq I(X;Z) + \epsilon$ and $D \leq \epsilon$, where $I(X;Z)$ is the mutual information between the random variables X and Z with joint distribution $P_X P_{Z|X}$. The quantity $\min_{P_X : \sum_x P_X W_{Z|X} = P_Z} I(X;Z)$ is called the channel resolvability since no lower such constant can be found.

The intuition behind Theorem 1.2 is that by exceeding the channel resolvability, the structure that exists within the coded sequences X^n gets obfuscated by the

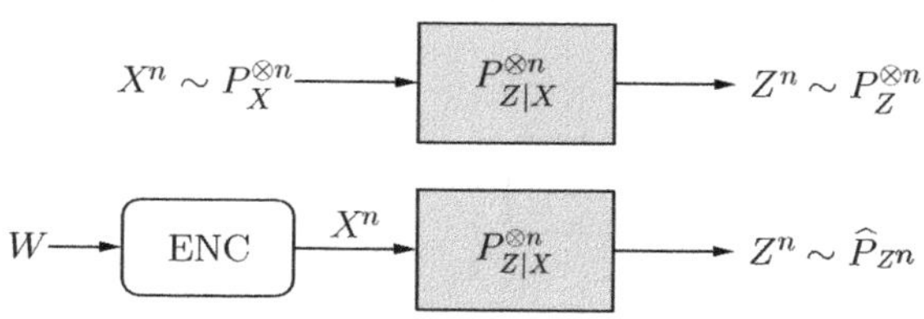

Figure 1.2 Soft-covering over a discrete memoryless channel.

channel noise and the distribution at the output of the channel resembles a product distribution. Note that Theorem 1.2 is not merely a consequence of Theorem 1.1 since Theorem 1.1 only ensures that reliable communication is impossible at rates above the channel capacity and some lingering structure could still be identified in the channel output. In contrast, Theorem 1.2 shows that some codes completely lose their structure at rates above the channel resolvability. The close resemblance in the expression of the channel resolvability and the channel capacity is slightly misleading and an artefact of the discrete memoryless nature of the channels considered here. In general, the channel resolvability exceeds the channel capacity for the same choice of input distribution [29].

The usefulness of soft covering for security shall be developed in Section 1.2; for now, one can interpret soft covering as a coding mechanism that obfuscates the existence of coded information. Although the design of soft-covering codes has been much less investigated than channel codes, specific instances can be built using polar codes or a combination of channel codes and invertible extractors [12, 18, 31, 60].

1.1.3 Source Coding with Side Information

The problem of source coding with side information is illustrated in Figure 1.3. Consider a discrete memoryless source with two components X and Y with known joint distribution P_{XY}. The objective is to encode n realizations of the source X^n into a message $W \in \{1, \dots, M\}$ to decode a reconstruction $\hat{X}^n$ using both W and the associated realizations Y^n as side information. The performance of source coding with side information may be measured in terms of the compression rate $R = \frac{\log_2 M}{n}$ and the probability of reconstruction error $P_e = P(\hat{X}^n \neq X^n)$.

The fundamental result of source coding with side information, discovered by Slepian and Wolf [59], is the following.

Theorem 1.3 Given a discrete memoryless source with known joint distribution P_{XY} and any $\epsilon > 0$, there exists a blocklength n and an encoder/decoder pair such that $R \leq H(X|Y) + \epsilon$ and $P_e \leq \epsilon$, where $H(X|Y)$ is conditional entropy of X given Y. In addition, $H(X|Y)$ is the lowest such constant that can be found.

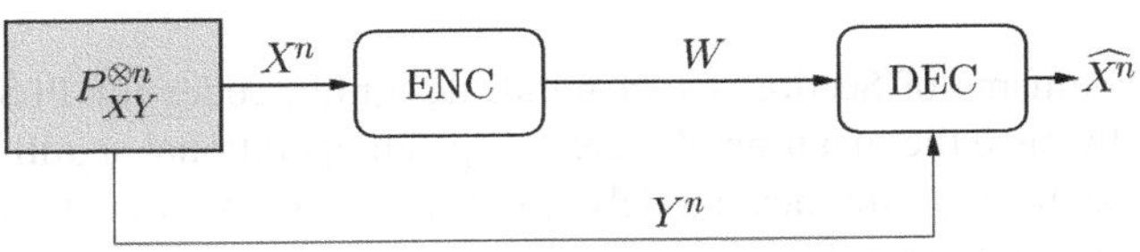

Figure 1.3 Source coding with side information for a discrete memoryless source.

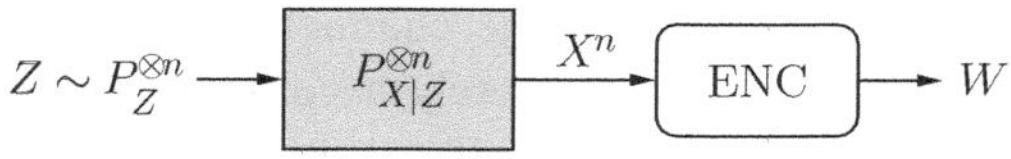

Figure 1.4 Privacy amplification from a discrete memoryless channel.

Again, specific instances of codes for source coding with side information can be constructed using low-density parity-check codes and polar codes [5, 37, 42].

1.1.4 Privacy Amplification

Privacy amplification plays a role dual to source coding with side information analogous to the role of soft covering with respect to channel coding. Instead of compressing a source while retaining enough dependence to enable reconstruction with side information, one can attempt to compress so much that all dependence against side information is lost. The operation of privacy amplification is illustrated in Figure 1.4. Consider a discrete memoryless source with two components X and Z with known joint distribution P_{XZ}. The objective is to encode n realizations of the source X^n into a message $W \in \{1, \dots, M\}$ that is uniformly distributed and nearly independent of the associated sequence Z^n. The performance of privacy amplification can be measured in terms of the extraction rate $R = \frac{\log_2 M}{n}$ and the relative entropy $D = D(P_{WZ^n} \| Q_M P_{Z^{\otimes n}})$ between the joint distribution P_{WZ^n} induced by the code and the source, and the product distribution $Q_M P_{Z^{\otimes n}}$ where Q_M is the uniform distribution on $\{1, \dots, M\}$.

The fundamental results of privacy amplification, which can be found in several forms [2, 9] including the so-called left-over hash lemma, is as follows.

Theorem 1.4 Given a discrete memoryless source with known joint distribution P_{XZ} and any $\epsilon > 0$, there exists a blocklength n and an encoder such that $R \geq H(X|Z) - \epsilon$ and $D \leq \epsilon$, where $H(X|Z)$ is conditional entropy of X given Z. In addition, $H(X|Z)$ is the highest such constant that can be found.

Intuitively, Theorem 1.4 states that the maximum randomness that can be extracted from the output of a channel $P_{X|Z}$ independently of the input P_Z is the entropy of the channel noise. The design of codes for privacy amplification is quite well understood, in particular is sufficed to use universal hash functions [14].

1.2 Coding for Physical-Layer Security

With the coding mechanisms of Section 1.1 at hand, designing codes for PLS follows almost naturally once the meaning of secrecy is given operational significance. We shall analyze the two canonical models of PLS: secure communication over the wiretap channel and secret-key generation from correlated observations of a source.

1.2.1 Secure Communication

The wiretap channel model is illustrated in Figure 1.5. The model captures a situation in which a transmitter attempts to reliably convey a message $W \in \{1, \dots, M\}$ to a legitimate receiver over discrete memoryless channel with known transition probability $P_{Y|X}$, while avoiding leaking information to an eavesdropper who obtains its observations through another discrete memoryless channel with known transition probability $P_{Z|X}$. These distributions may be the marginals of a transition probability $P_{YZ|X}$. The legitimate receiver attempts to construct a correct estimate $\widehat{W}$ of W while the eavesdropper attempts to infer information about W from its observations Z^n. The performance of the coding scheme is measured in terms of the secret communication rate $R_s = \frac{\log M}{n}$, the probability of decoding error $P_e = P(\widehat{W} \neq W)$, and the secret information leakage $S = \max_{P_W} I(W; Z^n)$. For reasons we detail next, it is convenient to introduce an auxiliary random message W' to randomize the encoding of the message W, and to allow the encoder to be stochastic. These additional assumptions do not trivialize the problem since the randomness introduced is local to the encoder and not shared with any other party.

The secrecy metric $\max_{P_W} I(W; Z^n)$ is called *semantic secrecy*, for a small value ensures that the eavesdropper cannot infer any information about any function of the message [7]. Other metrics, such as the strong secrecy $I(W; Z^n)$ and the weak secrecy $\frac{1}{n}I(W; Z^n)$, have found their use in the literature [20, 44] but have gathered skepticism in the cryptographic community. Fortunately, most results derived under weaker secrecy metrics have been shown to hold for semantic secrecy, and the coding techniques discussed next provide tools to directly analyze semantic secrecy.

The crucial idea to understand how the design of codes for secure communication is the following observation [20, 65]. For any distribution $Q_Z^{\otimes n}$, it holds that

$$\begin{aligned} I(W; Z^n) &= D(\widehat{P}_{WZ^n} \| Q_M P_{Z^n}) \leq D(\widehat{P}_{WZ^n} \| Q_M Q_{Z^n}) \\ &= \mathbb{E}_W \left(P_{Z^n|W} \| Q_Z^{\otimes n} \right). \end{aligned} \tag{1.1}$$

One interpretation of the upper bound (1.1) is that a sufficient condition to ensure semantic secrecy is to ensure that all messages W induce the same target

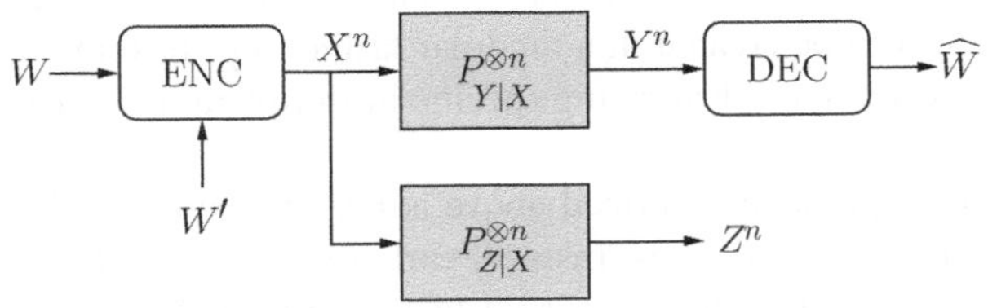

Figure 1.5 Secure communication over the wiretap channel.

distribution $Q_Z^{\otimes n}$. Intuitively, from the perspective of the eavesdropper, identifying the message is nearly impossible because all messages induce the same output distribution. This deceptively simple observation allows one to then make a conceptual connection with the mechanisms of Section 1.1. Specifically, the ability to control the output distribution is precisely the role of soft covering. Hence, one can construct a code for the wiretap channel by (i) associating a soft-covering code of rate R' to every message W, inducing the same target distribution at the output of the eavesdropper's channel $P_{Z|X}$; (ii) ensuring that the union of soft-covering codes forms a channel code of rate R for the main channel $P_{Y|X}$. This construction mandates the use of the auxiliary message W' alluded to earlier to randomize the encoding of every soft-covering code; hence, the encoding of every secrecy message is stochastic, associating a randomly chosen codeword within a codebook to every message.

Assuming a given distribution P_X, the rate R must not exceed but can approach $I(X;Y)$ according to Theorem 1.1. Similarly, the rate R' much exceed but can approach $I(X;Z)$ according to Theorem 1.2. Consequently, the rate $R_s = R - R'$ of secret messages must not exceed but can approach $I(X;Y) - I(X;Z)$, which is essentially the secrecy capacity.

An additional subtlety is that the transmitter could artificially introduce noise before transmission, in the form of a discrete memoryless channel with transition probability $P_{X|V}$ so that the effective channel is the concatenation of $P_{X|V}$ and $P_{YZ|X}$. This channel prefixing turns out to be required to achieve optimal secrecy rates [21, 63] and the secrecy capacity is given as follows.

Theorem 1.5 Given a discrete memoryless channel with known transition probability $P_{YZ|X}$, a distribution P_{XV}, and any $\epsilon > 0$, there exists a blocklength n and an encoder such that $R_s \geq I(V;Y) - I(V;Z) - \epsilon$ and $P_e \leq \epsilon$, $S \leq \epsilon$, where the random variables V, X, Y, Z have joint distribution $P_V P_{X|V} P_{YZ|X}$. The quantity $C_s = \sup_{P_V P_{X|V} P_{YZ|X}} I(V;Y) - I(V;Z)$ is the largest such constant that can be found and is called the secrecy capacity.

This conceptual approach to constructing codes for secure communication over the wiretap channel leads to concrete code instance, see for instance [8, 12, 17, 40, 57, 62]. The constructions [6, 17, 57] particularly stand out for they suggest that the soft-covering codes can be created in a modular fashion by "hashing" a good channel code into subcodes and leveraging specific hash function that can be stochastically inverted.

One should note that the approach described above is *not* the approach that was historically followed to construct codes. Instead, earlier constructions used an astute observation by Wyner [68] to tie secrecy to reliability [12]. Despite many code constructions, e.g., [36, 50], this approach is fundamentally limited to weak secrecy [10]. Furthermore, the approach outlined here based on soft covering generalizes to adversarial channels [27, 47, 53] and allows one to ensure secrecy under much weaker assumptions regarding channel knowledge than our presentation suggests.

1.2.2 Secret-Key Generation

The source model for secret-key generation is illustrated in Figure 1.6. This model captures a scenario in which three parties observe the correlated signals X^n, Y^n, Z^n, respectively, originating from a discrete memoryless source with distribution P_{XYZ}. The objective of the two parties observing X^n and Y^n is to distill a secret key $W \in \{1, \cdots, M\}$ from their observations, treating Z^n as an eavesdropper. The distillation of the secret key is enabled by a public-authenticated channel of unlimited capacity over which the two parties can transmit messages collectively denoted F. This assumption does not trivialize the problem because F is known to the eavesdropper. In addition, the cost of authenticating a channel is negligible compared to the size of the distilled key [64]. The performance of the coding scheme is measured in terms of the secret key distillation rate $R_{\text{k}} = \frac{\log M}{n}$, the probability of reconstruction error $P_{\text{e}} = P(\widehat{W} \neq W)$, and the secret information leakage $S = I(W; Z^n, F)$. The encoding operations that define the messages F can be generic, allowing for interactivity and randomization.

The design of secret-key generation codes follows an approach conceptually similar to that outlined in Section 1.2.1 for secure communication in an even more direct way. Specifically, secret keys may be distilled by exactly combining source coding with side information and privacy amplification so that (i) one associates a public index of rate R to every sequence X^n that enables the receiver observing Y^n to reconstruct X^n; (ii) one associates a secret key index of rate R_{k} to every sequence X^n that remains independent of the public index and independent of Z^n. The rate R must exceed but can approach $H(X|Y)$ by Theorem 1.3. The sum rate $R + R_{\text{k}}$ must not exceed but can approach $H(X|Z)$ by Theorem 1.4. Consequently, the secret-key

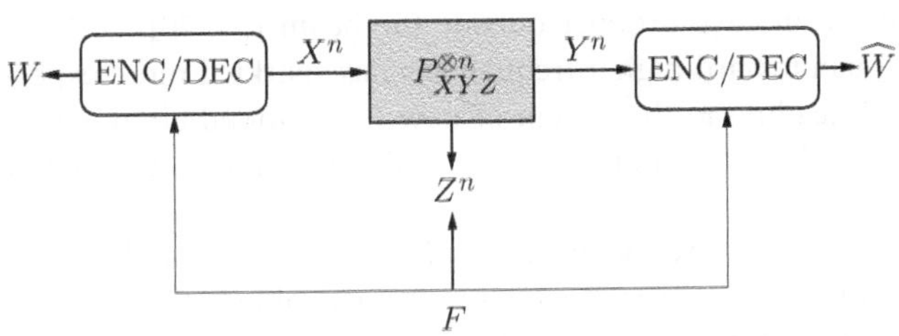

Figure 1.6 Secret-key generation from source model.

rate R_k must not exceed but can approach $H(X|Z) - H(X|Y)$, which is the essence of the secret-key capacity.

In general, the secret-key capacity is not known without placing additional restrictions on the source model. Nevertheless, upon remarking that the roles of X and Y are interchangeable, the following result holds.

Theorem 1.6 Given a source model with joint distribution P_{XYZ} and any $\epsilon > 0$, there exists a secret-key generation code such that

$$R_k \geq \max\left(H(X|Z) - H(X|Y), H(Y|Z) - H(Y|X)\right) - \epsilon, \tag{1.2}$$

and $P_e \leq \epsilon, S \leq \epsilon$. The secret key capacity is also known to be no greater $I(X;Y|Z)$.

This approach to generating secret keys has long been used in quantum key distribution [25] and is made much simpler than the construction of codes for the wiretap channel by the universal nature of privacy amplification [9] and the ability to ignore the exact structure of reconciliation [13] when applying privacy amplification.

Note that rewriting $H(X|Z) - H(X|Y)$ as $I(X;Y) - I(X;Z)$ suggests a connection between secure communication and secret-key generation. While this connection can be exploited [20, 52, 69], the instantiation of codes is more natural following the distinct coding mechanisms described earlier. In particular, the use of universal hash functions for privacy amplification offers some level of universality to the key generation process, such that minimal knowledge of the distribution $P_{Z|X}$ is actually required to extract a key [9, 23].

1.3 Engineering and Learning Channels

The push toward the use of higher frequencies of the wireless spectrum is naturally offering opportunities for PLS and 6G. For instance, the need for narrow beams to overcome the path loss at mmWave is a natural way to make communications more directional and effectively create channels that are naturally "better" for an intended receiver than for an eavesdropper outside the beam [39, 70]. Perhaps more importantly, the push toward higher frequencies has sparkled more experimental interest for PLS, leading to the design of new RF-frontend circuits and measurements [28, 38, 41] that offer more concrete evidence of the potential and limits of PLS.

These exciting experimental works open up possibilities for PLS and can be broadly described as solutions to *engineer* channels that will enable PLS. Nevertheless, a complete integration with the coding mechanisms described in Sections 1.1 and 1.2 is still largely missing. More precisely, secrecy capacity and secret-key capacity are still often used to measure the "secrecy level" of a link but without deploying the coding schemes required to transform the metric into an actual operational guarantee for a transmission. The main impediment toward this integration is perhaps not the coding mechanisms themselves, as solutions exist as described earlier. Rather, the main hurdle is the need to *learn* the wireless environment as some degree of channel knowledge is required to properly tune the parameters of the coding scheme and provide security guarantees.

References

1 3GPP. Study on the security for 5G URLLC (Release 16). Technical Specification (TR) 33.825, 3rd Generation Partnership Project (3GPP), Mar 2019.

2 R. Ahlswede and I. Csiszár. Common randomness in information theory and cryptography. I. Secret sharing. *IEEE Transactions on Information Theory*, 39(4):1121–1132, Jul 1993. doi: 10.1109/18.243431.

3 G. Alagic et al. Status report on the second round of the NIST post-quantum cryptography standardization process. Technical Report NISTIR 8309, National Institute of Standards and Technology, Jul 2020.

4 E. Arikan. Channel polarization: A method for constructing capacity-achieving codes for symmetric binary-input memoryless channels. *IEEE Transactions on Information Theory*, 55(7):3051–3073, 2009. doi: 10.1109/TIT.2009.2021379.

5 E. Arikan. Source polarization. In *Proceedings of IEEE International Symposium on Information Theory*, pages 899–903, Austin, TX, Jun 2010. doi: 10.1109/ISIT.2010.5513567.

6 M. Bellare and S. Tessaro. Polynomial-time, semantic-secure encryption achieving the secrecy capacity, Jan 2012.

7 M. Bellare, S. Tessaro, and A. Vardy. Semantic security for the wiretap channel. In R. Safavi-Naini and R. Canetti, editors, *Advances in Cryptology & CRYPTO 2012*, volume 7417 of *Lecture Notes in Computer Science*, pages 294–311. Springer-Verlag, Berlin Heidelberg, 2012. ISBN 978-3-642-32008-8. doi: 10.1007/978-3-642-32009-5i8. hard-copy.

8 M. Bellare, S. Tessaro, and A. Vardy. A cryptographic treatment of the wiretap channel, Jan 2012.

9 C. H. Bennett, G. Brassard, C. Crépeau, and U. Maurer. Generalized privacy amplification. *IEEE Transactions on Information Theory*, 41(6):1915–1923, Nov 1995. doi: 10.1109/18.476316.

10 M. R. Bloch. Achieving secrecy: Capacity vs. resolvability. In *Proceedings of IEEE International Symposium on Information Theory*, pages 632–636, Saint Petersburg, Russia, Aug 2011. doi: 10.1109/ISIT.2011.6034207.

11 M. Bloch and J. Barros. *Physical-Layer Security: From Information Theory to Security Engineering*. Cambridge University Press, Oct 2011. doi: 10.1017/CBO9780511977985.

12 M. R. Bloch, M. Hayashi, and A. Thangaraj. Error-control coding for physical-layer secrecy. *Proceedings of IEEE*, 103(10):1725–1746, Oct 2015. ISSN 0018-9219. doi: 10.1109/JPROC.2015.2463678.

13 C. Cachin and U. M. Maurer. Linking information reconciliation and privacy amplification. *Journal of Cryptology*, 10(2):97–110, Mar 1997.

14 J. L. Carter and M. N. Wegman. Universal classes of hash functions. *Journal of Computer and System Sciences*, 18:143–154, 1979.

15 R. Chen, C. Li, S. Yan, R. Malaney, and J. Yuan. Physical layer security for ultra-reliable and low-latency communications. *IEEE Wireless Communications*, 26(5):6–11, Oct 2019. doi: 10.1109/mwc.001.1900051.

16 A. Chorti, A. N. Barreto, S. Kopsell, M. Zoli, M. Chafii, P. Sehier, G. Fettweis, and H. V. Poor. Context-aware security for 6G wireless: The role of physical layer security. *IEEE Communications Standards Magazine*, 6(1):102–108, Mar 2022. doi: 10.1109/mcomstd.0001.2000082.

17 R. A. Chou. Explicit codes for the wiretap channel with uncertainty on the eavesdropper's channel. In *Proceedings of IEEE International Symposium on Information Theory*, pages 476–481, Vail, CO, Jun 2018. doi: 10.1109/ISIT.2018.8437777.

18 R. A. Chou, M. R. Bloch, and J. Kliewer. Low-complexity channel resolvability codes for the symmetric multiple-access channel. In *Proceedings of IEEE Information Theory Workshop*, pages 466–470, Hobart, Tasmania, Nov 2014. doi: 10.1109/ITW.2014.6970875.

19 Cisco. Cisco Annual Internet Report (2018–2023) White Paper, Mar 2020.

20 I. Csiszár. Almost independence and secrecy capacity. *Problems of Information Transmission*, 32(1):40–47, Jan-Mar 1996.

21 I. Csiszár and J. Körner. Broadcast channels with confidential messages. *IEEE Transactions on Information Theory*, 24(3):339–348, May 1978.

22 P. Cuff. Distributed channel synthesis. *IEEE Transactions on Information Theory*, 59(11):7071–7096, 2013. doi: 10.1109/TIT.2013.2279330.

23 Y. Dodis, R. Ostrovsky, L. Reyzin, and A. Smith. Fuzzy extractors: How to generate strong keys from biometrics and other noisy data. *SIAM Journal of Computing*, 38(1):97–139, Mar 2008. ISSN 0097-5397. doi: 10.1137/060651380.

24 R. G. Gallager. *Low Density Parity Check Codes*. PhD thesis, Massachusetts Institute of Technology, Cambridge, MA, 1963.

25 N. Gisin, G. Ribordy, W. Tittel, and H. Zbinden. Quantum cryptography. *Reviews of Modern Physics*, 74:145–195, Mar 2002. doi: 10.1103/RevModPhys.74.145.

26 Z. Goldfeld, P. Cuff, and H. Permuter. Semantic-security capacity for wiretap channels of type II. *IEEE Transactions on Information Theory*, 62(7):3863–3879, 2016. ISSN 0018-9448. doi: 10.1109/TIT.2016.2565483.

27 Z. Goldfeld, P. Cuff, and H. H. Permuter. Arbitrarily varying wiretap channels with type constrained states. *IEEE Transactions on Information Theory*, 62(12):7216–7244, Dec 2016. doi: 10.1109/TIT.2016.2619701.

28 J. Guo, L. Poli, M. A. Hannan, P. Rocca, S. Yang, and A. Massa. Time-modulated arrays for physical layer secure communications: Optimization-based synthesis and experimental assessment. *IEEE Transactions on Antennas and Propagation*, 66(12):6939–6949, Dec 2018. doi: 10.1109/tap.2018.2870381.

29 T. S. Han and S. Verdú. Approximation theory of output statistics. *IEEE Transactions on Information Theory*, 39(3):752–772, May 1993. ISSN 0018-9448. doi: 10.1109/18.256486.

30 M. Hayashi. General nonasymptotic and asymptotic formulas in channel resolvability and identification capacity and their application to the wiretap channels. *IEEE Transactions on Information Theory*, 52(4):1562–1575, Apr 2006. doi: 10.1109/TIT.2006.871040.

31 M. Hayashi. Tight exponential analysis of universally composable privacy amplification and its applications. *IEEE Transactions on Information Theory*, 59(11):7728–7746, Nov 2013. ISSN 0018-9448. doi: 10.1109/TIT.2013.2278971.

32 L. Jiao, N. Wang, P. Wang, A. Alipour-Fanid, J. Tang, and K. Zeng. Physical layer key generation in 5G wireless networks. *IEEE Wireless Communications*, 26(5):48–54, Oct 2019. ISSN 1536-1284, 1558-0687. doi: 10.1109/MWC.001.1900061.

33 S. Kudekar, T. Richardson, and R. L. Urbanke. Spatially coupled ensembles universally achieve capacity under belief propagation. *IEEE Transactions on Information Theory*, 59(12):7761–7813, Dec 2013. doi: 10.1109/tit.2013.2280915.

34 M. Lentmaier, A. Sridharan, D. J. Costello, and K. Sh. Zigangirov. Iterative decoding threshold analysis for LDPC convolutional codes. *IEEE Transactions on Information Theory*, 56(10):5274–5289, Oct 2010. doi: 10.1109/TIT.2010.2059490.

35 G. Li, C. Sun, J. Zhang, E. Jorswieck, B. Xiao, and A. Hu. Physical layer key generation in 5G and beyond wireless communications: Challenges and opportunities. *Entropy*, 21(5):497, May 2019. ISSN 1099-4300. doi: 10.3390/e21050497.

36 R. Liu, Y. Liang, H. V. Poor, and P. Spasojević. Secure nested codes for type ii wiretap channels. In *Proceedings of IEEE Information Theory Workshop*, pages 337–342, Lake Tahoe, CA, Sept 2007. doi: 10.1109/ITW.2007.4313097.

37 A. D. Liveris, Z. Xiong, and C. N. Georghiades. Compression of binary sources with side information at the decoder using LDPC codes. *IEEE Communications Letters*, 6(10):440–442, Oct 2002.

38 X. Lu, S. Venkatesh, B. Tang, and K. Sengupta. Space-time modulated 71-to-76ghz mm-wave transmitter array for physically secure directional wireless links. In *Proceedings of IEEE International Solid- State Circuits Conference*, pages 86–88, Feb 2020. doi: 10.1109/ISSCC19947.2020.9062929.

39 J. Ma et al. Security and eavesdropping in terahertz wireless links. *Nature*, 563(7729):89–93, Nov 2018. ISSN 1476-4687. doi: 10.1038/s41586-018-0609-x. URL https://www.nature.com/articles/s41586-018-0609-x.

40 H. Mahdavifar and A. Vardy. Achieving the secrecy capacity of wiretap channels using polar codes. *IEEE Transactions on Information Theory*, 57(10):6428–6443, 2011. doi: 10.1109/TIT.2011.2162275.

41 N. S. Mannem, T.-Y. Huang, E. Erfani, S. Li, and H. Wang. A mm-wave transmitter MIMO with constellation decomposition array (CDA) for keyless physically secured high-throughput links. In *2021 IEEE Radio Frequency Integrated Circuits Symposium (RFIC)*. IEEE, Jun 2021. doi: 10.1109/rfic51843.2021.9490442.

42 J. Martinez-Mateo, D. Elkouss, and V. Martin. Key reconciliation for high performance quantum key distribution. *Scientific Reports*, 3:1576, Apr 2013. doi: 10.1038/srep01576.

43 U. M. Maurer. Secret key agreement by public discussion from common information. *IEEE Transactions on Information Theory*, 39(3):733–742, May 1993. doi: 10.1109/18.256484.

44 U. M. Maurer and S. Wolf. Information-theoretic key agreement: From weak to strong secrecy for free. In *Advances in Cryptology - Eurocrypt 2000*, page 351. Lecture Notes in Computer Science, B. Preneel, 2000.

45 M. Mitev, A. Chorti, H. V. Poor, and G. P. Fettweis. What physical layer security can do for 6G security. *IEEE Open Journal of Vehicular Technology*, 4:375–388, 2023. doi: 10.1109/ojvt.2023.3245071.

46 A. Mukherjee. Physical-layer security in the internet of things: Sensing and communication confidentiality under resource constraints. *Proceedings of the IEEE*, 103(10):1747–1761, Oct 2015. ISSN 1558-2256. doi: 10.1109/JPROC.2015.2466548.

47 M. Nafea and A. Yener. A new wiretap channel model and its strong secrecy capacity. *IEEE Transactions on Information Theory*, 64(3):2077–2092, Mar 2018. doi: 10.1109/TIT.2017.2786541.

48 NIS Cooperation Group. EU coordinated risk assessment of the cybersecurity of 5G networks. Technical report, European Commission, 2019.

49 P. Porambage, G. Gur, D. P. M. Osorio, M. Liyanage, A. Gurtov, and M. Ylianttila. The roadmap to 6G security and privacy. *IEEE Open Journal of the Communications Society*, 2:1094–1122, 2021. doi: 10.1109/ojcoms.2021.3078081.

50 V. Rathi, M. Andersson, R. Thobaben, J. Kliewer, and M. Skoglund. Performance analysis and design of two edge-type LDPC codes for the BEC wiretap channel. *IEEE Transactions on Information Theory*, 59(2):1048–1064, Feb 2013. ISSN 0018-9448. doi: 10.1109/TIT.2012.2219577.

51 P. A. Regalia, A. Khisti, Y. Liang, and S. Tomasin. Secure communications via physical-layer and information-theoretic techniques [scanning the issue]. *Proceedings of the IEEE*, 103(10):1698–1701, Oct 2015. doi: 10.1109/jproc.2015.2473895

52 J. M. Renes and R. Renner. Noisy channel coding via privacy amplification and information reconciliation. *IEEE Transactions on Information Theory*, 57(11):7377–7385, 2011. doi: 10.1109/TIT.2011.2162226

53 R. F. Schaefer, H. Boche, and H. V. Poor. Secure communication under channel uncertainty and adversarial attacks. *Proceedings of the IEEE*, 103(10):1796–1813, Oct 2015. ISSN 0018-9219. doi: 10.1109/JPROC.2015.2459652

54 M. Shakiba-Herfeh, A. Chorti, and H. V. Poor. Physical layer security: Authentication, integrity, and confidentiality. In *Physical Layer Security*, pages 129–150. Springer International Publishing, 2021. doi: 10.1007/978-3-030-55366-16.

55 C. E. Shannon. A mathematical theory of communication. *The Bell System Technical Journal*, 27:379–423, 623–656, Jul, Oct 1948.

56 C. E. Shannon. Communication theory of secrecy systems. *The Bell System Technical Journal*, 28:656–715, 1948.

57 S. Sharifian, F. Lin, and R. Safavi-Naini. Hash-then-encode: A modular semantically secure wiretap code. In *Proceedings of the 2nd Workshop on Communication Security*, pages 49–63, Paris, France, Apr 2018. ISBN 978-3-319-59265-7

58 C. She, C. Sun, Z. Gu, Y. Li, C. Yang, H. V. Poor, and B. Vucetic. A tutorial on ultrareliable and low-latency communications in 6G: Integrating domain knowledge into deep learning. *Proceedings of the IEEE*, 109(3):204–246, Mar 2021. doi: 10.1109/jproc.2021.3053601

59 D. Slepian and J. K. Wolf. Noiseless coding of correlated information sources. *IEEE Transactions on Information Theory*, 19(4):471–480, Jul 1973.

60 R. Sultana and R. A. Chou. Explicit low-complexity codes for multiple access channel resolvability. In *Proceedings of 57th Annual Allerton Conference on Communication, Control, and Computing*, pages 116–123, Monticello, IL, Sept 2019.

61 M. S. Turan, K. McKay, Ç. Çalık, D. Chang, and L. Bassham. Status report on the first round of the NIST lightweight cryptography standardization process. Technical Report NISTIR 8268, National Institute of Standards and Technology, Oct 2019.

62 H. Tyagi and A. Vardy. Universal hashing for information-theoretic security. *Proceedings of the IEEE*, 103(10):1781–1795, Oct 2015. ISSN 0018-9219. doi: 10.1109/JPROC.2015.2462774

63 S. Watanabe and Y. Oohama. The optimal use of rate-limited randomness in broadcast channels with confidential messages. *IEEE Transactions on Information Theory*, 61(2):983–995, Feb 2015. ISSN 0018-9448. doi: 10.1109/TIT.2014.2382096

64 M. N. Wegman and J. L. Carter. New hash functions and their use in authentication and set equality. *Journal of Computer Sciences and Systems*, 22:265–279, 1981.

65 A. Winter, A. C. A. Nascimento, and H. Imai. Commitment capacity of discrete memoryless channels. In *Proceedings of 9th IMA International Conference*, pages 33–51, Cirencester, UK, 2003.

66 Y. Wu, A. Khisti, C. Xiao, G. Caire, K. Wong, and X. Gao. A survey of physical layer security techniques for 5G wireless networks and challenges ahead. *IEEE Journal on Selected Areas in Communications*, 36(4):679–695, Apr 2018. ISSN 1558-0008. doi: 10.1109/JSAC.2018.2825560

67 A. Wyner. The common information of two dependent random variables. *IEEE Transactions on Information Theory*, 21(2):163–179, Mar 1975. doi: 10.1109/TIT.1975.1055346

68 A. D. Wyner. The wire-tap channel. *The Bell System Technical Journal*, 54(8):1355–1367, Oct 1975.

69 M. H. Yassaee, M. R. Aref, and A. Gohari. Achievability proof via output statistics of random binning. *IEEE Transactions on Information Theory*, 60(11):6760–6786, Nov 2014. ISSN 0018-9448. doi: 10.1109/TIT.2014.2351812

70 C.-Y. Yeh, Y. Ghasempour, Y. Amarasinghe, D. M. Mittleman, and E. W. Knightly. Security in terahertz WLANs with leaky wave antennas. In *Proceedings of the 13th ACM Conference on Security and Privacy in Wireless and Mobile Networks*. ACM, Jul 2020. doi: 10.1145/3395351.3399365

71 A. Yener and S. Ulukus. Wireless physical-layer security: Lessons learned from information theory. *Proceedings of the IEEE*, 103(10):1814–1825, Oct 2015. ISSN 0018-9219. doi: 10.1109/JPROC.2015.2459592

72 M. Ylianttila, R. Kantola, A. Gurtov, L. Mucchi, and Oppermann I., editors. *6G White Paper: Research Challenges For Trust, Security And Privacy [White Paper]*. Number 9 in 6G Research Visions. University of Oulu, Jun 2020. doi: http://urn.fi/urn:isbn:9789526226804.

73 C. T. Zenger, M. Pietersz, and C. Paar. Preventing relay attacks and providing perfect forward secrecy using physec on 8-bit μc. In *2016 IEEE International Conference on Communications Workshops (ICC)*, pages 110–115, May 2016. doi: 10.1109/ICCW.2016.7503773

74 M. Zoli, A. N. Barreto, S. Köpsell, P. Sen, and G. Fettweis. Physical-layer-security box: A concept for time-frequency channel-reciprocity key generation. *EURASIP Journal on Wireless Communications and Networking*, 2020(1):114, Dec 2020. ISSN 1687-1499. doi: 10.1186/s13638-020-01712-6

2

Coding Theory Advances in Physical-Layer Secrecy

Laura Luzzi

ETIS, UMR 8051, CY Cergy Paris Université, ENSEA, CNRS, Cergy-Pontoise, France

Notation Given two (discrete or continuous) distributions p, q, we denote their variational distance by $\mathbb{V}(p, q)$. The finite field of size p is denoted by $\mathbb{F}_p$. The notation $[\![a, b]\!]$ refers to the set of integers between a and b, with extremes included. We will use the notation $\mathcal{U}_A$ for the uniform distribution over a set A.

2.1 Introduction

As seen in the previous chapter, the possibility of communicating securely over noisy channels without secret keys has been known since the pioneering work of Wyner [67]. However, these techniques have never been implemented in actual communication systems so far. One fundamental challenge is that the theoretical assumptions in Wyner's wiretap model, such as the fact that the eavesdropper's channel should be degraded with respect to the main channel, are often hard to verify in practice. Moreover, some information about the eavesdropper's channel is usually required in order to select the wiretap code parameters.

However, the technology advances envisioned for 6G networks, such as the deployment of massive multiple-input multiple-output (MIMO) systems with pencil-sharp beamforming in the sub 6 GHz frequency band, as well as of mmWave and THz wireless systems characterized by highly directional communications, could provide a practical application scenario for wiretap channels, where the signal is received with a high power inside a restricted target zone and with significantly lower power outside this area. The use of reconfigurable intelligent surfaces (RISs) for channel engineering could further improve the available secrecy capacity.

Physical-Layer Security for 6G, First Edition. Edited by Parthajit Mohapatra, Nikolaos Pappas, Arsenia Chorti, and Stefano Tomasin.

A challenging problem in order to implement wiretap coding is to translate the insight given by information-theoretic proofs into the design of explicit codes with low-complexity encoding and decoding for realistic channel models and to provide bounds for the information leakage in the finite blocklength regime.

In this chapter, we will review the most promising coding techniques in the literature. As we will see, most wiretap coding schemes can be divided into two main groups depending on their fundamental structure:

- the first group follows Wyner's classical approach, where the main codebook is partitioned into subcodebooks associated to each confidential message. In the case of linear codes, this nested structure is typically obtained through *coset coding*.
- the second group follows a modular approach, where secrecy coding is added as an additional preprocessing/postprocessing layer on top of an existing error-correcting code. These codes are typically based on *invertible extractors*.

One drawback of the first approach is that it requires a joint encoding design for reliability and secrecy. In particular, it is not known whether a suitable wiretap code partitioning always exists for any given error-correcting code. In contrast, the second approach allows to decouple the secrecy and reliability requirements and to use any "off the shelf" error-correcting code.

2.2 Wiretap Coding Schemes Based on Coset Coding

It can be shown that stochastic encoding is necessary to ensure information-theoretic security over wiretap channels [[12], Section 3.4.1]. Each confidential message m is then mapped to a subcodebook or "bin" rather than to a single codeword. An auxiliary message m', which carries no information of its own, is typically used to select a codeword inside the bin (see Figure 2.1).

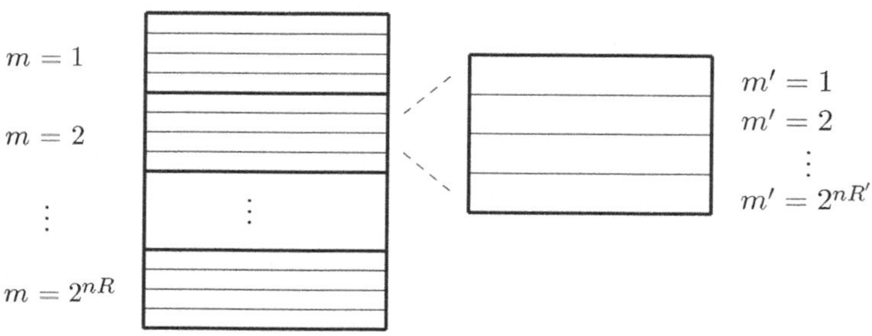

Figure 2.1 An illustration of the binning technique. Each message m is mapped to a "bin" of size $2^{nR'}$. A uniform auxiliary message $m' \in [\![1, 2^{nR'}]\!]$ is used to choose a codeword inside the bin.

The codebook should be chosen in such a way that Bob, who experiences better channel conditions, should decode the transmitted codeword with high probability, while Eve, who experiences worse channel conditions, should not be able to determine to which bin the codeword belongs.

In Wyner's achievability proof [67], the rate R' of each bin approaches the capacity C_e of Eve's channel *from below* as $n \to \infty$. These "capacity-based codes" achieve weak secrecy, but they fall short of achieving strong secrecy, since the leakage is lower bounded by a positive constant. In contrast, it has been shown that it is possible to design "resolvability-based codes" such that $R' > C_e$, which achieve strong secrecy, since the leakage tends to 0 with high probability [13, 14, 39]. The fundamental idea is that when the bin rate is higher than Eve's channel capacity, the output distributions induced by different confidential messages can become indistinguishable for the eavesdropper. As we will see in the next sections, similar findings apply when random codes are replaced by structured codes.

A natural choice when reproducing Wyner's random coding techniques using structured codes is to choose as "bins" the cosets of a linear code. In particular, low-density parity-check (LDPC) codes and polar codes are attractive solutions for secrecy coding due to their low complexity of encoding and decoding.

2.2.1 LDPC Codes for Binary Erasure Wiretap Channels

First, we review some properties of LDPC codes. We refer the reader to [25, 55] for more details. Recall that an (n, k) binary linear code can be represented by a set of k parity-check equations involving n variables. Equivalently, it can be represented by means of a *Tanner graph*, whose nodes are partitioned into two groups: *n variable nodes* $\{x_i\}_{\{i=1,\dots,n\}}$ and *k check nodes*. A variable node x_i is connected to a check node if the variable appears in the corresponding parity-check equation (see Figure 2.2). *Low-density parity-check* codes, or LDPC codes for short, are a class of binary linear codes whose Tanner graph is sparse, in the sense that the number of edges grows linearly with the codelength n.

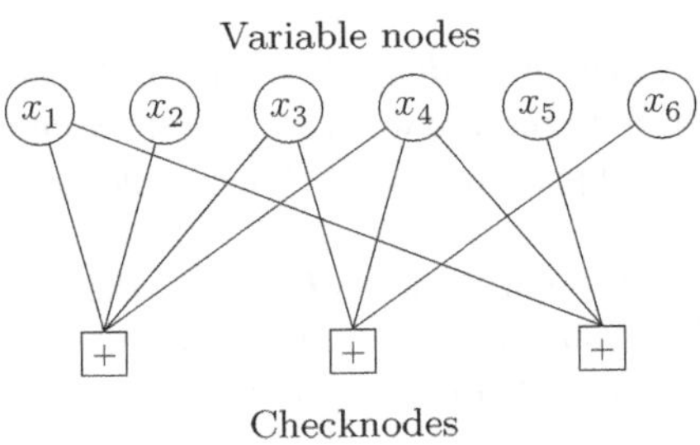

Figure 2.2 An example of factor graph for the (6, 3) binary linear code defined by the parity-check equations $x_1 \oplus x_2 \oplus x_3 \oplus x_4 = 0$, $x_3 \oplus x_4 \oplus x_6 = 0$, $x_1 \oplus x_4 \oplus x_5 = 0$.

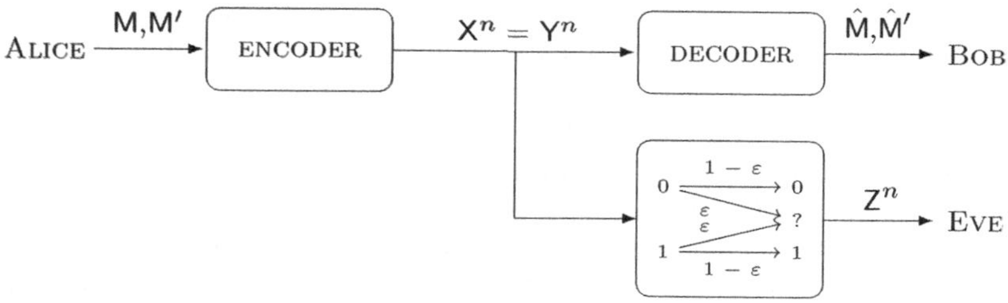

Figure 2.3 The binary erasure wiretap channel BEWC(ϵ).

The number of edges incident to a node is called the *degree* of the node. Let λ_i, $i = 1, \ldots, k$ be the fraction of edges connected to a variable node of degree i, and ρ_j, $j = 1, \ldots, n$ the fraction of edges connected to a check node of degree j. The polynomials $\lambda(x) = \sum_{i=1}^{k} \lambda_i x^{i-1}$ and $\rho(x) = \sum_{i=1}^{n} \rho_i x^{i-1}$ describe these "degree distributions from an edge perspective" [55]. Given a pair (λ, ρ) of degree distributions, an LDPC code ensemble $C_n(\lambda, \rho)$ is the set of all binary linear codes of length n obtained by selecting a bipartite graph uniformly at random according to these distributions.[1] An LDPC code ensemble is called *regular* if every variable node has a fixed degree l and every check node has a fixed degree r.

Thanks to the sparsity of their Tanner graph, LDPC codes can be decoded in linear time as a function of the codelength n using suboptimal iterative decoding algorithms, such as *message passing* [55].

The first wiretap coding constructions based on LDPC codes were proposed by Thangaraj et al. [63] for the binary erasur wiretap channel BEWC(ϵ), where the main channel is error-free and the eavesdropper's channel is a binary erasure channel with erasure probability ε (see Figure 2.3).

The first approach in [63] is to choose as "bins" the cosets of a good (i.e., capacity approaching) LDPC code. Unfortunately, these "capacity-based codes" can only achieve weak secrecy, similar to the random code case. Moreover, this scheme requires the construction of capacity-achieving LDPC codes, which are hard to design in practice.

A second approach is to use as "bins" the cosets of the *dual* of a good LDPC code. Although in [63] this scheme was only proven to achieve weak secrecy, it actually achieves strong secrecy for a restricted range of parameters [60]. We now describe the scheme in more detail.

Let C be an (n, k) binary LDPC code with parity-check matrix H, and $C^{\perp}$ the dual code of C, with generator matrix $G_{\perp} = H^T$, of size $n \times (n-k)$. Let G' be another binary matrix of size $n \times k$ chosen in such a way that the columns of $G_{\perp}$ and G' form a basis of $\mathbb{F}_2^n$.

1 More details on how to sample a code from the ensemble can be found in [54].

To encode the confidential message $\mathbf{m} \in \mathbb{F}_2^k$, Alice randomly draws an auxiliary message $\mathbf{m}' \in \mathbb{F}_2^{n-k}$ and transmits the codeword

$$\mathbf{x} = G'\mathbf{m} + G_{\perp}\mathbf{m}' \in \mathbb{F}_2^n. \tag{2.1}$$

Thus, the set of codewords $\mathbf{x}$ corresponding to the message $\mathbf{m}$ is a coset $\{G'\mathbf{m} + C^{\perp}\}$ of the dual code $C^{\perp}$.

Suppose that during the transmission on the eavesdropper's channel, the bits in the set $\mathcal{E} \subseteq \{1, 2, \dots, n\}$ are erased. The corresponding output $\mathbf{z} \in \{0, 1, ?\}^n$ is said to be *secured by* $C^{\perp}$ if all cosets of $C^{\perp}$ contain at least one element that matches with $\mathbf{z}$ in all the unerased positions. The following result was proven by Ozarow and Wyner [48].

Lemma 2.1 The output $\mathbf{z}$ is secured by $C^{\perp}$ if and only if the submatrix $G_{n-|\mathcal{E}|}$ of $G_{\perp}$ whose rows correspond to the *unerased* bits of $\mathbf{z}$ has rank $n - |\mathcal{E}|$.

Given an (n, k) parity check code C, let $P_{\text{e},C}^{\text{MAP}}(n, \epsilon)$ be the block error probability of the code under maximum a posteriori (MAP) decoding over a binary erasure channel (BEC(ε)) without secrecy constraints. Note that $P_{\text{e},C}^{\text{MAP}}(n, \varepsilon)$ is the probability that when selecting a subset of columns of H with probability ε, the corresponding submatrix $H_{\mathcal{E}}$ of H is not full rank. In fact, if the submatrix $H_{\mathcal{E}}$ associated to the erased bits is full rank, given the output $\mathbf{z}$ of the BEC(ε), the linear system $H\mathbf{x} = 0$ will have a unique solution $\mathbf{x} \in C$ which is consistent with $\mathbf{z}$.

An ensemble of LDPC codes is said to have *threshold* α if $P_{\text{e},C}^{\text{MAP}}(n, \varepsilon) \to 0$ as $n \to \infty$ if and only if $\varepsilon < \alpha$. Thus, the threshold property characterizes the error probability performance of the (primal) code C over a BEC (without secrecy constraints).

The following result highlights the relation between the reliability performance of C and the secrecy performance of the dual code $C^{\perp}$. In fact, the dual code achieves weak secrecy, provided that the erasure probability ϵ is larger than $1 - \alpha$, as well as strong secrecy under stronger hypotheses [60, 63].

Proposition 2.1 If the ensemble of LDPC codes $\{C_n\}$ has threshold α, then the dual wiretap code scheme (2.1) achieves weak secrecy over any BEWC(ε) with $\varepsilon > 1 - \alpha$. Moreover, if there exists a threshold $\alpha^* \le \alpha$ such that $P_{\text{e},C}^{\text{MAP}}(n, \varepsilon) = o\left(\frac{1}{n}\right)$ for all $\varepsilon < \alpha^*$, then the dual wiretap code construction achieves strong secrecy over any BEWC(ε) with $\varepsilon > 1 - \alpha^*$.

Proof: Consider the dual code construction over a BEWC(ε). Recall that $G_{\perp} = H^T$. So with probability $1 - P_{\text{e},C}^{\text{MAP}}(n, 1 - \varepsilon)$, the submatrix of $G_{\perp} = H^T$ obtained by selecting the *rows* corresponding to *unerased* bits with probability $1 - \varepsilon$ is full rank, and therefore $\mathbf{z}$ is secured by $C_\perp$. Consider the random variable Q defined as follows:

$$\mathsf{Q} = \begin{cases} 1 & \text{if } \mathsf{Z}^n \text{ is secured by } C_\perp, \\ 0 & \text{otherwise.} \end{cases}$$

Then, we can establish the following upper bound for the leakage:

$$\begin{aligned}\mathbb{I}(\mathsf{M};\mathsf{Z}^n) &= \mathbb{H}(\mathsf{M}) - \mathbb{H}(\mathsf{M}|\mathsf{Z}^n) \le \mathbb{H}(\mathsf{M}) - \mathbb{H}(\mathsf{M}|\mathsf{Z}^n,\mathsf{Q}) = \\ &= \mathbb{H}(\mathsf{M}) - \mathbb{H}(\mathsf{M}|\mathsf{Z}^n,\mathsf{Q}=1)\mathbb{P}(\mathsf{Q}=1) - \mathbb{H}(\mathsf{M}|\mathsf{Z}^n,\mathsf{Q}=0)\mathbb{P}(\mathsf{Q}=0) \\ &\le \mathbb{H}(\mathsf{M}) - \mathbb{H}(\mathsf{M})(1 - P_{\mathrm{e},C}^{\mathrm{MAP}}(n,1-\varepsilon)) = \mathbb{H}(\mathsf{M})P_{\mathrm{e},C}^{\mathrm{MAP}}(n,1-\varepsilon) \\ &= nRP_{\mathrm{e},C}^{\mathrm{MAP}}(n,1-\varepsilon).\end{aligned}$$

Recalling the definition of threshold, this concludes the proof of both statements. □

Note that a more practical bound on the leakage is obtained by replacing $P_{\mathrm{e},C}^{\mathrm{MAP}}$ by the block error probability $P_{\mathrm{e},C}^{\mathrm{MP}}$ under message passing decoding:

$$\mathbb{I}(\mathsf{M};\mathsf{Z}^n) \le nRP_{\mathrm{e},C}^{\mathrm{MP}}(n,1-\varepsilon).$$

The existence of suitable LDPC ensembles which achieve strong secrecy was proven in [60] through a stopping set analysis. A *stopping set* is a set of variable nodes such that the degree of all neighboring check nodes of the set in the Tanner graph is at least two. If the set of erased variable nodes contains a stopping set, then the iterative decoding will fail. The presence of stopping sets in an LDPC code is related to its *girth*, which is the length of the shortest cycle in its Tanner graph. In a graph with large girth, small stopping sets are less likely to exist, which results in better error performance. In [60], it was shown that for $k > 3$, the error probability under iterative decoding of certain ensembles of LDPC codes with girth greater or equal than $2k$ decays as $O\left(\frac{1}{n^2}\right)$ on a BEC(ε) for all $\varepsilon < \alpha^*$, where α^* is strictly smaller than the erasure threshold α of the ensemble. In particular, Proposition 2.1 guarantees that the dual wiretap code construction will achieve strong secrecy over a BEWC(ε) with $\varepsilon > 1 - \alpha^*$. Note that the strong and weak secrecy thresholds do not coincide, as shown in Figure 2.4.

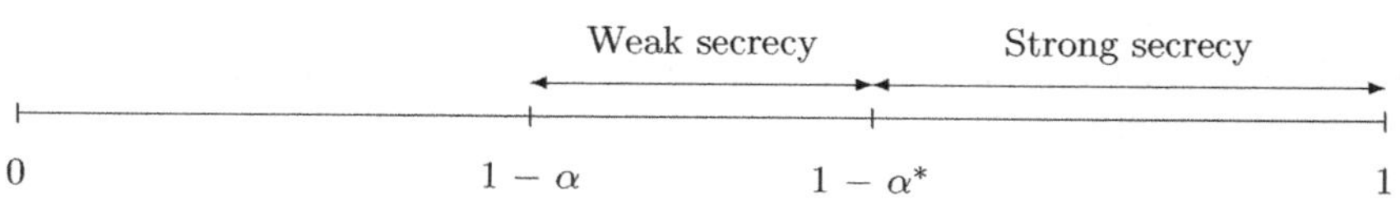

Figure 2.4 Range of eavesdropper's parameters ε such that the dual wiretap LDPC code construction in [60] achieves weak and strong secrecy. Source: Adapted from [60].

An improved scheme with no gap between these thresholds was proposed in [59], by considering the duals of LDPC codes whose girth grows logarithmically with the blocklength. Note that the block error probability $P^{\text{MP}}_{\text{e},C}$ is upper bounded by the bit error probability $P^{\text{MP}}_{\text{b},C}$ using the union bound:

$$P^{\text{MP}}_{\text{e},C}(n, 1-\varepsilon) \leq nP^{\text{MP}}_{\text{b},C}(n, 1-\varepsilon).$$

Consequently, we have the following Corollary of Proposition 2.1.

Corollary 2.1 If there exists a threshold $\alpha^* \leq \alpha$ such that $P^{\text{MP}}_{\text{b},C}(n, \varepsilon) = o\left(\frac{1}{n^2}\right)$ for all $\varepsilon < \alpha^*$, then the dual wiretap code construction achieves strong secrecy over any BEWC(ε) with $\varepsilon > 1 - \alpha^*$.

In order to prove the existence of LDPC codes whose bit error probability under message passing decays as $O\left(\frac{1}{n^3}\right)$ for the threshold $\alpha^* = \alpha$, the authors of [59] study the asymptotic behavior of $P^{\text{MP}}_{\text{b},C}$ using *density evolution*, by considering randomly chosen computation graphs with a fixed level t. The main challenge is that the density evolution approximation assumes that the decoding neighborhood is cycle free (the computation graph is a tree). This assumption can be made rigorous in the case of codes of sufficiently large girth compared to t. Moreover, the choice of *regular* LDPC ensembles ensures that the degrees of all nodes in the computation graph can be chosen independently. Thus, Subramanian et al. [59] prove the existence of ensembles of regular LDPC wiretap codes which achieve strong secrecy for $\varepsilon > 1 - \alpha$, where α is the erasure threshold of the ensemble under message passing decoding.

For *irregular* LDPC wiretap coset codes, it is only possible to establish the existence of codes achieving strong secrecy for $\varepsilon > 1 - \alpha$ with high probability over the code ensemble. However, for irregular LDPC codes, the value $1 - \alpha$ can be very close to the code rate, and so this construction allows to achieve secrecy rates close to the secrecy capacity.

We note that although the above schemes have been proposed for the BEWC setting, they can be extended to other settings. For example, wiretap channels where the main channel is error free and Eve's channel is binary symmetric or Gaussian can be modeled as degraded BEWCs. Thus, the dual LDPC wiretap coding schemes still guarantee positive strong secrecy rates for these scenarios. However, all the schemes reviewed so far require an error-free main channel, that is, they achieve secrecy but not reliability, due to the fact that the dual of a good code is typically not a good code.

In the case when Bob's channel is also a BEC, Rathi et al. [52] proposed an alternate construction using spatially coupled LDPC codes which achieves the whole weak secrecy rate-equivocation region.

In the next section, we review a different wiretap code construction using polar codes, which achieves the strong secrecy capacity of any symmetric-degraded wiretap channel.

2.2.2 Polar Codes for Binary Input Symmetric Channels

Polar codes [3] are the first codes achieving the capacity of any binary-input symmetric discrete memoryless channel with low complexity coding and decoding of the order $O(n \log n)$) as a function of code length. Polar codes are now part of the 5G New Radio standard and are used as the error correcting codes for the enhanced mobile broadband (eMBB) control channel.

The construction of polar codes is based on the technique of *channel polarization*, which allows to transform $n = 2^m$ copies of a discrete memoryless channel W into n sub-channels $W_n^{(i)} : \{0,1\} \to \mathcal{Y}^n \times \{0,1\}^{i-1}$, $i = 1, \ldots, n$, which are either almost noiseless or almost pure-noise channels when $n \to \infty$.

To measure the quality of a binary-input channel $W : \{0,1\} \to \mathcal{Y}$, one can consider the Bhattacharyya parameter defined as

$$Z(W) = \sum_{y \in \mathcal{Y}} \sqrt{W(y|0)W(y|1)} \in [0,1].$$

When $Z(W)$ is close to 1, the channel is very noisy, while for $Z(W)$ close to 0, the channel is almost error-free.

As an example, consider the case $n = 2$. The basic polarization step, illustrated in Figure 2.5, consists in transforming two copies of W into a pair of polarized channels $W' : \{0,1\} \to \mathcal{Y}^2$, $W'' : \{0,1\} \to \mathcal{Y}^2 \times \{0,1\}$ defined as follows:

$$W'(y_1, y_2|u_1) = \sum_{u'_2} \frac{1}{2} W(y_1|u_1 \oplus u'_2)W(y_2|u'_2),$$

$$W''(y_1, y_2, u_1|u_2) = \frac{1}{2} W(y_1|u_1 \oplus u_2)W(y_2|u_2).$$

In terms of Bhattacharyya parameters, one can show that

$$Z(W'') = Z(W)^2, \quad Z(W') \leq 2Z(W) - Z(W)^2, \quad Z(W') \leq Z(W) \leq Z(W''),$$

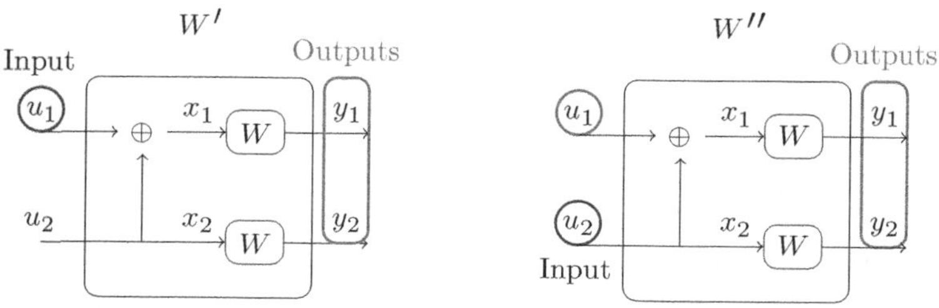

Figure 2.5 The basic polarization technique for $n = 2$.

that is, W' is more noisy than the original channel, since the second input u_2 plays the role of random noise, while W'' is less noisy, thanks to the side information u_1 which is included in the output.

Given the polarization kernel

$$G = \begin{pmatrix} 1 & 0 \\ 1 & 1 \end{pmatrix},$$

the *bit reversal permutation* P_n such that $P_n(u_1, \dots, u_n)^T = (u_n, \dots, u_1)^T$, and defining $G_n = G^{\otimes m} P_n$, the general polarization transform for length $n = 2^m$ proceeds in two steps:

1. *Channel combining*: define a vector channel
$$W_n(\mathbf{y}|\mathbf{u}) = W^n(\mathbf{y}|\mathbf{x}), \quad \mathbf{x} = G_n \mathbf{u}.$$
2. *Channel splitting*: define the *bit channels* $W_n^{(i)} : \{0,1\} \to \mathcal{Y}^n \times \{0,1\}^{i-1}$, $i = 1, \dots, n$ as
$$W_n^{(i)}(\mathbf{y}, (u_1, \dots, u_{i-1})|u_i) = \sum_{u_{i+1},\dots,u_n} \frac{1}{2^{n-1}} W_n(\mathbf{y}|\mathbf{u}).$$

The combined channel W_n in the case $n = 4$ is illustrated in Figure 2.6.

For $0 < \beta < \frac{1}{2}$, define the set of *good bit channels*

$$\mathcal{G}_n(W, \beta) = \left\{ i \in [\![1, n]\!] \mid Z(W_n^{(i)}) < 2^{-n^\beta}/n \right\} \tag{2.2}$$

and the set of *bad bit channels*

$$\mathcal{B}_n(W, \beta) = \left\{ i \in [\![1, n]\!] \mid Z(W_n^{(i)}) \geq 2^{-n^\beta}/n \right\}. \tag{2.3}$$

Then, the following theorem holds [3].

Theorem 2.1 ***(Channel polarization)*** When $n \to \infty$, the ratio $\frac{|\mathcal{G}_n(W,\beta)|}{n}$ of good bit channels converges to $C(W)$ and the ratio $\frac{|\mathcal{B}_n(W,\beta)|}{n}$ of bad bit channels converges to $1 - C(W)$.

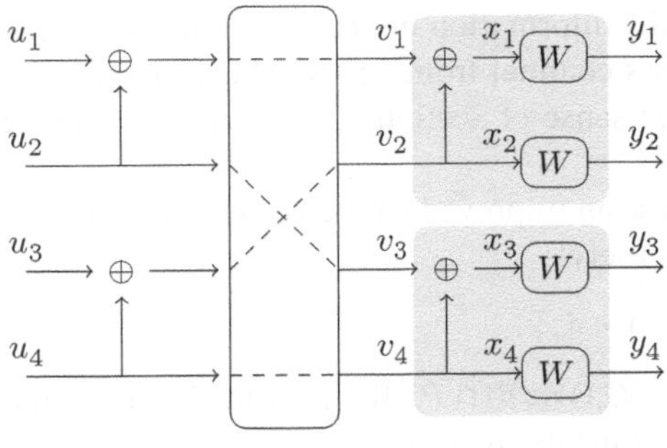

Figure 2.6 The channel polarization construction for $n = 4$.

The channel polarization phenomenon can be exploited to build capacity-achieving polar codes for channel coding simply by transmitting information bits through the good bit channels and zeros (or a fixed choice of *frozen bits*) through the bad bit channels. Given a set of information bits $\mathcal{A} \subset [\![1, n]\!]$, and assuming that the frozen bits $\mathbf{u}_{\mathcal{A}^c}$ are known at both transmitter and receiver, the *successive cancelation* (SC) decoder decides on the value of the bits $i \in \mathcal{A}$ in increasing order, according to the likelihood $L_n^{(i)}(\mathbf{y}, \mathbf{u}_1^{i-1}) = \frac{W_n^{(i)}(\mathbf{y},\mathbf{u}_1^{i-1}|0)}{W_n^{(i)}(\mathbf{y},\mathbf{u}_1^{i-1}|1)}$, which is a function of the output $\mathbf{y}$ and the previously decoded bits, as follows:

$$\hat{u}_i = \begin{cases} 0 & \text{if} \quad L_n^{(i)}(\mathbf{y}, \mathbf{u}_1^{i-1}) \geq 1, \\ 1 & \text{if} \quad L_n^{(i)}(\mathbf{y}, \mathbf{u}_1^{i-1}) < 1. \end{cases}$$

The following bound [5] characterizes the block error probability for polar codes under SC decoding:

$$P_e \leq \sum_{i \in \mathcal{A}} Z(W_n^{(i)}).$$

Polar codes are very versatile and have been used for a wide range of applications. Their structure makes them particularly suitable for secrecy coding. Mahdavifar and Vardy [41] and Hof and Shamai [34] proposed a wiretap coding construction based on polar codes that achieves the weak secrecy capacity, provided that the main channel W_b and the eavesdropper's channel W_e are symmetric discrete memoryless channels, and W_e is degraded with respect to W_b. As in (2.2) and (2.3), one can consider the sets of good and bad bit channels for W_b and W_b. In the degraded case, one can show that $\forall \beta \in \left(0, \frac{1}{2}\right)$, $\mathcal{G}_n(W_e, \beta) \subseteq \mathcal{G}_n(W_b, \beta)$, that is, the good bit channels for Eve are also good for Bob.

The key strategy of the wiretap coding construction in [41] is to send *information bits* on the set $\mathcal{A} = \mathcal{G}_n(W_b) \backslash \mathcal{G}_n(W_e)$ of bit channels which are good for Bob but bad for Eve, *zeros* on the set $\mathcal{B} = \mathcal{B}_n(W_b)$ of bit channels that are bad for both Bob and Eve, and *random bits* on the set $\mathcal{R} = \mathcal{G}_n(W_e, \beta)$ of bit channels which are good for both. Note that in this scheme, the rate of each "bin" (i.e., the set of possible codewords for a fixed choice of the information bits) corresponds to $|\mathcal{R}|/n$, which approaches the capacity of Eve's channel from below. Thus, this wiretap code is a "capacity-based code" in the sense of Section 2.2 and cannot achieve strong secrecy.

This motivated the authors to propose an improved strong secrecy scheme [42] by introducing the set of "poor bit channels"

$$\mathcal{P}_n(W, \delta_n) = \left\{ i \in [\![1, n]\!] \mid C(W^{(i)}) < \delta_n \right\}.$$

Information bits are then sent on $\mathcal{A} = \mathcal{G}_n(W_b, \beta) \cap \mathcal{P}_n(W_e, \delta_n)$ (bit channels that are poor for Eve, but good for Bob); random bits are sent on $\mathcal{R} = [\![1, n]\!] \backslash \mathcal{P}_n(W_e, \delta_n)$, and zeros are sent on $\mathcal{B} = \mathcal{P}_n(W_e, \delta_n) \backslash \mathcal{G}_n(W_m, \beta)$.

Note that this is a "resolvability-based code" according to Section 2.2, since it can be shown that $|\mathcal{R}|/n \searrow C(W_e)$, that is, the size of the bins is now strictly greater than Eve's channel capacity.

To prove that this new scheme achieves the strong secrecy capacity for a suitable sequence δ_n, the authors consider an induced channel which comprises the polar code and the average over the random bit channels in $\mathcal{R}$. More precisely, given two vectors $\mathbf{x} \in \{0,1\}^{n-r}$ and $\mathbf{e} \in \{0,1\}^r$, we denote by $(\mathbf{x};\mathbf{e})$ the vector $\mathbf{v} \in \{0,1\}^n$ such that $\mathbf{v}_{\mathcal{R}^c} = \mathbf{x}$ and $\mathbf{v}_{\mathcal{R}} = \mathbf{e}$. Consider the channel

$$Q_n(\mathbf{z}|\mathbf{x}) = \frac{1}{2^r} \sum_{\mathbf{e} \in \{0,1\}^r} W_e^n \left(\mathbf{z} \middle| G_n(\mathbf{x};\mathbf{e})^T \right), \tag{2.4}$$

which takes as input the bits that are not in $\mathcal{R}$. One can show that Q_n is a symmetric channel, therefore the uniform input $\overline{\mathsf{X}}^{n-r}$ on the nonrandom bit channels maximizes the mutual information and for any other input distribution X^{n-r},

$$\mathbb{I}(\mathsf{X}^{n-r}; \mathsf{Z}^n) \leq C(Q_n) = \sum_{i \in \mathcal{R}^c} C(W_e^{(i)}) \leq \delta_n \left| \mathcal{P}_n(W_e, \delta_n) \right|. \tag{2.5}$$

Finally, to prove that the mutual information tends to 0, it is enough to show that one can choose $\delta_n = o\left(\frac{1}{n}\right)$. Observe that Mahdavifar and Vardy's codes work for *any a priori distribution of the source messages*, while the other secrecy schemes usually require a uniform message distribution, a condition that is unrealistic for most information sources. That is, the scheme achieves *semantic security* as well [10].

Unfortunately, in [42], the authors were not able to prove that the legitimate receiver's probability of error vanishes; therefore, the applicability of the strong secrecy scheme is limited to the case of an error-free main channel. The main obstacle to guaranteeing both strong secrecy and reliability is the presence of a small set of bit channels that are at the same time unreliable and insecure. This problem was addressed in [56] using a multi-block polar coding technique, which however has an increased complexity and requires Alice and Bob to share a small random seed. A more general polar coding scheme, which also applies to non-degraded wiretap channels and broadcast channels with confidential messages, was proposed in [26].

2.2.3 Lattice Codes for Gaussian and Fading Wiretap Channels

Wireless systems are particularly vulnerable to eavesdropping since every transmission can potentially be overheard by all the neighboring nodes in the network. Moreover, the wireless medium is inherently a source of randomness, which can be exploited to provide security. Therefore, the design of wiretap codes over continuous channels such as Gaussian and fading channels is of particular interest for the practical integration of PLS techniques into future communication systems.

Lattice codes are particularly well-suited to wiretap coding over continuous channels. In general, lattice signal constellations, carved from a discrete subgroup Λ of $\mathbb{R}^n$, are a natural counterpart of linear codes when the Hamming metric is replaced by Euclidean distance. In the presence of an eavesdropper, coset coding constructions using two nested lattices $\Lambda_e \subset \Lambda_b$ can be seen as the structured equivalent of random binning. More precisely, the set of confidential messages $\mathcal{M}$ can be identified with the quotient group Λ_b/Λ_e through a bijection $\phi : \mathcal{M} \to \Lambda_b/\Lambda_e$ which maps each message $m \in \mathcal{M}$ to a coset leader $\lambda_m \in \Lambda_b \cap \mathcal{R}(\Lambda_e)$, where $\mathcal{R}(\Lambda_e)$ is some fundamental region of the lattice Λ_e.[2] In order to encode the message $m \in \mathcal{M}$, Alice transmits $\lambda_m + \lambda$, where $\lambda \in \Lambda_e$ is a random vector.

The earliest wiretap code designs for continuous channels aimed to maximize Eve's error probability.[3] Oggier et al. [6, 46] first considered nested lattice codes for the Gaussian wiretap channel, introduced the concept of *secrecy gain* and showed that Eve's error probability tends to 1 for even unimodular lattices when the dimension tends to ∞. These lattices are further investigated in [23]. This secrecy coding framework under an error probability criterion is extended to the case of fading and multiple antenna channels using ideal lattices from number fields [7, 8, 35, 47].

Follow-up works focused on information-theoretic metrics and introduced new techniques to bound the leakage of nested lattice constructions. Choo et al. [17] showed the existence of sequences of lattices achieving the weak secret key capacity. Lattice codes were also used to provide weak/strong secrecy in the presence of a cooperative jammer and of interference [31–33].

Ling et al. [37] showed that lattice codes achieve strong secrecy and semantic security over the Gaussian wiretap channel. Following the approach by Csiszár [14, 20], the information leakage vanishes if the output distributions of the eavesdropper's channel associated to different messages can be made arbitrarily close in variational distance. Moreover, they proposed the *flatness factor* of a lattice, corresponding to the L^∞ distance of Gaussian noise-(folded modulo the lattice) from the uniform distribution, as a new design criterion to bound the variational distance of output distributions.[4] They showed the existence of sequences of "secrecy-good" lattices whose flatness factor vanishes when the dimension of the lattice tends

---

2 A fundamental region $\mathcal{R}(\Lambda)$ of the lattice Λ is a bounded set such that its translates with respect to the lattice are disjoint and their union is $\mathbb{R}^n$.

3 Some of these were not based on lattices, such as the work [36] using punctured LDPC codes.

4 We note that the flatness factor is a reformulation of the *smoothing parameter* used in lattice-based cryptography [53].

to infinity. These codes achieve semantic security for rates up to $1/(2\ln 2)$ bits from the secrecy capacity, provided that the codeword corresponding to each message m is sampled according to a *discrete Gaussian distribution* over the coset $\lambda_m + \Lambda_e$.

The flatness factor criterion can be extended to multiple-antenna fading wiretap channels [15, 40]. In particular, Luzzi et al. [40] propose a design criterion for lattice wiretap codes over fading channels: the normalized product distance (in the single antenna case) or normalized minimum determinant (in the multiple antenna case) of the lattice and of its dual should be maximized simultaneously.

The problem of finding good wiretap lattice codes in fixed dimension was considered in [21], where it was shown that one can restrict the search to the set of *well-rounded lattices*, that is, lattices such that all the successive minima are equal to the first minimum. Algebraic constructions of ideal lattices with small average flatness factor were also provided in [21].

A different lattice scheme using *polar lattices* [38] built from polar codes using Construction D [19] was shown to achieve semantic security over Gaussian wiretap channels with encoding/decoding complexity $O(N \log N \log(\log N))$.

Finally, in the next section, we will discuss an alternative approach to achieve the strong secrecy capacity of the Gaussian wiretap channel, which is based on 2-universal hash functions [64].

2.3 Wiretap Coding Schemes Based on Invertible Extractors

In this section, we review wiretap code constructions based on extractors [10, 16, 29, 30, 64]. As mentioned before, the advantage of this approach is that it allows to address the reliability and secrecy constraints separately.

Informally speaking, an *extractor* is a function that takes as input a sequence X and a seed S and outputs a shorter sequence K which is almost uniform and almost independent of the seed, provided that X has enough randomness. The latter condition is typically expressed as a lower bound on the *min-entropy* of X, which is given by

$$H_\infty(\mathsf{X}) = -\log\left(\max_{x\in\mathcal{X}} p_{\mathsf{X}}(x)\right).$$

We note that the min-entropy is a stronger measure of unpredictability than the Shannon entropy, since $H_\infty(\mathsf{X}) \leq H(\mathsf{X})$ for any discrete random variable X.

More generally, given two random variables (X, Z), where X represents the source of randomness and Z the adversary's side information, it is possible to extract from X a uniform key which is almost independent of Z, provided that the *average*

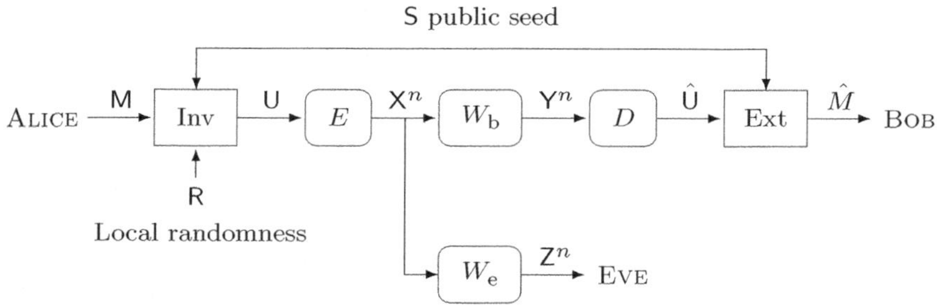

Figure 2.7 Modular wiretap codes based on invertible extractors.

conditional min-entropy

$$\tilde{H}_\infty(\mathsf{X}|\mathsf{Z}) = -\log\left(\mathbb{E}_\mathsf{Z}\left[\max_{x\in\mathcal{X}} p_{\mathsf{X}|\mathsf{Z}}(x|\mathsf{Z})\right]\right)$$

is large enough.[5]

Definition 2.1 ***(Strong extractor)*** A function $\text{Ext} : \{0,1\}^l \times \{0,1\}^s \to \{0,1\}^k$ is a *strong* (h, ϵ)*-extractor* if for any $(\mathsf{X},\mathsf{Z}) \in \{0,1\}^l \times \{0,1\}^*$ such that $\tilde{H}_\infty(\mathsf{X}|\mathsf{Z}) \geq h$, and for S uniformly distributed over $\{0,1\}^s$, we have

$$\mathbb{V}(p_{\text{Ext}(\mathsf{X},\mathsf{S}),\mathsf{Z},\mathsf{S}}, \mathcal{U}_{\{0,1\}^k} p_{\mathsf{ZS}}) \leq \epsilon.$$

An extractor Ext is called *regular* if $\forall \mathbf{s} \in \{0,1\}^s$ all the points in the range of Ext have the same number of pre-images.

We are interested in extractors that can be inverted, in the following sense.

Definition 2.2 ***(Invertible extractor)*** The extractor Ext is *invertible* if there exists an (efficient) function $\text{Inv} : \{0,1\}^s \times \{0,1\}^r \times \{0,1\}^k \to \{0,1\}^l$ such that, for all $\mathbf{s} \in \{0,1\}^s$, for all $\mathbf{m} \in \{0,1\}^k$, and for R uniform over $\{0,1\}^r$, $\text{Inv}(\mathbf{s}, \mathsf{R}, \mathbf{m})$ is uniform on the set of pre-images $\{\mathbf{x} \in \{0,1\}^l : \text{Ext}(\mathbf{x}, \mathbf{s}) = \mathbf{m}\}$.

We now describe how to use an invertible extractor to transform a good error-correcting code into a wiretap code [9, 10, 29, 66]. This kind of protocol is called *Invert-then-Encode* [10] and is illustrated in Figure 2.7. Let (E, D) be an encoder–decoder pair for the main channel W_b, with $E : \{0,1\}^l \to \mathcal{X}^n$, $D : \mathcal{Y}^n \to \{0,1\}^l$. Suppose that $\text{Ext} : \{0,1\}^l \times \{0,1\}^s \to \{0,1\}^k$ is an invertible extractor with inverter Inv. Then, preprocessing and postprocessing security layers can be added as follows.

5 See [22] for an explanation of why the average conditional min-entropy is the most relevant definition in this context.

Let us first suppose that Alice and Bob share a uniform *public* seed $\mathsf{S} \in \mathcal{S}$. To transmit the confidential message $\mathsf{M} \in \mathcal{M}$, given the public seed S and local randomness R, Alice computes $\mathsf{U} = \text{Inv}(\mathsf{S}, \mathsf{R}, \mathsf{M}) \in \{0,1\}^l$, encodes $\mathsf{X}^n = E(\mathsf{U}) \in \mathcal{S}^n$, and sends it over the wiretap channel. By construction, U is chosen uniformly in the set of pre-images $\{\mathbf{u} \in \{0,1\}^l : \text{Ext}(\mathbf{u}, \mathsf{S}) = \mathsf{M}\}$. Since (E, D) is a good error-correcting code for the main channel, Bob can recover U from the channel output Y^n with small probability of error and can reconstruct $\mathsf{M} = \text{Ext}(\mathsf{U}, \mathsf{S})$.

The strong secrecy and semantic security for the Invert-then-Encode construction can be proven in two steps [10]. First, the authors show that it achieves a weaker definition of security called *distinguishing security for random messages* (RDS), provided that the extractor is regular and that the eavesdropper's channel is symmetric. Then, they show that random message security implies semantic security if the encryption function is separable and message-linear.

Up to now, we have assumed that Alice and Bob share a randomness source. In practice, if this is not the case, the seed S can be transmitted reliably over the main channel. The same seed can be "recycled" over several consecutive blocks so that the rate loss due to this transmission becomes negligible [9, 10]. This version of the protocol is called *Repeat Invert-then-Encode.*

One way to obtain efficiently computable extractors is through *universal hash functions* (UHFs).

Definition 2.3 A family $\{f_\mathbf{s}\}_{\mathbf{s}\in\mathcal{S}}$ of *hash functions* $f_\mathbf{s} : \{0,1\}^l \to \{0,1\}^k$ is *2-universal* if for every pair of distinct strings $\mathbf{x}, \mathbf{x}' \in \{0,1\}^l$, for S uniformly random in $\mathcal{S}$,

$$\mathbb{P}\{f_\mathsf{S}(\mathbf{x}) = f_\mathsf{S}(\mathbf{x}')\} \leq 2^{-k}.$$

The following lemma [22] shows that given $\epsilon > 0$, the function $\text{Ext}(\mathbf{x}, \mathbf{s}) = f_\mathbf{s}(\mathbf{x})$ is an (h, ϵ)-extractor for $h = k - 2 - 2\log_2 \epsilon$.

Lemma 2.2 ***(Leftover Hash Lemma)*** Let $\{f_s\}_{s\in\mathcal{S}}$ be a family of 2-universal hash functions. Then, for any pair of random variables (X, Z),

$$\mathbb{V}(p_{f_\mathsf{S}(\mathsf{X})\mathsf{ZS}}, \mathcal{U}_{\{0,1\}^k} p_{\mathsf{ZS}}) \leq \frac{1}{2}\sqrt{2^{k-\tilde{H}_\infty(\mathsf{X}|\mathsf{Z})}}.$$

We note that there exist many versions of the Leftover Hash Lemma which replace the conditional min-entropy by different metrics, see e.g., [66] for more references.

An explicit family of 2-universal hash functions $f_\mathbf{s} : \{0,1\}^l \to \{0,1\}^k$ that can be implemented efficiently for well-chosen values of the parameters can be obtained by identifying $\{0,1\}^l$ with the Galois field $\mathbb{F}_{2^l}$ and by defining $\forall \mathbf{s} \in \mathcal{S} = \{0,1\}^l$

$$f_{\mathbf{s}}(\mathbf{x}) = (\mathbf{x} \cdot \mathbf{s})_k, \tag{2.6}$$

where $(\mathbf{x})_k$ denotes the subvector of the k most significant bits of $\mathbf{x}$. An explicit inverter for this family can be defined as

$$\text{Inv}(\mathbf{s}, \mathbf{r}, \mathbf{m}) = \mathbf{s}^{-1} \cdot (\mathbf{m}||\mathbf{r}),$$

where $\mathbf{r} \in \{0,1\}^{l-k}$.

The follow-up work [58] proposes a new modular wiretap code construction such that the cost of inversion is the same as the cost of hashing. First, they define a new notion of XOR-universal hash functions (XOR-UHFs).

Definition 2.4 A family of hash functions $f_{\mathbf{s}} : \{0,1\}^l \to \{0,1\}^k$ is *XOR-universal* if for every pair of distinct strings $\mathbf{x}, \mathbf{x}' \in \{0,1\}^l$, for S uniformly random in $\mathcal{S}$,

$$\mathbb{P}\left\{f_{\mathsf{S}}(\mathbf{x}) \oplus f_{\mathsf{S}}(\mathbf{x}') = \mathbf{a}\right\} \leq 2^{-k} \quad \forall \mathbf{a} \in \{0,1\}^k.$$

In particular, XOR-UHFs are also 2-universal. If one relaxes the constraint that the domain and the set of seeds are the same, a variant of the UHF family (2.6) satisfies this definition. Let $\mathcal{S} = \{0,1\}^l$ if $l \geq k$, and $\mathcal{S} = \{0,1\}^k$ if $l < k$, and define $f_{\mathbf{s}} : \{0,1\}^l \to \{0,1\}^k$ as follows:

$$f_{\mathbf{s}}(\mathbf{x}) = \begin{cases} (\mathbf{x} \cdot \mathbf{s})_k & \text{if } l \geq k, \\ ((\mathbf{x}||\mathbf{0}^{k-l}) \cdot \mathbf{s}) & \text{if } l < k. \end{cases}$$

Given a family $f_{\mathbf{s}} : \{0,1\}^l \to \{0,1\}^k$ of XOR-UHFs, one can obtain a family $g_{\mathbf{s}} : \{0,1\}^l \times \{0,1\}^k \to \{0,1\}^k$ of regular and invertible hash functions by defining

$$g_{\mathbf{s}}(\mathbf{x}, \mathbf{y}) = f_{\mathbf{s}}(\mathbf{x}) \oplus \mathbf{y} \quad \forall \mathbf{x} \in \{0,1\}^l, \forall \mathbf{y} \in \{0,1\}^k,$$

together with the inverter function

$$g_{\mathbf{s}}^{\text{Inv}}(\mathbf{y}, \mathbf{r}) = (\mathbf{r}, f_{\mathbf{s}}(\mathbf{r}) \oplus \mathbf{y}) \quad \forall \mathbf{y} \in \{0,1\}^k, \forall \mathbf{r} \in \{0,1\}^l.$$

Given an encoder–decoder pair (E, D), a variant of the Invert-then-Encode protocol, called *Hash-then-Encode*, is defined as follows. To send the confidential message $\mathsf{M} \in \mathcal{M}$, using local randomness R, Alice computes $\mathsf{U} = (\mathsf{R}||f_{\mathsf{S}}(\mathsf{R} \oplus \mathsf{M}))$, encodes $\mathsf{X}^n = E(\mathsf{U})$ and transmits it over the channel. Using the decoder D, Bob recovers $\hat{\mathsf{U}} = (\mathsf{U}_1||\mathsf{U}_2)$ from the channel output, which is equal to $(\mathsf{R}||f_{\mathsf{S}}(\mathsf{R} \oplus \mathsf{M}))$ with high probability, and obtains $\hat{\mathsf{M}} = f_{\mathsf{S}}(\mathsf{U}_1) \oplus \mathsf{U}_2$. This construction is more efficient than Invert-then-Encode and has a better effective rate. It achieves the secrecy capacity for all symmetric degraded discrete wiretap channels.

A modular construction has also been used in [18] to design wiretap codes that do not require any symmetry or degradation assumption, and can deal with uncertainties about the statistics of Eve's channel, which is modeled as a *compound* or *arbitrarily varying* channel. This construction achieves the best known communication rates for these channel models. It employs the UHF family (2.6) for secrecy and multi-block polar codes for reliability, but unlike [42], it relies on source polarization [4] rather than channel polarization.

2.3.1 Secrecy Capacity-Achieving Codes for the Gaussian Channel

The approach based on invertible extractors described in the previous section extends to continuous channels [61]. In particular, the work [64] proposes a construction of codes achieving the strong secrecy capacity of the Gaussian wiretap channel by using the universal hash family (2.6) together with a power-constrained error-correcting code. This was shown to be semantically secure in [65].

2.4 Finite-Length Results

The potential use cases for PLS technologies in sixth-generation communication systems include Internet of Things (IoT) networks featuring short packet payloads and ultra-reliable low-latency communications (URLLCs). For both applications, it is essential to obtain precise bounds for the leakage in finite blocklength. The first non-asymptotic bounds for the leakage in wiretap channels were obtained by Hayashi [28]. Building on the recent theoretical breakthrough in the analysis of finite-length channel coding rates [50], Yang et al. [68] proved tight bounds on the second-order secrecy rate for discrete memoryless and Gaussian wiretap channels. Rather than the mutual information, they consider the average total variation distance $S(\mathsf{M}|\mathsf{Z}^n)$ between the message M and the Eve's observation Z^n as a measure of leakage[6]:

$$S(\mathsf{M}|\mathsf{Z}^n) = \mathbb{V}(p_{\mathsf{MZ}^n}, \mathcal{U}_{\mathcal{M}} p_{\mathsf{Z}^n}).$$

They show that the maximal secrecy rate $R^*(n, \epsilon, \delta)$ for block length n, error probability ϵ, and under the constraint $S(\mathsf{M}|\mathsf{Z}^n) \leq \delta$ is bounded by

$$C_s - \sqrt{\frac{V_1}{n}} Q^{-1}(\epsilon) - \sqrt{\frac{V_2}{n}} Q^{-1}(\delta) + \mathcal{O}\left(\frac{\log n}{n}\right) \leq R^*(n, \epsilon, \delta)$$
$$\leq C_s - \sqrt{\frac{V_3}{n}} Q^{-1}(\epsilon + \delta) + \mathcal{O}\left(\frac{\log n}{n}\right), \tag{2.7}$$

6 An alternative metric considered in [68] is the maximum total variation distance $S_{\max}(\mathsf{M}|\mathsf{Z}^n) = \max_{m \in \mathcal{M}} \mathbb{V}(p_{\mathsf{Z}^n|\mathsf{M}=m}, p_{\mathsf{Z}^n|\mathsf{M}} \circ \mathcal{U}_{\mathcal{M}})$, which is related to distinguishing security and semantic security.

where C_s is the secrecy capacity, Q is the tail probability of the standard Gaussian distribution, and V_1, V_2, V_3 are constants depending on the distributions of the main channel and eavesdropper's channel. For semi-deterministic wiretap channels (in which the main channel is deterministic), they show that for $\epsilon + \delta < 1$, the maximal secrecy rate is of the form

$$R^*(n, \epsilon, \delta) = C_s - \sqrt{\frac{V_S}{n}} Q^{-1}\left(\frac{\delta}{1-\epsilon}\right) + \mathcal{O}\left(\frac{\log n}{n}\right), \tag{2.8}$$

where V_S is the conditional variance of the information density of p_{XZ}. In particular, Eq. (2.8) evidences a trade-off between reliability and secrecy.

In spite of these theoretical advances, to this date, only few works have considered practical wiretap code constructions in finite blocklength.

Coset coding schemes for binary symmetric and Gaussian channels based on turbo codes and LDPC codes under an eavesdropper error probability criterion were considered in [43, 44]. The follow-up work [45] considered randomized Reed–Muller codes for Gaussian wiretap channels under a mutual information criterion for very short blocklengths. The limitation in blocklength is due to the difficulty of providing a tight estimate of the leakage, which is done using neural networks. Other works [49] use Monte-Carlo techniques to quantify the equivocation of short wiretap codes over a BEC. Algebraic criteria to compute exact equivocation expressions have been found in [27].

In the case of the polar coding construction of [42], finite-length bounds for the leakage can be obtained directly in terms of the sum of the capacities of the bit channels, as in Eq. (2.5). The same bound holds for other types of codes which have polarization properties, such as Reed–Muller codes [1].

In [57], the bounds on secrecy performance of polar and Reed–Muller codes are compared to the second-order secrecy rate (2.8) over a BEWC. When Eve's channel W_e is a BEC, these bounds are particularly simple since all the bit channels $W_e^{(i)}$ are also BECs with some parameter p_i, and their average total variation distance is $\frac{1}{2}(1-p_i) = \frac{1}{2}C(W_e^{(i)})$. The parameters p_i can be estimated through Monte-Carlo simulations (or, for polar codes, directly using the closed form expression for the bit channel capacity in [3]). As shown in Figure 2.8, the secrecy rate of Reed–Muller codes is close to the second-order coding rate, while the achievable secrecy rate of polar codes appears to be suboptimal. This might be due to the faster polarization properties of Reed–Muller codes compared to polar codes [1]. However, Reed–Muller codes are still unsuitable for practical applications due to the lack of an efficient decoding algorithm, except in the low code rate and the high code rate regime [2].

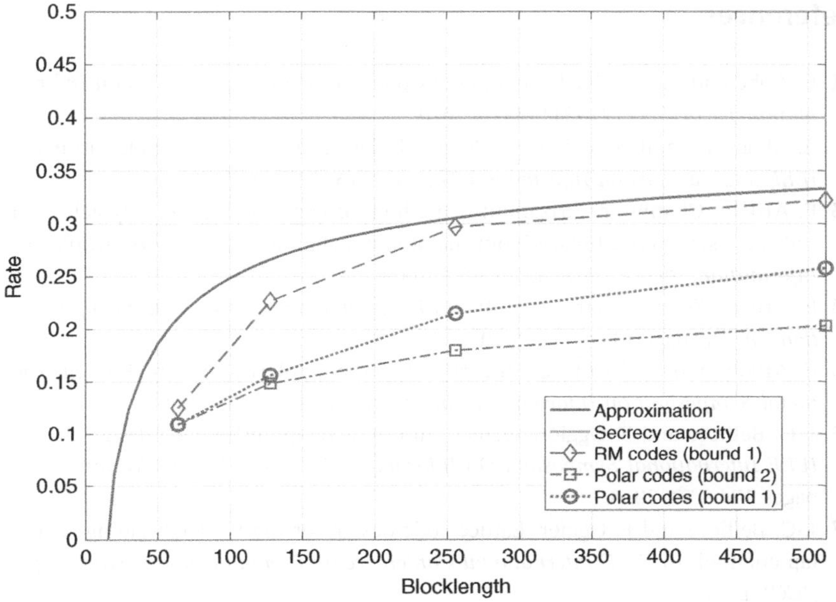

Figure 2.8 Comparison of the lower bound on secrecy rates of Reed–Muller and polar codes for $p = 0.4$ and $\delta = 0.001$, with the second-order approximation in (2.8). Bound 1 is computed by Monte-Carlo simulations and bound 2 is obtained using the expression for the bit-channel capacity.

A different approach to obtain a tight lower bound on the leakage for finite-blocklength polar codes using a modified SC decoder was proposed in [62].

Other recent works have used deep learning to design wiretap codes. In [24], a clustering technique is used to obtain a structure that resembles coset coding, under a secrecy criterion based on cross entropy. In [11], a neural network autoencoder is trained to optimize the secrecy and reliability constraints, under a mutual information criterion. The work [51] proposes a modular construction where hash functions guarantee secrecy, while the reliability layer is optimized through an autoencoder. These codes show promising performance compared to the state-of-the-art constructions from coding theory.

The advantage of deep-learning techniques is that they do not require the knowledge of the channel statistics, but only a sufficient amount of input and output channel samples to be used for training. However, the complexity of estimating the leakage remains a significant challenge, which currently limits their use to short blocklengths or low rates.

References

1 E. Abbe and M. Ye. Reed-Muller codes polarize. *IEEE Transactions on Information Theory*, 66(12):7311–7332, 2020.

2 E. Abbe, A. Shpilka, and M. Ye. Reed–Muller codes: Theory and algorithms. *IEEE Transactions on Information Theory*, 67(6):3251–3277, 2020.

3 E. Arikan. Channel polarization: A method for constructing capacity-achieving codes for symmetric binary-input memoryless channels. *IEEE Transactions on Information Theory*, 55(7):3051–3073, 2009.

4 E. Arikan. Source polarization. In *IEEE International Symposium on Information Theory*, pages 899–903, 2010.

5 E. Arikan and E. Telatar. On the rate of channel polarization. In *IEEE International Symposium on Information Theory*, pages 1493–1495, 2009.

6 J.-C. Belfiore and F. Oggier. Secrecy gain: A wiretap lattice code design. In *IEEE International Symposium On Information Theory & Its Applications*, pages 174–178, 2010.

7 J.-C. Belfiore and F. Oggier. Lattice code design for the Rayleigh fading wiretap channel. In *IEEE International Conference on Communications Workshops (ICC)*, pages 1–5, 2011.

8 J.-C. Belfiore and F. Oggier. An error probability approach to MIMO wiretap channels. *IEEE Transactions on Communications*, 61(8):3396–3403, 2013.

9 M. Bellare, S. Tessaro, and A. Vardy. A cryptographic treatment of the wiretap channel. *arXiv preprint arXiv:1201.2205*, 2012.

10 M. Bellare, S. Tessaro, and A. Vardy. Semantic security for the wiretap channel. In *Advances in Cryptology–CRYPTO 2012: 32nd Annual Cryptology Conference*, pages 294–311. Springer, Aug 2012.

11 K.-L. Besser, P.-H. Lin, C. R. Janda, and E. A. Jorswieck. Wiretap code design by neural network autoencoders. *IEEE Transactions on Information Forensics and Security*, 15:3374–3386, 2020. doi: 10.1109/TIFS.2019.2945619.

12 M. Bloch and J. Barros. *Physical-Layer Security: From Information Theory to Security Engineering*. Cambridge University Press, 2011.

13 M. Bloch and J. N. Laneman. On the secrecy capacity of arbitrary wiretap channels. In *Allerton Conference on Communication, Control and Computing*, pages 818–825, Sept 2008.

14 M. R. Bloch and J. N. Laneman. Strong secrecy from channel resolvability. *IEEE Transactions on Information Theory*, 59(12):8077–8098, Dec 2013.

15 A. Campello, C. Ling, and J.-C. Belfiore. Semantically secure lattice codes for compound MIMO channels. *IEEE Transactions on Information Theory*, 66(3):1572–1584, 2019.

16 M. Cheraghchi, F. Didier, and A. Shokrollahi. Invertible extractors and wiretap protocols. *IEEE Transactions on Information Theory*, 58(2):1254–1274, 2011.

17 L. Choo, C. Ling, and K. Wong. Achievable rates for lattice coding over the Gaussian wiretap channel. In *ICC Physical Layer Security Workshop*, 2011.

18 R. A. Chou. Explicit wiretap channel codes via source coding, universal hashing, and distribution approximation, when the channels' statistics are uncertain. *IEEE Transactions on Information Forensics and Security*, 18:117–132, 2022.

19 J. Conway and N. Sloane. *Sphere Packings, Lattices and Groups*, volume 290. Springer Science & Business Media, 2013.

20 I. Csiszár. Almost independence and secrecy capacity. *Problems of Information Transmission*, 32:40–47, 1996.

21 M. T. Damir, A. Karrila, L. Amoros, O. W. Gnilke, D. Karpuk, and C. Hollanti. Well-rounded lattices: Towards optimal coset codes for Gaussian and fading wiretap channels. *IEEE Transactions on Information Theory*, 67(6):3645–3663, 2021.

22 Y. Dodis, R. Ostrovsky, L. Reyzin, and A. Smith. Fuzzy extractors: How to generate strong keys from biometrics and other noisy data. *SIAM Journal on Computing*, 38(1):97–139, 2008.

23 A.-M. Ernvall-Hytönen and C. Hollanti. On the eavesdropper's correct decision in Gaussian and fading wiretap channels using lattice codes. In *IEEE Information Theory Workshop*, pages 210–214, 2011.

24 R. Fritschek, R. F. Schaefer, and G. Wunder. Deep learning based wiretap coding via mutual information estimation. In *Proceedings of the 2nd ACM Workshop on Wireless Security and Machine Learning*, pages 74–79, 2020.

25 R. Gallager. Low-density parity-check codes. *IRE Transactions on Information Theory*, 8(1):21–28, 1962.

26 T. C. Gulcu and A. Barg. Achieving secrecy capacity of the wiretap channel and broadcast channel with a confidential component. *IEEE Transactions on Information Theory*, 63(2):1311–1324, 2017.

27 W. K. Harrison. Exact equivocation expressions for wiretap coding over erasure channel models. *IEEE Communications Letters*, 24(12):2687–2691, 2020.

28 M. Hayashi. General nonasymptotic and asymptotic formulas in channel resolvability and identification capacity and their application to the wiretap channel. *IEEE Transactions on Information Theory*, 52(4):1562–1575, 2006.

29 M. Hayashi. Exponential decreasing rate of leaked information in universal random privacy amplification. *IEEE Transactions on Information Theory*, 57(6):3989–4001, 2011.

30 M. Hayashi and R. Matsumoto. Construction of wiretap codes from ordinary channel codes. In *IEEE International Symposium on Information Theory*, pages 2538–2542, 2010.

31 X. He and A. Yener. Providing secrecy with lattice codes. In *46th annual Allerton Conference on Communication, Control, and Computing*, pages 1199–1206. IEEE, 2008.

32 X. He and A. Yener. The Gaussian many-to-one interference channel with confidential messages. *IEEE Transactions on Information Theory*, 57(5):2730–2745, 2011.

33 X. He and A. Yener. Strong secrecy and reliable Byzantine detection in the presence of an untrusted relay. *IEEE Transactions on Information Theory*, 59(1):177–192, 2012.

34 E. Hof and S. Shamai. Secrecy-achieving polar-coding. In *Information Theory Workshop (ITW), 2010 IEEE*, pages 1–5, 2010. doi: 10.1109/CIG.2010.5592878.

35 D. Karpuk, A.-M. Ernvall-Hytönen, C. Hollanti, and E. Viterbo. Probability estimates for fading and wiretap channels from ideal class zeta functions. *Advances in Mathematics of Communications*, 9(4):391–413, 2015.

36 D. Klinc, J. Ha, S. McLaughlin, J. Barros, and B. Kwak. LDPC codes for the Gaussian wiretap channel. *IEEE Transactions on Information Forensics and Security*, 6(3):532–540, 2011.

37 C. Ling, L. Luzzi, J.-C. Belfiore, and D. Stehlé. Semantically secure lattice codes for the Gaussian wiretap channel. *IEEE Transactions on Information Theory*, 60(10):6399–6416, Oct 2014.

38 L. Liu, Y. Yan, and C. Ling. Achieving secrecy capacity of the Gaussian wiretap channel with polar lattices. *IEEE Transactions on Information Theory*, 64(3):1647–1665, 2018.

39 L. Luzzi and M. Bloch. Capacity-based random codes cannot achieve strong secrecy over symmetric wiretap channels. In *1st International ICST Workshop on Secure Wireless Networks (Securenets)*, May 2011.

40 L. Luzzi, R. Vehkalahti, and C. Ling. Almost universal codes for MIMO wiretap channels. *IEEE Transactions on Information Theory*, 64(11):7218–7241, 2018.

41 H. Mahdavifar and A. Vardy. Achieving the secrecy capacity of wiretap channels using polar codes. In *IEEE International Symposium on Information Theory*, pages 913–917, 2010.

42 H. Mahdavifar and A. Vardy. Achieving the secrecy capacity of wiretap channels using polar codes. *IEEE Transactions on Information Theory*, 57(10):6428–6443, Oct 2011.

43 A. Nooraiepour and T. M. Duman. Randomized turbo codes for the wiretap channel. In *GLOBECOM 2017-2017 IEEE Global Communications Conference*, pages 1–6. IEEE, 2017.

44 A. Nooraiepour and T. M. Duman. Randomized serially concatenated LDGM codes for the Gaussian wiretap channel. *IEEE Communications Letters*, 22(4):680–683, 2018.

45 A. Nooraiepour, S. R. Aghdam, and T. M Duman. On secure communications over Gaussian wiretap channels via finite-length codes. *IEEE Communications Letters*, 24(9):1904–1908, 2020.

46 F. Oggier, P. Solé, and J.-C. Belfiore. Lattice codes for the wiretap Gaussian channel: Construction and analysis. *IEEE Transactions on Information Theory*, 62(10):5690–5708, Oct 2016.

47 S. S. Ong and F. Oggier. Wiretap lattice codes from number fields with no small norm elements. *Designs, Codes and Cryptography*, 73:425–440, 2014.

48 L. Ozarow and A. Wyner. Wire-tap channel II. *AT&T Bell Laboratories Technical Journal*, 63(10):2135–2157, 1984.

49 J. Pfister, M. A. C. Gomes, J. P. Vilela, and W. K. Harrison. Quantifying equivocation for finite blocklength wiretap codes. In *IEEE International Conference on Communications (ICC)*, pages 1–6, 2017.

50 Y. Polyanskiy, H. V. Poor, and S. Verdú. Channel coding rate in the finite blocklength regime. *IEEE Transactions on Information Theory*, 56(5):2307–2359, 2010.

51 V. Rana and R. A. Chou. Short blocklength wiretap channel codes via deep learning: Design and performance evaluation. *IEEE Transactions on Communications*, 71(3):1462–1474, 2023. doi: 10.1109/TCOMM.2023.3237259.

52 V. Rathi, R. Urbanke, M. Andersson, and M. Skoglund. Rate-equivocation optimal spatially coupled LDPC codes for the BEC wiretap channel. In *IEEE International Symposium on Information Theory Proceedings*, pages 2393–2397, 2011.

53 O. Regev. On lattices, learning with errors, random linear codes, and cryptography. *Journal of the ACM*, 56(6):34:1–34:40, 2009.

54 T. Richardson and R. Urbanke. Efficient encoding of low-density parity-check codes. *IEEE Transactions on Information Theory*, 47(2):638–656, 2001.

55 T. Richardson and R. Urbanke. *Modern Coding Theory*. Cambridge University Press, 2008.

56 E. Şaşoğlu and A. Vardy. A new polar coding scheme for strong security on wiretap channels. In *IEEE International Symposium on Information Theory*, pages 1117–1121, 2013.

57 M. Shakiba-Herfeh, L. Luzzi, and A. Chorti. Finite blocklength secrecy analysis of polar and Reed-Muller codes in BEC semi-deterministic wiretap channels. In *IEEE Information Theory Workshop (ITW)*, pages 1–6, 2021.

58 S. Sharifian, F. Lin, and R. Safavi-Naini. Hash-then-encode: A modular semantically secure wiretap code. In *Proceedings of the 2nd Workshop on Communication Security: Cryptography and Physical Layer Security 2*, pages 49–63. Springer, 2018.

59 A. Subramanian, A. Thangaraj, M. Bloch, and S. W. McLaughlin. Strong secrecy on the binary erasure wiretap channel using large-girth LDPC codes. *IEEE Transactions on Information Forensics and Security*, 6(3):585–594, 2011.

60 A. Suresh, A. Subramanian, A. Thangaraj, M. Bloch, and S. McLaughlin. Strong secrecy for erasure wiretap channels. In *IEEE Information Theory Workshop*, pages 1–5, 2010.

61 I. Tal and A. Vardy. Channel upgrading for semantically-secure encryption on wiretap channels. In *IEEE International Symposium on Information Theory*, pages 1561–1565, 2013.

62 K. Taleb and M. Benammar. On the information leakage of finite block-length wiretap polar codes. In *IEEE International Symposium on Information Theory (ISIT)*, pages 61–65, 2021. doi: 10.1109/ISIT45174.2021.9518180.

63 A. Thangaraj, S. Dihidar, R. Calderbank, S. McLaughlin, and J.-M. Merolla. Applications of LDPC codes to the wiretap channel. *IEEE Transactions on Information Theory*, 53(8):2933–2945, 2007.

64 H. Tyagi and A. Vardy. Explicit capacity-achieving coding scheme for the Gaussian wiretap channel. In *IEEE International Symposium on Information Theory*, pages 956–960, 2014.

65 H. Tyagi and A. Vardy. Semantically-secure coding scheme achieving the capacity of a Gaussian wiretap channel. *arXiv preprint arXiv:1412.4958*, 2014.

66 H. Tyagi and A. Vardy. Universal hashing for information-theoretic security. *Proceedings of the IEEE*, 103(10):1781–1795, 2015.

67 A. D. Wyner. The wire-tap channel. *Bell System Technical Journal*, 54:1355–1387, Oct 1975.

68 W. Yang, R. F. Schaefer, and H. V. Poor. Wiretap channels: Nonasymptotic fundamental limits. *IEEE Transactions on Information Theory*, 65(7):4069–4093, 2019.

Part II

Physical-Layer Security in Emerging Scenarios

3

Beamforming Design for Secure IRS-Assisted Multiuser MISO Systems

Dongfang Xu[1], Derrick Wing Kwan Ng[2], and Robert Schober[3]

[1]*Academy of Interdisciplinary Studies, The Hong Kong University of Science and Technology, Kowloon, Hong Kong, China*
[2]*School of Electrical Engineering and Telecommunications, University of New South Wales, Sydney, New South Wales, Australia*
[3]*Institute for Digital Communications, Friedrich-Alexander-University Erlangen Nürnberg, Erlangen, Germany*

3.1 Introduction

For almost a century, secure communication has been intensively studied when designing wireless networks [14]. Due to the free propagation property of radio frequency signals, the information transmitted by base stations (BSs) can be leaked to undesired users. While upper-layer encryption techniques, including data encoding, are widely used to facilitate secure communication, they have limitations. These techniques require codebooks to be stored in advance by the receivers, which becomes invalid once the codebooks are acquired by eavesdroppers. Additionally, new computing technologies can easily decipher the encryption, making these techniques obsolete. To overcome these limitations, researchers are exploring more flexible and effective means to improve security. By developing innovative solutions, we can ensure that our communication remains secure and protected from potential threats. In the literature, numerous works have investigated system design to enhance physical-layer security (PLS) [4, 25, 37]. The main goal of PLS is to utilize the wireless channel state information (CSI) available to ensure a positive secrecy rate at the intended user. However, this requires that the channel gain between the BS and the intended user is greater than that of the eavesdropper, which can be challenging to accomplish due to the unpredictable nature of channel fading. Although multiple antenna techniques can potentially reduce information signal leakage by employing beamforming, it still has its own limitations. In fact, since the radio propagation environment is

Physical-Layer Security for 6G, First Edition. Edited by Parthajit Mohapatra, Nikolaos Pappas, Arsenia Chorti, and Stefano Tomasin.

largely uncontrollable and the eavesdropper may intentionally create a favorable condition for wiretapping in the vicinity of legitimate users, the multiple antenna techniques may not always be effective in achieving PLS. As such, researchers are exploring alternative approaches to achieve PLS, taking into account the dynamic nature of wireless channels and potential threats. By developing innovative solutions, we can ensure that our communication remains secure and protected from potential eavesdropping attacks.

Last few years, the revolutionary intelligent reflecting surfaces (IRSs) have popped up as an effective lever to enhance PLS, see [10] and [21]. In particular, IRSs comprise a group of economic low-power consumption components (e.g., positive-intrinsic-negative diodes, varactor-based micro-strip patches, or integrated phase shifters) and each element can be flexibly tuned to enable the reflection of the incident signal toward an intended direction, see [11, 15], and [29]. Hence, by exploiting their programmability, IRSs can be adaptively configured to create a favorable wireless channel condition, which can potentially improve the system performance, see [42]. Additionally, IRS is resizable, allowing it to be seamlessly integrated into current wireless systems to expand service coverage, see [28]. Thus, we can smoothly deploy IRS into existing wireless networks and tune their reflection patterns to constructively combine desired signals at legitimate users and destructively mitigate informational signals at eavesdroppers. The aforementioned advantages have motivated numerous works to address the resource allocation design of IRS-enabled wireless networks, see, e.g., [16, 26, 27, 39], and [31]. For example, Wu and Zhang [26] initiated to employ a passive IRS in a wireless system to establish a high-quality reflected link for a desired user and studied the BS precoding and IRS reflection pattern optimization problem to achieve the maximum transmit rate for the focused user. Also, two novel polynomial-time suboptimal algorithms were designed in [39], where the throughput of a multiple-input single-output (MISO) wireless network is maximized exploiting an IRS. Moreover, a few works have attempted to employ IRS to achieve secure communication. For instance, a joint active BS beamforming and IRS reflection pattern design algorithm was developed in [43] to enhance the PLS in an MISO wireless network. In [22], an efficient BS precoding and IRS reflection configuration optimization framework was developed to guarantee PLS. Yet, the schemes developed in [22, 43], and [38] may not guarantee secure communication, especially when the eavesdropper enjoys much better channel conditions compared to the legitimate users. In fact, there is a more flexible and effective technique that can further enhance PLS, namely, artificial noise (AN). This technique involves intentionally generating noise that impairs the quality of eavesdroppers' channels and has been proposed in [19] as a means of improving PLS. The primary concept behind AN is to generate a noise signal and transmit it toward the eavesdroppers to intentionally impair their signal-to-noise ratio [17, 20]. As a result, by carefully designing the AN, we can effectively impair the quality of the channel of the eavesdropper without significantly impairing the legitimate users. In addition to the above observations, Shen et al. [22] and Yu et al. [38] adopted a simple wireless network comprising only one legitimate user which rarely happens in practice. Besides, Chen et al. [7] considered an IRS-assisted multiuser system and developed corresponding optimization framework for security provisioning. However, Chen et al. [7] proposed to replace the IRS-induced non-convex unit modulus constraint by an excessively relaxed convex constraint, which may result in infeasible solutions for the original problem and cause considerable performance loss in practical systems. Based on the above discussions, in this chapter, we concentrate on an IRS-enhanced multiuser wireless network and investigate the efficient IRS reflection configuration, AN, and downlink (DL) beamforming policy to achieve the maximum secrecy rate of the considered network.

This chapter is structured as follows: Section 3.2 introduces the IRS-enhanced multiuser system model. Section 3.3 presents the optimization problem, while Section 3.4 describes the resource allocation optimization algorithm. The proposed scheme's performance is evaluated through simulation in Section 3.5. Finally, Sections 3.6 and 3.7 provide conclusions and future research directions, respectively.

Notations: In this chapter, lowercase boldface letters represent vectors, while uppercase boldface letters represent matrices. The operation $[b]^+$ represents the maximum of zero and b. The function $\text{diag}(\mathbf{b})$ creates an N-dimensional diagonal matrix with the main diagonal elements being $b_1, \ldots, b_N$. The function $\Re\{\cdot\}$ extracts the real component of a complex-valued variable. The function $\mathcal{E}\{\cdot\}$ denotes the statistical expectation. Lastly, the notation $\nabla_{\mathbf{b}} f(\mathbf{b})$ represents the gradient of the function $f(\mathbf{b})$ with respect to a vector argument $\mathbf{b}$.

3.2 System Model

This chapter adopts a DL transmission scenario where a BS installed with $N_\text{A} > 1$ antennas is utilized to provide satisfactory information transmission for K single-antenna legitimate users while inhibiting the wiretapping of a single-antenna eavesdropper, cf. Figure 3.1. We consider unfavorable radio propagation conditions, where the line-of-sight components of both BS-to-legitimate users links and BS-to-eavesdropper links are absent because of blockage. To facilitate secure transmission, a passive IRS is applied to collaborate with the BS. Specifically, the IRS consists of M phase shift elements, which can be flexibly controlled and reconfigured via a controller. Furthermore, to reveal the theoretical system performance benchmark of the focused network, we follow

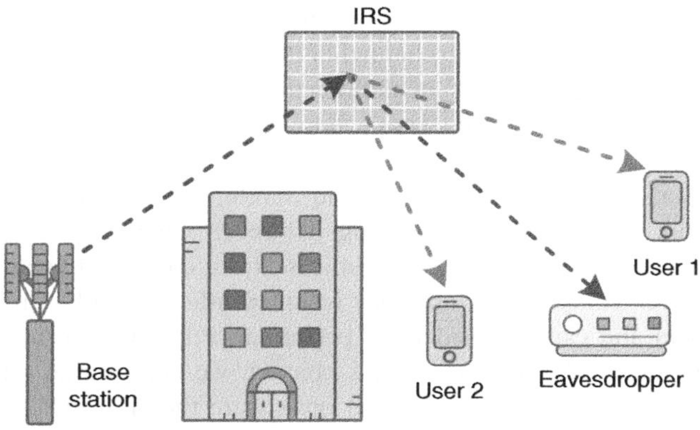

Figure 3.1 Illustration of the considered IRS-enhanced wireless networks involving two users and one eavesdropper.

the widely adopted assumption where the CSI of the network is accurately estimated and frequently updated by the BS.[1] Two sets, i.e., $\mathcal{M} = \{1, \dots, M\}$ and $\mathcal{K} = \{1, \dots, K\}$, are defined to indexify the elements of the IRS and the legitimate users, respectively.

A vector signal $\mathbf{x} \in \mathbb{C}^{N_A \times 1}$ is transmitted scheduling time slot by the BS. Specifically, $\mathbf{x}$ contains K independent data streams for all legitimate users and AN. Specifically, vector $\mathbf{x}$ is given by

$$\mathbf{x} = \sum_{k \in \mathcal{K}} \mathbf{w}_k b_k + \mathbf{s}. \tag{3.1}$$

Here, vector $\mathbf{w}_k \in \mathbb{C}^{N_A \times 1}$ denotes the beamformers dedicated to user k. Accordingly, scalar variable $b_k \in \mathbb{C}$ represents the data symbol and without loss of generality, we assume $\mathcal{E}\{|b_k|^2\} = 1$, $\forall k \in \mathcal{K}$, and $\mathcal{E}\{b_k^* b_r\} = 0$, $\forall k \neq r$. The AN signal $\mathbf{s} \in \mathbb{C}^{N_A \times 1}$ is produced and transmitted by the BS for security provisioning. In this chapter, vector $\mathbf{s}$ is characterized by a zero-mean CSCG random variable with a covariance matrix denoted by $\mathbf{S}$.

Via the reflected links established by the IRS, the received signal per time slot at legitimate user k and the eavesdropper are expressed as follows, respectively,

$$y_k = \mathbf{h}_k^H \boldsymbol{\Theta} \mathbf{G} \left(\sum_{k \in \mathcal{K}} \mathbf{w}_k b_k + \mathbf{s} \right) + n_k, \tag{3.2}$$

$$y_e = \mathbf{l}^H \boldsymbol{\Theta} \mathbf{G} \left(\sum_{k \in \mathcal{K}} \mathbf{w}_k b_k + \mathbf{s} \right) + n_e. \tag{3.3}$$

1 Extension to the case where imperfect CSI is available at the BS has been investigated in our work [41].

Here, vectors $\mathbf{h}_k \in \mathbb{C}^{M\times 1}$ and $\mathbf{l} \in \mathbb{C}^{M\times 1}$ characterize the IRS reflecting links of for user k and for the eavesdropper, respectively. Moreover, an M-dimensional diagonal matrix, i.e., $\boldsymbol{\Theta} = \text{diag}\left(e^{j\theta_1}, \ldots, e^{j\theta_M}\right)$, is adopted to characterize the IRS reflection pattern.[2] The main diagonal elements of $\boldsymbol{\Theta}$, i.e., the scalar $\theta_m, \forall m \in \mathcal{M}$, indicates the reflection coefficients of the mth phase shifters of the IRS, see [26]. The BS-IRS link is characterized by channel matrix $\mathbf{G} \in \mathbb{C}^{M\times N_A}$. Besides, CSGC random variables n_k and n_e represent the additive white Gaussian noises at legitimate user k and the eavesdropper with zero means and variances $\sigma_{n_k}^2$ and $\sigma_{n_e}^2$, respectively.

3.3 Resource Allocation Optimization Problem

This section presents the adopted secure communication performance metric and the optimization problem formulation for the focused network.

3.3.1 Performance Metrics of Secure Communication

The achievable rate (bits/s/Hz) of user is calculated based on k $R_k = \log_2(1 + \Gamma_k)$. Here, the signal-to-interference-plus-noise ratio (SINR) Γ_k is given by

$$\Gamma_k = \frac{\left|\mathbf{h}_k^H \boldsymbol{\Theta}\mathbf{G}\mathbf{w}_k\right|^2}{\sum\limits_{r\in\mathcal{K}\setminus\{k\}} \left|\mathbf{h}_k^H \boldsymbol{\Theta}\mathbf{G}\mathbf{w}_r\right|^2 + \text{Tr}(\mathbf{G}^H\boldsymbol{\Theta}^H\mathbf{h}_k\mathbf{h}_k^H\boldsymbol{\Theta}\mathbf{G}\mathbf{S}) + \sigma_{n_k}^2}. \tag{3.4}$$

Here, we impose a worst-case assumption for secure resource allocation algorithm design, see [40]. In particular, the eavesdropper is assumed to have the ability in canceling the undesired multiuser interference while keeping the desired signal for information decoding. As a result, assuming the eavesdropper wiretaps user k, the corresponding channel capacity is given by [35]

$$C_k^{\text{E}} = \log_2\left(1 + \frac{\left|\mathbf{l}^H\boldsymbol{\Theta}\mathbf{G}\mathbf{w}_k\right|^2}{\text{Tr}(\mathbf{G}^H\boldsymbol{\Theta}^H\mathbf{l}\mathbf{l}^H\boldsymbol{\Theta}\mathbf{G}\mathbf{S}) + \sigma_{n_e}^2}\right). \tag{3.5}$$

Hence, for user k, we calculate the achievable secrecy rate based on $R_k^{\text{Sec}} = \left[R_k - C_k^{\text{E}}\right]^+$ [9]. Note that by imposing a non-negative achievable secrecy rate for each user, we can guarantee the PLS of the considered system.

2 In this chapter, we adopt the passive IRS model where each IRS element has a unit gain. In fact, integrated with an additional reflection-type amplifier, a novel IRS structure, namely, active IRSs, is capable of amplifying the incident signal when powered by an energy supply. The corresponding active IRS design for achieving green communication is studied in our works [33] and [44].

3.3.2 Problem Formulation

The design objective is to promote the SSR of the focused network. Specifically, we jointly optimize the IRS reflection configuration $\boldsymbol{\Theta}$, beamformers $\mathbf{w}_k$, and AN covariance matrix $\mathbf{S}$. As such, we target to tackle the following optimization problem:

$$\begin{aligned}
\underset{\mathbf{S}\in\mathbb{H}^{N_A},\mathbf{w}_k,\boldsymbol{\Theta}}{\text{maximize}} \quad & \sum_{k\in\mathcal{K}} \left[R_k - C_k^{\mathrm{E}}\right]^+ \\
\text{subject to} \quad & \text{C1: } \sum_{k\in\mathcal{K}} \|\mathbf{w}_k\|^2 + \mathrm{Tr}(\mathbf{S}) \le P_{\max}, \\
& \text{C2: } \mathbf{S} \succeq \mathbf{0}, \\
& \text{C3: } \left|[\boldsymbol{\Theta}]_{m,m}\right| = 1, \forall m \in \mathcal{M}.
\end{aligned} \tag{3.6}$$

In this context, constraint C1 restricts the maximum BS transmit power to $P_{\max}$, while $\mathbf{S} \in \mathbb{H}^{N_A}$ and constraint C2 are applied to ensure that $\mathbf{S}$ is a covariance matrix. Additionally, constraint C3 enforces the unit-modulus constraint for each element of the considered passive IRS. Furthermore, if a user has a nonpositive secrecy rate, the corresponding information transmission should be terminated. As such, the operator $[\cdot]^+$ in (3.6) can be omitted while preserving the optimality.

The coupling between $\mathbf{S}$, $\mathbf{w}_k$, and $\boldsymbol{\Theta}$ and the non-convexity originated from the unit-modulus constraint C3 make it very challenging to optimally solve (3.6) in polynomial time. As a compromise, in the next section, we propose an optimization framework by exploiting AO theory and obtain a suboptimal solution of (3.6), see [3].

3.4 Solution of the Optimization Problem

This section expounds on the details of the AO-based low-complexity suboptimal algorithm for handling (3.6). To enable the application of AO, we commence by restructuring the initial non-convex optimization problem into a more manageable and equivalent form. We then break down the resulting problem into two subproblems, which we address using successive convex approximation (SCA) and manifold optimization methods, respectively. This optimization framework allows us to efficiently tackle the problem and achieve satisfactory performance.

3.4.1 Problem Reformulation

To start with, we define the equivalent channel matrices $\mathbf{H}_k = \mathrm{diag}(\mathbf{h}_k^H)\mathbf{G}$, $\mathbf{L} = \mathrm{diag}(\mathbf{l}^H)\mathbf{G}$, and the beamforming matrix $\mathbf{W}_k = \mathbf{w}_k\mathbf{w}_k^H$. Accordingly, we represent the matrix $\boldsymbol{\Theta}$ as vector variable $\mathbf{p} = \left[e^{j\theta_1}, \dots, e^{j\theta_M}\right]^H$ to equivalently characterize the reflection configuration of the IRS. As such, the received SINR at user k is expressed as

$$\Gamma_k = \frac{\mathrm{Tr}(\mathbf{W}_k\mathbf{H}_k^H\mathbf{p}\mathbf{p}^H\mathbf{H}_k)}{\sum\limits_{r\in\mathcal{K}\setminus\{k\}} \mathrm{Tr}(\mathbf{W}_r\mathbf{H}_k^H\mathbf{p}\mathbf{p}^H\mathbf{H}_k) + \mathrm{Tr}(\mathbf{S}\mathbf{H}_k^H\mathbf{p}\mathbf{p}^H\mathbf{H}_k) + \sigma_{n_k}^2}. \tag{3.7}$$

Also, C_k^{E} in (3.5) can be rewritten as

$$C_k^{\mathrm{E}} = \log_2\left(1 + \frac{\mathrm{Tr}(\mathbf{W}_k\mathbf{L}^H\mathbf{p}\mathbf{p}^H\mathbf{L})}{\mathrm{Tr}(\mathbf{S}\mathbf{L}^H\mathbf{p}\mathbf{p}^H\mathbf{L}) + \sigma_{n_e}^2}\right). \tag{3.8}$$

Next, to facilitate efficient algorithm design, we reformulate (3.6) into the following equivalent form:

$$\begin{aligned}
&\underset{\mathbf{W}_k,\mathbf{S}\in\mathbb{H}^{N_A},\mathbf{p}}{\text{minimize}} && F \triangleq g_1 + g_2 - h_1 - h_2\\
&\text{subject to} && \mathrm{C1}: \mathrm{Tr}(\mathbf{S}) + \sum_{k\in\mathcal{K}} \mathrm{Tr}(\mathbf{W}_k) \le P_{\max},\\
&&& \mathrm{C2}: \mathbf{S} \succeq \mathbf{0},\\
&&& \mathrm{C3}: |p_m| = 1,\ \forall m \in \mathcal{M},\\
&&& \mathrm{C4}: \mathbf{W}_k \succeq \mathbf{0},\ \forall k \in \mathcal{K},\\
&&& \mathrm{C5}: \mathrm{rank}(\mathbf{W}_k) \le 1,\ \forall k \in \mathcal{K}.
\end{aligned} \tag{3.9}$$

Here, the complex variable p_m in constraint C3 represents the mth component of $\mathbf{p}$. Moreover, two additional constraints, i.e., C4 and C5, are imposed on the problem formulation to ensure the successful recovery of the vector $\mathbf{w}_k$ after obtaining $\mathbf{W}_k$. Furthermore, the objective function F in (3.9) is equivalently expressed as the linear combination of functions g_1, g_2, h_1, and h_2, which are defined as follows, respectively,

$$g_1 = -\sum_{k\in\mathcal{K}} \log_2\left(\sum_{r\in\mathcal{K}} \mathrm{Tr}(\mathbf{W}_r\mathbf{H}_k^H\mathbf{p}\mathbf{p}^H\mathbf{H}_k) + \mathrm{Tr}(\mathbf{S}\mathbf{H}_k^H\mathbf{p}\mathbf{p}^H\mathbf{H}_k) + \sigma_{n_k}^2\right), \tag{3.10}$$

$$g_2 = -K\log_2\left(\mathrm{Tr}(\mathbf{S}\mathbf{L}^H\mathbf{p}\mathbf{p}^H\mathbf{L}) + \sigma_{n_e}^2\right), \tag{3.11}$$

$$h_1 = -\sum_{k\in\mathcal{K}} \log_2\left(\sum_{r\in\mathcal{K}\setminus\{k\}} \mathrm{Tr}(\mathbf{W}_r\mathbf{H}_k^H\mathbf{p}\mathbf{p}^H\mathbf{H}_k) + \mathrm{Tr}(\mathbf{S}\mathbf{H}_k^H\mathbf{p}\mathbf{p}^H\mathbf{H}_k) + \sigma_{n_k}^2\right), \tag{3.12}$$

$$h_2 = -\sum_{k\in\mathcal{K}} \log_2\left(\mathrm{Tr}(\mathbf{W}_k\mathbf{L}^H\mathbf{p}\mathbf{p}^H\mathbf{L}) + \mathrm{Tr}(\mathbf{S}\mathbf{L}^H\mathbf{p}\mathbf{p}^H\mathbf{L}) + \sigma_{n_e}^2\right). \tag{3.13}$$

By exploiting AO theory, see [3], we construct two optimization variable blocks, i.e., $\{\mathbf{W}_k, \mathbf{S}\}$ and $\{\mathbf{p}\}$, each of which is associated with a subproblem. Then, we iteratively optimize the two subproblems associated respectively with two blocks in an alternating manner until convergence. Specifically, for a given $\mathbf{p}$, we solve the corresponding subproblem by capitalizing on SCA, see [12], and semidefinite relaxation (SDR), see [30]. Subsequently, by fixing $\mathbf{W}_k$ and $\mathbf{S}$, we solve the resulting subproblem for $\mathbf{p}$ by exploiting manifold optimization theory, see [1].

3.4.2 Successive Convex Approximation

By fixing optimization variable $\mathbf{p}$, we can rewrite (3.9) as the following subproblem:

$$\begin{aligned} &\underset{\mathbf{W}_k, \mathbf{S} \in \mathbb{H}^{N_A}}{\text{minimize}} \quad g_1 + g_2 - h_1 - h_2 \\ &\text{subject to} \quad \text{C1, C2, C4, C5.} \end{aligned} \tag{3.14}$$

The term $g_1 + g_2 - h_1 - h_2$ of (3.14) can be interpreted as the difference of two convex functions. We note that such a form can be efficiently tackled by SCA. As a result, we first construct global underestimators for h_1 and h_2 by employing the first-order Taylor expansion, see [12]. Specifically, given any feasible points $\mathbf{W}_k^t$ and $\mathbf{S}^t$, we can derive an inequality for the differentiable convex function $h_1(\mathbf{W}_k, \mathbf{S})$ as follows[3]:

$$\begin{aligned} h_1(\mathbf{W}_k, \mathbf{S}) \geq\ & h_1(\mathbf{W}_k^t, \mathbf{S}^t) \\ & + \sum_{k \in \mathcal{K}} \text{Tr}\left(\left(\nabla_{\mathbf{W}_k} h_1(\mathbf{W}_k^t, \mathbf{S}^t) \right)^H (\mathbf{W}_k - \mathbf{W}_k^t) \right) \\ & + \text{Tr}\left((\nabla_{\mathbf{S}} h_1(\mathbf{W}_k^t, \mathbf{S}^t))^H (\mathbf{S} - \mathbf{S}^t) \right) \\ \triangleq\ & \widetilde{h}_1(\mathbf{W}_k, \mathbf{S}, \mathbf{W}_k^t, \mathbf{S}^t). \end{aligned} \tag{3.15}$$

The function $\widetilde{h}_1(\mathbf{W}_k, \mathbf{S}, \mathbf{W}_k^t, \mathbf{S}^t)$ in (3.15) is a global underestimator of function $h_1(\mathbf{W}_k, \mathbf{S})$ [36]. Also, a global underestimator for function $h_2(\mathbf{W}_k, \mathbf{S})$ can be derived at the feasible point $\mathbf{W}_k^t$ and $\mathbf{S}^t$ in a similar manner as follows:

$$\begin{aligned} \widetilde{h}_2(\mathbf{W}_k, \mathbf{S}, \mathbf{W}_k^t, \mathbf{S}^t) \triangleq\ & h_2(\mathbf{W}_k^t, \mathbf{S}^t) \\ & + \sum_{k \in \mathcal{K}} \text{Tr}\left(\left(\nabla_{\mathbf{W}_k} h_2(\mathbf{W}_k^t, \mathbf{S}^t) \right)^H (\mathbf{W}_k - \mathbf{W}_k^t) \right) \\ & + \text{Tr}\left((\nabla_{\mathbf{S}} h_2(\mathbf{W}_k^t, \mathbf{S}^t))^H (\mathbf{S} - \mathbf{S}^t) \right). \end{aligned} \tag{3.16}$$

3 The superscript t indicates the iteration index of the SCA algorithm.

Thus, an upper bound of the value of $g_1 + g_2 - h_1 - h_2$ can be acquired when tackling the problem below

$$\begin{aligned} \underset{\mathbf{W}_k, \mathbf{S} \in \mathbb{H}^{N_A}}{\text{minimize}} \quad & r(\mathbf{W}_k, \mathbf{S}) \triangleq g_1 + g_2 - \widetilde{h}_1 - \widetilde{h}_2 \\ \text{subject to} \quad & \text{C1, C2, C4, C5.} \end{aligned} \tag{3.17}$$

Now, the only obstacle to efficiently solve problem (3.17) is the challenging rank constraint C5. For handing this obstacle, we apply SDR to eliminate constraint C5. By relaxing the rank constraint in (3.17), we can solve the problem optimally in polynomial time using convex solvers like CVX [13] and YALMIP [18]. In the following, we present a theorem to verify the optimality of SDR.

Theorem 3.1 We can always obtain an optimal $\mathbf{W}_k$ where $\mathbf{W}_k$ can be equivalently represented as $\mathbf{w}_k \mathbf{w}_k^H$.

Proof: The details of the proof can be found in the Appendix of this chapter. □

Algorithm 3.1 illustrates the key steps of the proposed SCA algorithm. We recall that the minimum objective function value of (3.17) serves as an upper bound of (3.14). As such, by continuously solving (3.17) over iterations, we can constantly narrow the gap between the objective value of (3.14) and the upper bound, producing a locally optimal solution $\mathbf{W}_k$ and $\mathbf{S}$ for (3.14), see [12].

Algorithm 3.1 Successive convex approximation-based algorithm

```
Initialize the algorithm by setting loop index t = 1 and convergence tolerance factor ε_SCA ≪ 1. The optimization variables W_k^1 and S^1 are initialized by randomly and uniformly producing their components from the interval (0, P_max/((K+1)N_A)).
repeat
   Solve problem (3.17) with W_k^t and S^t
   Update t = t + 1 and W_k^t = W_k and S^t = S
until |r(W_k^t, S^t) − r(W_k^{t−1}, S^{t−1})| / r(W_k^{t−1}, S^{t−1}) ≤ ε_SCA, where r(·,·) is defined in (3.17)
W_k^* = W_k^t and S^* = S^t
```

3.4.3 Complex Circle Optimization

For given fixed $\mathbf{W}_k$ and $\mathbf{S}$ obtained with Algorithm 3.1, the optimization problem in (3.9) can be equivalently recast as follows:

$$\begin{aligned} \underset{\mathbf{p}}{\text{minimize}} \quad & f(\mathbf{p}) \triangleq g_1 + g_2 - h_1 - h_2 \\ \text{subject to} \quad & \text{C3: } |p_m| = 1, \ \forall m. \end{aligned} \tag{3.18}$$

The unit modulus constraint C3 is an obstacle for efficiently obtaining the optimal solution to (3.18). To overcome this obstacle, the unit modulus constraint is usually relaxed to the convex hull of the original non-convex set, see [7]. Yet, this method is very likely to result in an infeasible solution to the original problem with the unit modulus constraint. Alternatively, one can successively employ SDR and Gaussian randomization, e.g., [26]. However, this approach may cause divergence of the algorithm as the objective function does not preserve monotonicity over different iterations. To address this issue, we exploit manifold optimization theory and propose an effective algorithm that provides a high-quality feasible solution while ensuring guaranteed convergence. This approach is motivated by the observations made and allows us to achieve optimal results efficiently.

According to Riemannian manifold optimization theory, see [1], constraint C3 defines a complex circle manifold that can be characterized as follows:

$$\mathcal{O} = \left\{ \mathbf{p} \in \mathbb{C}^M \,\middle|\, |p_1| = |p_2| = \cdots = |p_m| = 1 \right\}. \tag{3.19}$$

It is important to note that any vector $\mathbf{p}$ optimized on the complex circle manifold automatically satisfies constraint C3. Specifically, the search space is a Riemannian submanifold of $\mathbb{C}^M$. To develop the corresponding algorithm, we introduce some mathematical preliminaries necessary for manifold optimization. This allows us to efficiently optimize the vectors and achieve optimal results.

3.4.3.1 Tangent Space

Given a point $\mathbf{p}_j$ on a complex circle manifold $\mathcal{O}$, the corresponding tangent space can be regarded as a set that comprises all tangent vectors of the complex circle manifold $\mathcal{O}$ originating from $\mathbf{p}_j$. A toy example of the tangent space in the complex plane $\mathbb{C}$, i.e., $M = 1$, is demonstrated in Figure 3.2. As such, the tangent space for $\mathcal{O}$ at vector $\mathbf{p}_j$ can be expressed [[1], Sec. 3.5]

$$T_{\mathbf{p}_j}\mathcal{O} = \left\{ \mathbf{v} \in \mathbb{C}^M \,\middle|\, \Re\left\{ \mathbf{v} \circ \mathbf{p}_j^* \right\} = \mathbf{0}_M \right\}, \tag{3.20}$$

where $\mathbf{v}$ is a tangent vector for point $\mathbf{p}_j$.

3.4.3.2 Riemannian Gradient

Among all elements in set $T_{\mathbf{p}_j}\mathcal{O}$, we select the one associated with the fastest decrease of the function value at the Riemannian space gradient on $\mathcal{O}$, referred to as $\text{grad}_{\mathbf{p}_j} f$. Accordingly, the value of $\text{grad}_{\mathbf{p}_j} f$ can be computed orthogonally projecting the Euclidean gradient $\nabla_{\mathbf{p}_j} f$ onto the set $T_{\mathbf{p}_j}\mathcal{O}$, see [[1], Sec. 3.6.1]. Specifically, $\text{grad}_{\mathbf{p}_j} f$ is given by [[1], Sec. 3.6]

$$\text{grad}_{\mathbf{p}_j} f = \nabla_{\mathbf{p}_j} f - \Re\left\{ \nabla_{\mathbf{p}_j} f \circ \mathbf{p}_j^* \right\} \circ \mathbf{p}_j, \tag{3.21}$$

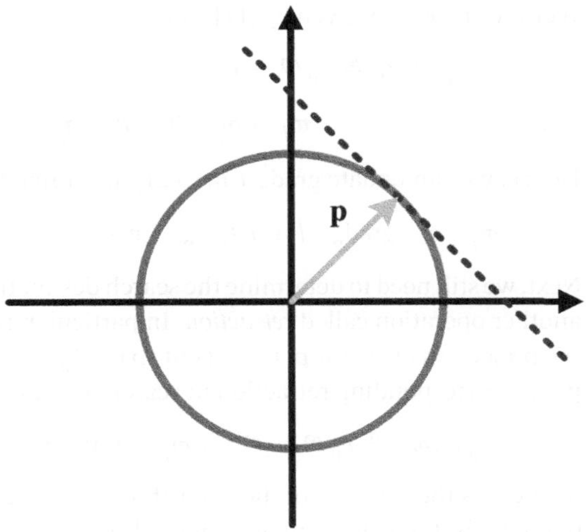

Figure 3.2 Illustration of the tangent space for a point **p** at a unit circle in the complex space ℂ. The circle denotes the unit circle in ℂ. The dashed line represents the tangent space with respect to **p** (denoted by the arrow).

where the gradient $\nabla_{\mathbf{p}_j} f$ is given by, see [6]

$$\nabla_{\mathbf{p}_j} f = -\frac{\sum_{k\in\mathcal{K}} \mathbf{L}(\mathbf{W}_k^H + \mathbf{W}_k + \mathbf{S}^H + \mathbf{S})\mathbf{L}^H\mathbf{p}_j}{(\ln 2)h_2(\mathbf{p}_j)} + \frac{K\mathbf{L}(\mathbf{S}^H + \mathbf{S})\mathbf{L}^H\mathbf{p}_j}{(\ln 2)g_2(\mathbf{p}_j)}$$
$$-\frac{\sum_{k\in\mathcal{K}}\sum_{r\in\mathcal{K}\setminus\{k\}} \left[\mathbf{H}_k(\mathbf{W}_r^H + \mathbf{W}_r + \mathbf{S}^H + \mathbf{S})\mathbf{H}_k^H\mathbf{p}_j\right]}{(\ln 2)h_1(\mathbf{p}_j)}$$
$$+\frac{\sum_{k\in\mathcal{K}}\sum_{r\in\mathcal{K}} \left[\mathbf{H}_k(\mathbf{W}_r^H + \mathbf{W}_r + \mathbf{S}^H + \mathbf{S})\mathbf{H}_k^H\mathbf{p}_j\right]}{(\ln 2)g_1(\mathbf{p}_j)}. \tag{3.22}$$

We borrow the Euclidean space-oriented optimization methods to solve the formulated optimization problem into Riemannian space after obtaining the gradient $\mathrm{grad}_{\mathbf{p}_j} f$. To update the next search direction in Euclidean space, we use the conjugate gradient (CG) method, see [2], as demonstrated below:

$$\varpi_{j+1} = -\nabla_{\mathbf{p}_{j+1}} f + \alpha_j \varpi_j. \tag{3.23}$$

Here, variable ϖ_j indicates the search direction of function f at point $\mathbf{p}_j$. Moreover, scalar α_j refers to the Polak–Ribiére parameter which facilitates a fast convergence speed of the algorithm, see [2]. Yet, because ϖ_j and ϖ_{j+1} in (3.23) in general belong to two different tangent spaces $T_{\mathbf{p}_j}\mathcal{O}$ and $T_{\mathbf{p}_{j+1}}\mathcal{O}$, we cannot integrate them directly in Riemannian space. To overcome this difficulty, we exploit an operation referred to as *transport* which establishes a mapping for ϖ_j between $T_{\mathbf{p}_j}\mathcal{O}$ and $T_{\mathbf{p}_{j+1}}\mathcal{O}$. Specifically, for complex circle manifold $\mathcal{O}$, ϖ_j under the operation of vector transport, is given by [[1], Sec. 5.4]

$$\mathcal{T}_{\mathbf{p}_j \to \mathbf{p}_{j+1}}(\boldsymbol{\varpi}_j) \triangleq T_{\mathbf{p}_j}\mathcal{O} \to T_{\mathbf{p}_{j+1}}\mathcal{O} :$$
$$\boldsymbol{\varpi}_j \to \boldsymbol{\varpi}_j - \Re\left\{\boldsymbol{\varpi}_j \circ \mathbf{p}^*_{j+1}\right\} \circ \mathbf{p}_{j+1}. \tag{3.24}$$

Hence, we can update $\text{grad}_{\mathbf{p}_j} f$ in (3.21) according to the following equation:

$$\boldsymbol{\varpi}_{j+1} = -\text{grad}_{\mathbf{p}_{j+1}} f + \alpha_j \mathcal{T}_{\mathbf{p}_j \to \mathbf{p}_{j+1}}(\boldsymbol{\varpi}_j). \tag{3.25}$$

Next, we still need to determine the search destination on $\mathcal{O}$. To this end, we exploit another operation called *retraction*. In particular, retraction allows us to ascertain the projection of vector $\mathbf{p}_j$ in tangent space $T_{\mathbf{p}_j}\mathcal{O}$ onto manifold $\mathcal{O}$. Specifically, for $\mathbf{p}_j$, the corresponding retraction in search direction $\boldsymbol{\varpi}_j$ is given as

$$\mathcal{R}_{\mathbf{p}_j}(\delta_j \boldsymbol{\varpi}_j) \triangleq T_{\mathbf{p}_j}\mathcal{O} \to \mathcal{O} : \delta_j \boldsymbol{\varpi}_j \to \text{unt}(\delta_j \boldsymbol{\varpi}_j), \tag{3.26}$$

where δ_j is the step size of the algorithm. For more basic properties and computational tools, the readers might refer to [5].

Employing Algorithm 3.2, the problem (3.18) can be tackled in polynomial time. Considering the fact that Algorithm 3.2 exploits the gradient in each iteration, the objective function f in (3.18) is guaranteed to constantly decreasing. It has been shown in [2] that Algorithm 3.2 can always produce a stationary point of (3.18) with guaranteed convergence.

Algorithm 3.2 Complex circle manifold optimization-based algorithm

1: Initialize convergence tolerance $\epsilon_{\text{MO}} \ll 1$ and starting point $\mathbf{p}_j$, set loop index $j = 1$
2: **repeat**
3: Compute $\text{grad}_{\mathbf{p}_j} f$ based on (3.21)
4: Obtain the step interval δ_j according to [1]
5: Implement retraction in (3.26) and update $\mathbf{p}_{j+1}$
6: Update $\text{grad}_{\mathbf{p}_{j+1}} f$ according to (3.21)
7: Compute $\mathcal{T}_{\mathbf{p}_j \to \mathbf{p}_{j+1}}(\boldsymbol{\varpi}_j)$ according to (3.24)
8: Obtain α_j by using (Eq. 8.24) in [1]
9: Update $\boldsymbol{\varpi}_{j+1}$ according to (3.25) and the loop index $j = j + 1$
10: **until** $\left\|\text{grad}_{\mathbf{p}_j} f\right\|_2 \leq \epsilon_{\text{MO}}$
11: Recover $\boldsymbol{\Phi}$ based on $\boldsymbol{\Phi} = \text{diag}\left((\mathbf{p}_j^T)^H\right)$

Algorithm 3.3 describes the key steps for solving (3.9). For both Algorithms 3.1 and 3.2, the objective functions preserve monotonicity over iterations. As such, the overall algorithm produces a suboptimal solution of (3.9) with guaranteed convergence, see [32]. The flow of the overall algorithm is demonstrated in Figure 3.3. Moreover, the worst-case complexities of Algorithms 3.1 and 3.2 can

Algorithm 3.3 Overall algorithm

1: Generate the initial point $\mathbf{W}_k^{(t)}$, $\mathbf{S}^{(t)}$, and $\mathbf{p}^{(t)}$, set algorithm termination factor $\epsilon_{\text{AO}} \ll 1$, and loop index $t = 1$.
2: **repeat**
3: Tackle (3.14) by employing **Algorithm 3.1** with $\mathbf{p}^{(t)}$ and obtain $\mathbf{W}_k^{(t+1)}$ and $\mathbf{S}^{(t+1)}$
4: Solve (3.18) via **Algorithm 3.2** with $\mathbf{W}_k = \mathbf{W}_k^{(t+1)}$ and $\mathbf{S} = \mathbf{S}^{(t+1)}$ and update $\mathbf{p}^{(t+1)}$ and loop index $t = t + 1$
5: **until** $\frac{|F^{(t)} - F^{(t-1)}|}{F^{(t-1)}} \leq \epsilon_{\text{AO}}$, where F is defined in (3.9)
6: Acquire $\mathbf{W}_k^* = \mathbf{W}_k^{(t)}$, $\mathbf{S}^* = \mathbf{S}^{(t)}$, and $\mathbf{p}^* = \mathbf{p}^{(t)}$

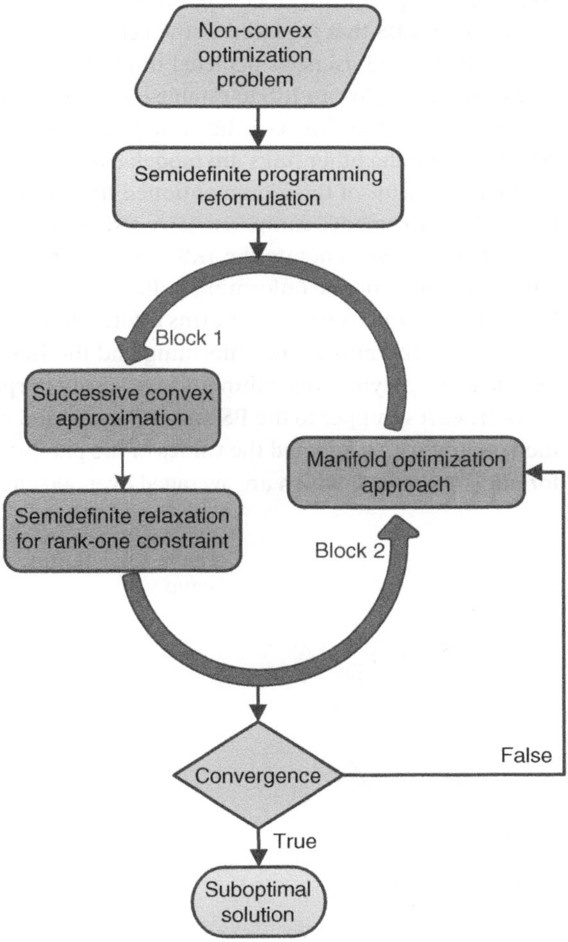

Figure 3.3 Illustration of the flow and the key optimization steps of the overall algorithm.

be calculated based on $C_1 = \mathcal{O}\left(\log_2(\frac{1}{\epsilon_{\text{SCA}}})\left((K+1)N_{\text{A}}^3 + (K+1)^2N_{\text{A}}^2 + (K+1)^3\right)\right)$, see [8], and $C_2 = \mathcal{O}\left(\log_2(\frac{1}{\epsilon_{\text{MO}}})M^{\frac{3}{2}}\right)$, see [24], respectively, where $\mathcal{O}(\cdot)$ is the big-O notation. Therefore, the worst-case computational complexity of Algorithm 3.3 can be expressed as $\mathcal{O}\left(\log(1/\epsilon_{\text{AO}})\right)(C_1 + C_2)$.

3.5 Experimental Results

In this section, we assess the performance of the proposed scheme via computer simulations. Specifically, focusing on a 120° sector of a cell of a 500-m radius, we consider a BS that is placed at the cell's center to serve K randomly and uniformly distributed users, as depicted in Figure 3.4. Since the sector is blocked by buildings, we deploy an IRS to establish reflecting links for the users. We characterize the BS-to-IRS link as a Rician fading channel with the Rician factor being set as 3, while the other links are modeled as Rayleigh fading channels. Also, the attenuation factor of the aforementioned links is assumed to be 2 and 3, respectively. For performance comparison, we also study two baseline schemes. For baseline scheme 1, we generate the IRS reflection pattern in a random manner and jointly optimize the beamforming policy using the SDR method [34]. For baseline scheme 2, the BS barely performs information beamforming without injecting AN, and the BS transmit beamforming and the IRS reflection pattern are jointly designed by applying Algorithm 3.3 for security purposes. We denote the distance from the eavesdropper to the BS and IRS by r_{Be} and r_{Re}, respectively. Unless specified otherwise, we adopted the values of the parameters summarized in Table 3.1 for our simulations, which are averaged over 500 channel realizations.

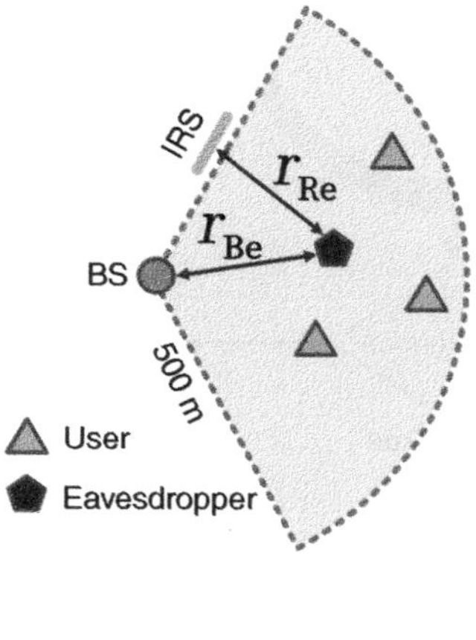

Figure 3.4 Illustration of the adopted simulation setup.

Table 3.1 System parameters.

System bandwidth	200 kHz
Noise power at the users, $\sigma_{n_k}^2$	−110 dBm
Noise power at the eavesdropper, $\sigma_{n_e}^2$	−110 dBm
Maximum BS transmit power allowance, $P_{\max}$	40 dBm
Algorithm termination factors, ϵ_{SCA}, ϵ_{MO}, and ϵ_{AO}	10^{-3}

3.5.1 Average SSR Versus BS Power Budget

Figure 3.5 displays the average SSR versus the maximum BS transmit power for different schemes. In particular, it can be seen that all schemes achieve higher system SSRs as $P_{\max}$ increases, since a higher transmit power budget allows for more flexible power allocation. Notably, the proposed scheme shows significantly better performance than the two baseline schemes. The joint design of $\mathbf{w}_k$, $\mathbf{S}$, and $\mathbf{\Theta}$ in the proposed scheme can customize appropriate wireless channel conditions

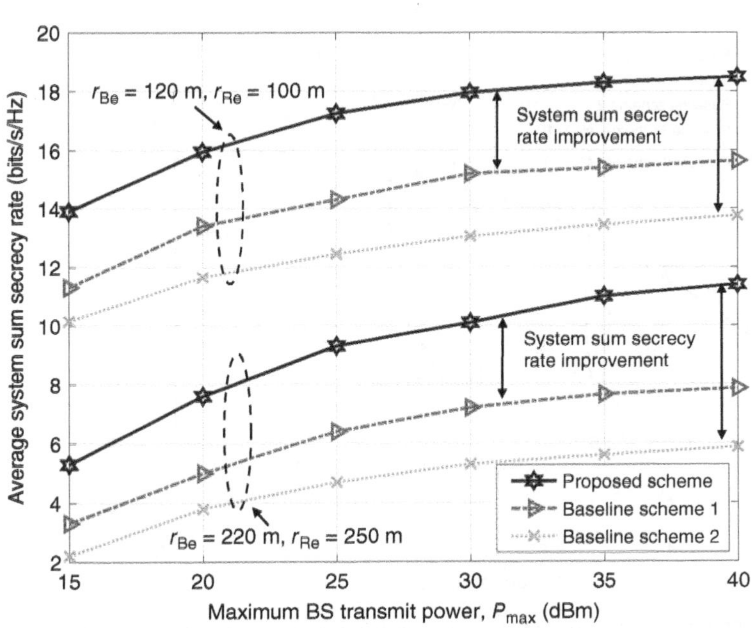

Figure 3.5 System SSR (bits/s/Hz) versus the BS maximum transmit power (dBm) averaged over different channel realizations with $N_A = 6$, $M = 6$, and $K = 3$.

for the legitimate users while degrading the wiretapping rate. In comparison, the two baseline schemes sacrifice performance to facilitate simpler implementations. Additionally, Figure 3.5 highlights that the system security is considerably affected by the network geometry, i.e., r_{Be} and r_{Re}, emphasizing the importance of employing a suitable IRS deployment policy based on the application scenario and system design objective.

3.5.2 Average SSR Versus Number of Legitimate Users

In Figure 3.6, we demonstrate the SSR versus the cardinality of legitimate users. We notice that all considered schemes achieve higher system SSRs when there are more legitimate users in the system, benefiting from the exploitation of multiuser diversity and optimized beamforming policies. To demonstrate the advantages of deploying IRSs in wireless networks, we also assess the effectiveness of the developed scheme for different setups in Figure 3.6: Case 1 ($M = 6$ and $N_{\text{A}} = 10$) and Case 2 ($M = 10$ and $N_{\text{A}} = 6$). We observe that Case 2 achieves a higher SSR than Case 1. This can be attributed to two factors. First, a larger IRS can reflect more

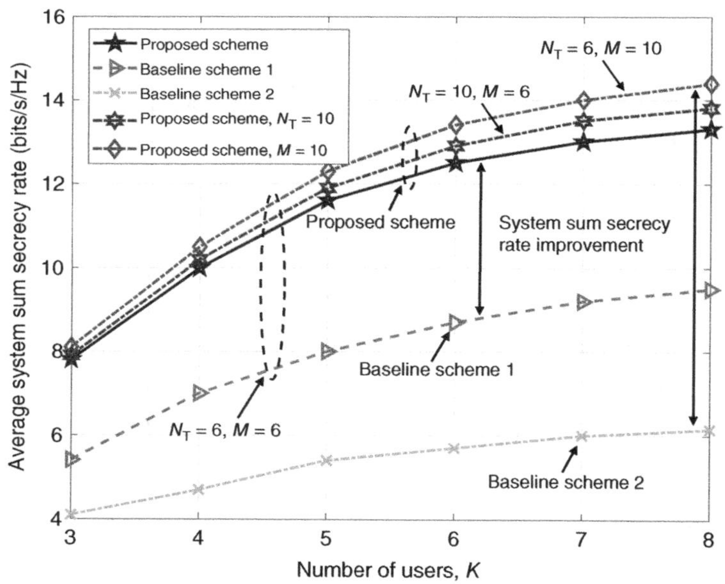

Figure 3.6 SSR (bits/s/Hz) versus the number of users averaged over different channel realizations and the geometry of the network being $r_{\text{Be}} = 200$ m and $r_{\text{Re}} = 250$ m.

signal power from the BS to the legitimate users, resulting in a higher power gain. Second, the inclusion of extra IRS elements offer greater versatility in assigning resources, resulting in an improved beamforming gain for the intended reflected connections.

3.6 Conclusion

This chapter presents an algorithm for resource allocation that effectively enhances the PLS in multiuser MISO systems with the aid of IRSs. To further improve secure communications, we have incorporated appropriately designed AN into the data stream transmission from the BS. The joint design of both BS and IRS was studied by solving an optimization problem, where the SSR of the considered system was maximized. To efficiently address the formulated non-convex optimization problem, we have utilized AO theory and developed a two-layer optimization framework that delivers a suboptimal solution. The experimental results demonstrate that the proposed scheme is a valid means to enhance the PLS compared to the two low-complexity baseline schemes. Furthermore, our findings validate that integrating IRS and AN is a predominant approach to promote the PLS of real-world wireless networks.

3.7 Future Extension

This chapter has focused on demonstrating the effectiveness of the proposed optimization framework for security provisioning in a system with a single eavesdropper, assuming perfect CSI of the wireless network is obtainable. However, the developed optimization algorithm can be modified to handle a more comprehensive and practical scenario where there are multiple eavesdroppers and their CSI is only partially known at the BS, as shown in [40]. Another model, the impedance network (INW)-based IRS model, assumes that all IRS elements can be jointly controlled via an impedance network, as proposed in [23]. Diverse models of IRSs often result in distinct constraints, which consequently give rise to various types of optimization problems for resource allocation. In order to thoroughly unloose the capabilities of IRSs and ensure secure communication in practical wireless networks, it is imperative to create a comprehensive optimization framework that can accommodate different models of IRSs. These observations highlight the importance of further research in this area to develop more sophisticated and adaptable optimization frameworks.

3.A Appendix – Proof of Theorem 3.1

We start by reformulating the rank constraint relaxed version of (3.17) equivalently as follows:

$$\begin{aligned}
\underset{\mathbf{W}_k,\mathbf{S}\in\mathbb{H}^{N_\mathrm{A}},\tau,\eta_k,\vartheta}{\text{minimize}} \quad & \tau \\
\text{subject to} \quad & \text{C1, C2, C4,} \\
& \text{C6:} \overline{g_1} + \overline{g_2} - \widetilde{h}_1 - \widetilde{h}_2 \le \tau, \\
& \text{C7: } \eta_k \le \sum_{r\in\mathcal{K}} \mathrm{Tr}(\mathbf{W}_r\mathbf{H}_k^H\mathbf{p}\mathbf{p}^H\mathbf{H}_k) + \mathrm{Tr}(\mathbf{S}\mathbf{H}_k^H\mathbf{p}\mathbf{p}^H\mathbf{H}_k), \\
& \text{C8: } \vartheta \le \mathrm{Tr}(\mathbf{S}\mathbf{L}^H\mathbf{p}\mathbf{p}^H\mathbf{L}),
\end{aligned} \tag{3.A.1}$$

where $\eta_k \in \mathbb{R}$ and $\vartheta \in \mathbb{R}$ are auxiliary optimization variables. Moreover, functions $\overline{g_1} \in \mathbb{R}$ and $\overline{g_2} \in \mathbb{R}$ are defined as follows:

$$\overline{g_1} = -\sum_{k\in\mathcal{K}} \log_2(\eta_k + \sigma_{n_k}^2), \tag{3.A.2}$$

$$\overline{g_2} = -\sum_{k\in\mathcal{K}} \log_2(\vartheta + \sigma_{n_e}^2). \tag{3.A.3}$$

We note that problem (3.A.1) preserves joint convexity for all the optimization variables and the Slater's condition is also fulfilled, which implies the strong duality between (3.A.1) and its dual problem holds. As such, instead of focusing on the primal problem in (3.A.1), we solve the dual counterpart of (3.A.1) for the optimal solution, cf. [6]. Motivated by this observation, we express the Lagrangian of (3.A.1) with respect to $\mathbf{W}_k$ as follows:

$$\begin{aligned}
\mathcal{L} = \sum_{k\in\mathcal{K}} \Bigg(& \varsigma\mathrm{Tr}(\mathbf{W}_k) - \chi_k \sum_{r\in\mathcal{K}} \mathrm{Tr}\left(\mathbf{W}_r\mathbf{H}_k^H\mathbf{p}\mathbf{p}^H\mathbf{H}_k\right) - \mathrm{Tr}(\mathbf{W}_k\mathbf{Z}_k) \\
& -\kappa\mathrm{Tr}\left(\left[\nabla_{\mathbf{W}_k} h_1(\mathbf{W}_k^t) + \nabla_{\mathbf{W}_k} h_2(\mathbf{W}_k^t)\right]^H (\mathbf{W}_k - \mathbf{W}_k^t)\right)\Bigg) + \Xi.
\end{aligned} \tag{3.A.4}$$

Here, Ξ collects all proof-irrelevant terms of (3.A.1). Real-valued variables ς, κ, and χ_k refer to the Lagrangian variables related to constraints C1, C6, and C7 in (3.A.1), respectively. Matrix $\mathbf{Z}_k \in \mathbb{C}^{N_\mathrm{A}\times N_\mathrm{A}}$ denotes the Lagrangian variables related to the semidefinite constraint C4. Then, the dual problem of (3.17) is formulated as follows:

$$\underset{\substack{\mathbf{Z}_k\succeq\mathbf{0},\\ \varsigma,\kappa,\chi_k\ge 0}}{\text{maximize}} \ \underset{\substack{\mathbf{W}_k,\mathbf{S}\in\mathbb{H}^{N_\mathrm{A}},\\ \tau,\eta_k,\vartheta}}{\text{minimize}} \ \mathcal{L}(\mathbf{W}_k, \mathbf{S}, \tau, \mathbf{Z}_k, \varsigma, \kappa, \chi_k). \tag{3.A.5}$$

Next, we explore the structure of $\mathbf{W}_k^*$ by inspecting its Karush–Kuhn–Tucker (KKT) condition demonstrated as follows:

$$\text{KKT1: } \mathbf{Z}_k^* \succeq \mathbf{0}, \kappa^*, \varsigma^*, \chi_k^* \geq 0, \qquad \text{KKT2: } \mathbf{Z}_k^*\mathbf{W}_k^* = \mathbf{0}, \tag{3.A.6}$$

$$\text{KKT3: } \mathbf{Z}_k^* = \varsigma^*\mathbf{I}_{N_A} - \kappa^*\nabla_{\mathbf{W}_k}\left(h_1(\mathbf{W}_k^t) + h_2(\mathbf{W}_k^t)\right) - \sum_{r\in\mathcal{K}}\chi_r^*\mathbf{H}_r^H\mathbf{p}\mathbf{p}^H\mathbf{H}_r, \tag{3.A.7}$$

where ς^*, κ^*, χ_r^*, and $\mathbf{Z}_k^*$ are the optimal Lagrange multipliers of (3.A.5) with respect to $\mathbf{W}_k^*$. Subsequently, we define a matrix variable $\mathbf{\Lambda}$ as follows:

$$\mathbf{\Lambda} = \kappa^*\nabla_{\mathbf{W}_k}\left(h_1(\mathbf{W}_k^t) + h_2(\mathbf{W}_k^t)\right) + \sum_{r\in\mathcal{K}}\chi_r^*\mathbf{H}_r^H\mathbf{p}\mathbf{p}^H\mathbf{H}_r, \tag{3.A.8}$$

and rewrite K3 as $\mathbf{Z}_k^* = \varsigma^*\mathbf{I}_{N_A} - \mathbf{\Lambda}$.

Next, we discuss the eigenvalues of $\mathbf{\Lambda}$ and show that $\text{rank}(\mathbf{W}^*) \leq 1$ always holds. For brevity, we denote the maximum eigenvalue of matrix $\mathbf{\Lambda}$ by $\lambda_{\mathbf{\Lambda}}^{\max} \in \mathbb{R}$. Considering $\mathbf{Z}_k^* = \varsigma^*\mathbf{I}_{N_A} - \mathbf{\Lambda}$, if $\varsigma^* < \lambda_{\mathbf{\Lambda}}^{\max}$, then $\mathbf{Z}_k^*$ is supposed to be positive semidefinite which violates $\mathbf{Z}_k^* \succeq \mathbf{0}$ in K1. Moreover, if $\varsigma^* > \lambda_{\mathbf{\Lambda}}^{\max}$, this leads to $\mathbf{Z}_k^* \succ \mathbf{0}$, which implies $\text{rank}(\mathbf{Z}_k^*) = N_A$. Under this condition, $\mathbf{W}_k^*$ is supposed to be $\mathbf{0}$ which affirms $\text{rank}(\mathbf{W}^*) \leq 1$. Furthermore, for the case where $\varsigma^* = \lambda_{\mathbf{\Lambda}}^{\max}$, we have $\text{rank}(\mathbf{Z}_k^*) = N_A - 1$. This results in the conclusion $\text{rank}(\mathbf{W}_k^*) = 1$, which folds up the proof.

References

1 P.-A. Absil, R. Mahony, and R. Sepulchre. *Optimization Algorithms on Matrix Manifolds*. Princeton University Press, 2009.

2 M. Avriel. *Nonlinear Programming: Analysis and Methods*. Courier Corporation, 2003.

3 J. C. Bezdek and R. J. Hathaway. Some notes on alternating optimization. In *AFSS International Conference on Fuzzy Systems*, pages 288–300. Springer, 2002.

4 M. Bloch and J. Barros. *Physical-Layer Security: From Information Theory to Security Engineering*. Cambridge University Press, 2011.

5 N. Boumal, B. Mishra, P. A. Absil, and R. Sepulchre. Manopt, a Matlab toolbox for optimization on manifolds. *Journal of Machine Learning Research*, 15(42):1455–1459, 2014. URL https://www.manopt.org.

6 S. P. Boyd and L. Vandenberghe. *Convex Optimization*. Cambridge University Press, 2004.

7 J. Chen, Y. C. Liang, Y. Pei, and H. Guo. Intelligent reflecting surface: A programmable wireless environment for physical layer security. *IEEE Access*, 7:82599–82612, 2019.

8 T. H. Cormen, C. E. Leiserson, R. L. Rivest, and C. Stein. *Introduction to Algorithms*. MIT Press, 2009.

9 I. Csiszar and J. Korner. Broadcast channels with confidential messages. *IEEE Transactions on Information Theory*, 24(3):339–348, May 1978.

10 T. J. Cui, M. Q. Qi, X. Wan, J. Zhao, and Q. Cheng. Coding metamaterials, digital metamaterials and programmable metamaterials. *Light: Science & Applications*, 3(10):e218, 2014.

11 M. Di Renzo et al. Smart radio environments empowered by reconfigurable AI meta-surfaces: An idea whose time has come. *EURASIP Journal on Wireless Communications and Networking*, 2019(1):1–20, May 2019.

12 Q. T. Dinh and M. Diehl. *Local Convergence of Sequential Convex Programming for Nonconvex Optimization*. Springer, 2010.

13 M. Grant and S. P. Boyd. CVX: Matlab software for disciplined convex programming, version 2.1. http://cvxr.com/cvx, Jan 2020.

14 R. Hartley. Transmission of information. *Bell Labs Technical Journal*, 7(3):535–563, 1928.

15 S. Hu, Z. Wei, Y. Cai, D. W. K. Ng, and J. Yuan. Sum-rate maximization for multiuser MISO downlink systems with self-sustainable IRS. In *Proceedings of IEEE Global Communications Conference (GLOBECOM)*, pages 1–7, Taipei, Taiwan, Dec 2020.

16 S. Hu, Z. Wei, Y. Cai, C. Liu, D. W. K. Ng, and J. Yuan. Robust and secure sum-rate maximization for multiuser MISO downlink systems with self-sustainable IRS. *IEEE Transactions on Communications*, 69(10):7032–7049, Oct 2021.

17 P. H. Lin, S. H. Lai, S. C. Lin, and H. J. Su. On secrecy rate of the generalized artificial-noise assisted secure beamforming for wiretap channels. *IEEE Journal on Selected Areas in Communications*, 31(9):1728–1740, Sept 2013.

18 J. Lofberg. YALMIP : A toolbox for modeling and optimization in MATLAB. In *Proceedings of International Conference on Robotics and Automation*, pages 284–289, Taipei, Taiwan, 2004.

19 R. Negi and S. Goel. Secret communication using artificial noise. In *Proceedings of IEEE Vehicular Technology Conference*, pages 1906–1910, Dallas, TX, USA, Sept 2005.

20 D. W. K. Ng, E. S. Lo, and R. Schober. Secure resource allocation and scheduling for OFDMA decode-and-forward relay networks. *IEEE Transactions on Wireless Communications*, 10(10):3528–3540, Oct 2011.

21 Md. Noor-A-Rahim, F. Firyaguna, J. John, M. O. Khyam, D. Pesch, E. Armstrong, H. Claussen, and H. V. Poor. Toward industry 5.0: Intelligent reflecting surface in smart manufacturing. *IEEE Communications Magazine*, 60(10):72–78, Oct 2022.

22 H. Shen, W. Xu, S. Gong, Z. He, and C. Zhao. Secrecy rate maximization for intelligent reflecting surface assisted multi-antenna communications. *IEEE Communications Letters*, 23(9):1488–1492, Sept 2019.

23 S. Shen, B. Clerckx, and R. Murch. Modeling and architecture design of reconfigurable intelligent surfaces using scattering parameter network analysis. *IEEE Transactions on Wireless Communications*, 21(2):1229–1243, Feb 2022.

24 J. R. Shewchuk. *An Introduction to the Conjugate Gradient Method without the Agonizing Pain*. Carnegie-Mellon University. Department of Computer Science Pittsburgh, 1994.

25 Y. S. Shiu, S. Y. Chang, H.-C. Wu, S. C.-H. Huang, and H.-H. Chen. Physical layer security in wireless networks: A tutorial. *IEEE Wireless Communications*, 18(2):66–74, Apr 2011.

26 Q. Wu and R. Zhang. Intelligent reflecting surface enhanced wireless network: Joint active and passive beamforming design. In *IEEE Global Communications Conference (GLOBECOM)*, pages 1–6, Dec 2018.

27 Q. Wu and R. Zhang. Beamforming optimization for intelligent reflecting surface with discrete phase shifts. In *IEEE International Conference on Acoustics, Speech and Signal Processing (ICASSP)*, pages 7830–7833, Brighton, UK, May 2019.

28 Q. Wu and R. Zhang. Towards smart and reconfigurable environment: Intelligent reflecting surface aided wireless network. *IEEE Communications Magazine*, 58(1):106–112, Jan 2020.

29 Y. Wu, D. Xu, D. W. K. Ng, R. Schober, and W. Gerstacker. Globally optimal resource allocation design for IRS-assisted multiuser networks with discrete phase shifts. In *IEEE International Conference on Communications Workshops (ICC Workshops)*, 2023.

30 D. Xu, Y. Sun, D. W. K. Ng, and R. Schober. Multiuser MISO UAV communications in uncertain environments with no-fly zones: Robust trajectory and resource allocation design. *IEEE Transactions on Communications*, 68(5):3153–3172, May 2020.

31 D. Xu, X. Yu, and R. Schober. Resource allocation for intelligent reflecting surface-assisted cognitive radio networks. In *Proceedings of IEEE International Workshop on Signal Processing Advances in Wireless Communications (SPAWC)*, pages 1–5, Atlanta, GA, USA, May 2020.

32 D. Xu, X. Yu, Y. Sun, D. W. K. Ng, and R. Schober. Resource allocation for IRS-assisted full-duplex cognitive radio systems. *IEEE Transactions on Communications*, 68(12):7376–7394, Dec 2020.

33 D. Xu, X. Yu, D. W. K. Ng, and R. Schober. Resource allocation for active IRS-assisted multiuser communication systems. In *Proceedings of the Fifty-Fifth*

Asilomar Conf. Signals, Systems and Computers (ASILOMAR), pages 113–119, Pacific Grove, CA, USA, Oct 2021.

34 D. Xu, V. Jamali, X. Yu, D. W. K. Ng, and R. Schober. Optimal resource allocation design for large IRS-assisted SWIPT systems: A scalable optimization framework. *IEEE Transactions on Communications*, 70(2):1423–1441, Feb 2022.

35 D. Xu, X. Yu, D. W. K. Ng, A. Schmeink, and R. Schober. Robust and secure resource allocation for ISAC systems: A novel optimization framework for variable-length snapshots. *IEEE Transactions on Communications*, 70(12):8196–8214, Dec 2022.

36 D. Xu, A. Khalili, X. Yu, D. W. K. Ng, and R. Schober. Integrated sensing and communication in distributed antenna networks. In *IEEE International Conference on Communications Workshops (ICC Workshops)*, 2023.

37 N. Yang, L. Wang, G. Geraci, M. Elkashlan, J. Yuan, and M. Di Renzo. Safeguarding 5G wireless communication networks using physical layer security. *IEEE Communications Magazine*, 53(4):20–27, Apr 2015.

38 X. Yu, D. Xu, and R. Schober. Enabling secure wireless communications via intelligent reflecting surfaces. In *Proceedings of IEEE Global Communications Conference (GLOBECOM)*, pages 1–6, Waikoloa, HI, USA, Dec 2019.

39 X. Yu, D. Xu, and R. Schober. MISO wireless communication systems via intelligent reflecting surfaces. In *Proceedings of IEEE/CIC International Conference on Communications in China (ICCC)*, pages 1–6, Changchun, China, May 2019.

40 X. Yu, D. Xu, Y. Sun, D. W. K. Ng, and R. Schober. Robust and secure wireless communications via intelligent reflecting surfaces. *IEEE Journal on Selected Areas in Communications*, 38(11):2637–2652, Nov 2020.

41 X. Yu, D. Xu, D. W. K. Ng, and R. Schober. IRS-assisted green communication systems: Provable convergence and robust optimization. *IEEE Transactions on Communications*, 69(9):6313–6329, Sept 2021.

42 X. Yu, V. Jamali, D. Xu, D. W. K. Ng, and R. Schober. Smart and reconfigurable wireless communications: From IRS modeling to algorithm design. *IEEE Wireless Communications*, 28(6):118–125, 2021.

43 F. Zhu, F. Gao, T. Zhang, K. Sun, and M. Yao. Physical-layer security for full duplex communications with self-interference mitigation. *IEEE Transactions on Wireless Communications*, 15(1):329–340, Jan 2016.

44 Z. Zhuang, X. Zhang, D. Xu, and S. Song. Active IRS-aided MIMO systems: How much gain can we get? *arXiv:2305.05277*, 2023.

4

Physical-Layer Security for Optical Wireless Communications

Shenjie Huang, Mohammad Dehghani Soltani, and Majid Safari

School of Engineering, The University of Edinburgh, Edinburgh, UK

4.1 Introduction

With the dramatic increase in global data traffic, future wireless networks should provide high data rates and energy efficiency, massive device connectivity, ultralow latency, and also high-security level. Smartphones, laptops, tablets, wireless sensors, and other wireless devices struggle to get their required share of the radio frequency (RF) spectrum. As a consequence, the data throughput of the devices is affected severely, and the established connection may be poor. In the next generations (6G and beyond) of wireless networks, 1000-fold capacity growth is anticipated within the next 10 years with respect to the long-term evolution (LTE) wireless networks [28]. This rapidly increasing requirement for wireless connectivity can be fulfilled through optical wireless communications (OWCs), which are feasible across the ultra-violet, visible light, and infrared (IR) spectra.

The primary uses of OWC include free-space optical (FSO) communication, employed for long-distance high-speed data transmission, and IR and visible light communication (VLC), utilized for short-range wireless networks. In particular, VLC is considered as a promising solution for future indoor wireless communications. The main advantages of VLC over RF include the very large and license-free spectrum, surpassing the entire 10^3, accurate indoor localization, high data rates, low cost, and rather straightforward deployment involving the use of commercially available light-emitting diodes (LEDs) and photodiodes (PDs) at the respective transmitters and receivers. Moreover, VLC finds potential applicability in RF-constrained environments including hospitals and airplane cabins [24]. Light-fidelity (LiFi) is a bidirectional and high-speed wireless networked system

Physical-Layer Security for 6G, First Edition. Edited by Parthajit Mohapatra, Nikolaos Pappas, Arsenia Chorti, and Stefano Tomasin.

that uses visible light for communication and illumination. LiFi can serve as a supplementary infrastructure in conjunction with current RF networks to form a hybrid network to support high-data rates and seamless connectivity for mobile users.

Since optical beams have limited divergence angles and can be blocked by opaque materials like walls, the OWC is intrinsically more secure than its RF counterpart in nature. Nevertheless, the open and broadcast property of the VLC system can still make it prone to eavesdropping attacks and bring security issues, e.g., when eavesdroppers (EDs) are present within the area covered by the VLC access point (AP). A number of physical-layer security (PLS) measures have been proposed to further enhance the security level ensuring that confidential messages can only be detected by legitimate users. Through the utilization of the randomness of the channels, resources, and noise, PLS can effectively reduce the amount of decipherable information accessible to potential EDs without resorting to modifications to layers above the physical layer [56].

Extensive prior research has been dedicated to exploring PLS for RF systems [29]. However, due to the fundamental distinction in transmission protocols and modulation schemes employed by RF and VLC systems, the developed PLS measures for RF systems are not directly transferable to VLC systems. In fact, compared to its RF counterpart, VLC normally employs intensity-modulation direct detection (IM/DD) because of the utilization of low-cost LEDs and PDs; hence, the transmitted signal should be both real and non-negative. In addition, since the practical light sources have a limited dynamic range, the extra peak transmitted power constraint should be considered. Therefore, the PLS techniques specifically designed for VLC are highly demanded. In this chapter, the PLS in VLC is discussed and extensive prior relevant works are reviewed. The PLS for VLC systems with emerging technologies is also introduced.

4.2 PLS for SISO VLC

4.2.1 PLS Performance Metrics

The commonly used performance metrics for the secrecy analysis include the secrecy capacity, secrecy outage probability (SOP), and average secrecy capacity. Secrecy capacity is characterized as the maximal data rate that can be decoded by the legitimate receivers (LRs) while ensuring that the EDs remain unaware of the transmitted information [38]. Let us consider a simple three-node model in which a source (Alice) is transmitting confidential messages to an LR (Bob) in the presence of an ED (Eve). The secrecy capacity can be calculated as

$$C_{\mathrm{s}} = \max_{p_X(x)} \left\{ I(X;Y_{\mathrm{B}}) - I(X;Y_{\mathrm{E}}) \right\}, \tag{4.1}$$

where $I(\cdot;\cdot)$ refers to the mutual information. The signal X denotes the transmitted signal with a distribution of $p_X(x)$, and $Y_{\rm B}$ and $Y_{\rm E}$ present the signals received by LR and ED, respectively. If the mobility or the randomness of the LD and ED locations is considered, the SOP and average secrecy capacity are more appropriate metrics. The SOP is characterized by the likelihood that the secrecy capacity falls below a specified capacity threshold where a secure outage occurs [20], i.e.,

$$P_{\rm SOP} = \Pr\left\{C_{\rm s} \leq C_{\rm th}\right\}, \tag{4.2}$$

where $C_{\rm th}$ denotes the threshold value. The average capacity is defined as $\overline{C}_{\rm s} = \mathbb{E}[C_{\rm s}]$, where $\mathbb{E}[\cdot]$ denotes the expectation operation.

4.2.2 SISO VLC Secrecy Analysis

In this section, let us concentrate on the PLS of the SISO VLC systems where only one LED is employed at the transmitter and both LRs and EDs are with a single PD, as illustrated in Figure 4.1. Because of the substantial detection area compared to the optical wavelength, the influence of small-scale fading effects in VLC is minimal [25]. As a result, the VLC channel is mainly determined by the geometric properties of the transceivers. A VLC channel normally contains both the line-of-sight (LoS) and non-line-of-sight (NLoS) components. The NLoS components are introduced by the diffusive reflections. For a typical indoor scenario, the NLoS components are normally weaker than the LoS ones, particularly when the receiver is situated at a distance from reflective surfaces like walls. As a result, the NLoS signal can be ignored to simplify the performance analysis [53]. Considering that the employed LEDs are with Lambertian radiation profile, the channel gain between an LED and a PD is given by [25]

$$h = \frac{(m+1)A_{\rm d}R}{2\pi d^2} g(\psi)\cos^m(\phi)\cos(\psi), \tag{4.3}$$

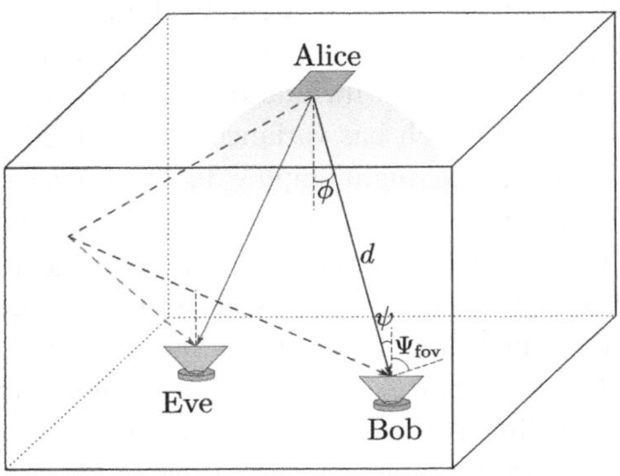

Figure 4.1 The indoor SISO VLC wiretap system.

where $m = -\ln(2)/\ln\left(\cos(\phi_{1/2})\right)$ denotes the Lambertian emission order with the half-power semi-angle of LED $\phi_{1/2}$, A_d is the detection area of the PD, R represents the PD responsivity, d is the distance between transceivers, and ϕ and ψ are the light radiance and incidence angles, respectively. The expression for the gain of the optical concentrator, denoted as $g(\psi)$, is given by

$$g(\psi) = \begin{cases} \dfrac{n_c^2}{\sin^2(\Psi_{fov})}, & 0 \le \psi \le \Psi_{fov}, \\ 0, & \psi > \Psi_{fov}, \end{cases} \tag{4.4}$$

with n_c denotes the optical concentrator's refractive index and Ψ_{fov} refers to the receiver field-of-view (FOV).

The presence of device orientation, link blockage, and mobility in indoor OWC networks necessitate more realistic and measurement-based models [47]. This feature in fact makes optical wireless networks inherently more secure [2]. It is shown that the secrecy performance of VLC, when accounting for the real-world statistics of receiver orientation and link blockage, is 30% higher compared to the limited assumption of a fixed orientation [3]. Typically, the locations of receivers in a VLC system are dynamic, and their orientations are random. From (4.3), the channel gain of a VLC receiver can be rewritten as

$$h = C\frac{g(\psi)\cos^m(\phi)\cos(\psi)}{d^2}, \tag{4.5}$$

where $C = \frac{(m+1)A_d R}{2\pi}$. The mobility of the user can introduce variability in the distance d and radiance and incidence angles ϕ and ψ, but the device orientation can only introduce randomness in the incidence angle ψ.

For an SISO VLC wiretap channel, the signals received by the legitimate receiver (Bob) and eavesdropper (Eve) are given by

$$Y_B = \alpha_c h_B X + n_B, \quad Y_E = \alpha_c h_E X + n_E, \tag{4.6}$$

respectively, where α_c denotes the transmitter electro-optical conversion factor, X is the zero-mean channel input signal with a distribution of $p_X(x)$, and n_B and n_E represent the additive white Gaussian noise (AWGN) affecting both Bob and Eve which has variances represented by σ_B^2 and σ_E^2, respectively. Note that Bob and Eve might employ different receiver designs, hence, could have distinct receiver noise variances. The channel gains h_B and h_E can be calculated by (4.3) which are related to the corresponding geometric properties. Practical LEDs have limited dynamic range as illustrated in Figure 4.2. To avoid excessive nonlinear distortion, the output optical power should keep in the linear regime, i.e., $[P_{min}, P_{max}]$, where P_{min} and P_{max} refer to the minimal and maximal optical power of the LED linear dynamic range. This means that the input electrical current should be within a range $[I_{min}, I_{max}]$, where $I_{min} = P_{min}/\alpha_c$ and $I_{max} = P_{max}/\alpha_c$. Denote

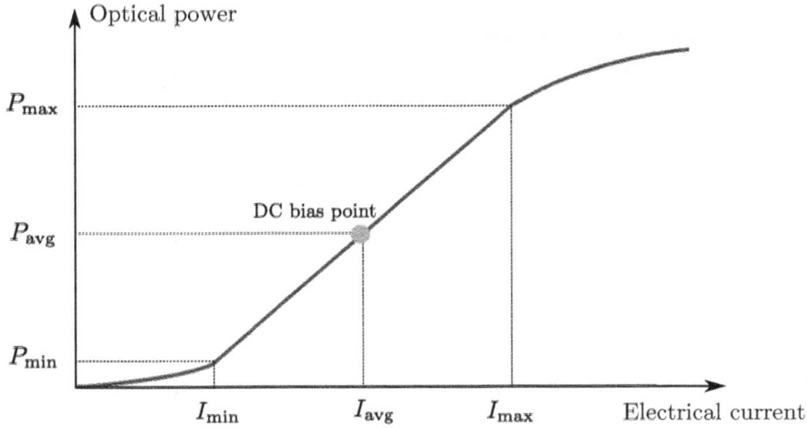

Figure 4.2 Limited dynamic range of a typical LED.

that the DC bias current, i.e., $I_{\text{avg}} = \frac{1}{2}(I_{\text{min}} + I_{\text{max}})$. The electrical signal used to drive the LED can be expressed as $X + I_{\text{avg}}$. Therefore, the constraint induced by limited LED dynamic range is given by $I_{\text{min}} \leq X + I_{\text{avg}} \leq I_{\text{max}}$, which corresponds to an amplitude constraint on X as $|X| \leq A_c$ where $A_c = \frac{1}{2}(I_{\text{max}} - I_{\text{min}})$.

While for the SISO VLC wiretap channel, the closed-form secrecy capacity given in (4.1) is still unknown, some works have been conducted to explore the lower bounds of the secrecy capacity or equivalently the achievable secrecy. It has been concluded that for such a channel achieving, the secrecy capacity requires using a discrete input distribution with a limited support set [39]. With this assumption, the analytical expression for a good secrecy capacity lower bound, as a function of the input distribution $p_X(x)$, can be expressed as $\max(R_s, 0)$ with [7]

$$R_s(p_X(x)) = \frac{1}{2}\log\left[\frac{1 + 2\gamma_B^2\sigma_X^2}{1 + \gamma_B^2\sigma_X^2}\right] - \log\left[\sum_{i=1}^{K}\sum_{j=1}^{K} p_i p_j e^{d_{i,j}}\right] - \frac{1}{2}\log\left[1 + \gamma_E^2\sigma_X^2\right], \tag{4.7}$$

where

$$d_{i,j} = \frac{-\gamma_B^2}{4\gamma_B^2\sigma_X^2 + 2}\left[\gamma_B^2\sigma_X^2(x_i - x_j)^2 - 2x_i x_j\right], \tag{4.8}$$

$\gamma_B^2 = h_B^2/\sigma_B^2$, $\gamma_E^2 = h_E^2/\sigma_E^2$, σ_X^2 denotes the variance of the input signal, and the probability mass function of X is given by $p_X(x) = \sum_{i=1}^{K} p_i \delta(x - x_i)$ with x_i denotes the ith mass point and p_i refers to the corresponding probability. The attainable secrecy rate (4.7) holds true for any specified bounded discrete distribution of X. Considering that VLC systems normally have an amplitude constraint on X,

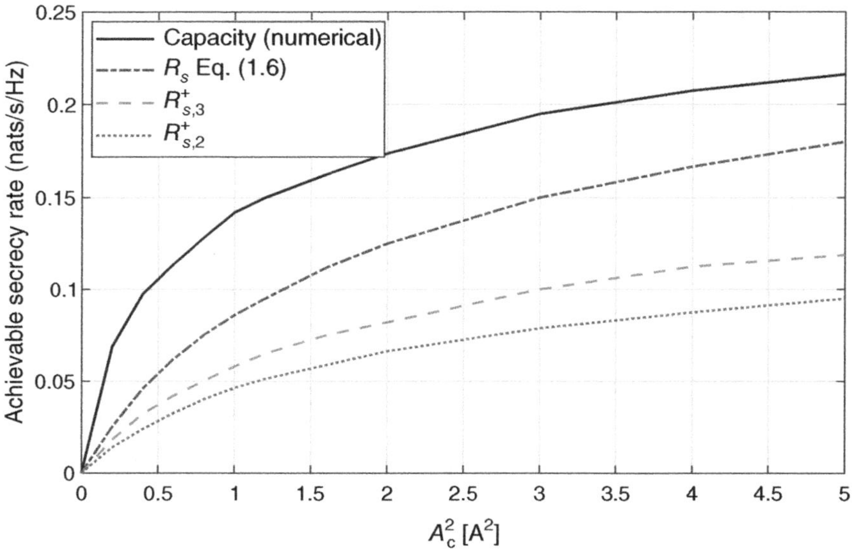

Figure 4.3 Secrecy capacity and different achievable secrecy rates. Source: Adapted from [11].

Arfaoui et al. [7] proposed to utilize the truncated discrete generalized normal (TDGN) distribution over the dynamic range of LED to optimize the secrecy rate, which results in a relatively tight lower bound on the secrecy capacity. Figure 4.3 demonstrates the various secrecy rates versus the peak constraint A_c^2. The secrecy capacity is achieved through the numerical method introduced in [39]. The curve of R_s is obtained by using (4.7) when TDGN distribution with the optimal parameters is adopted. The results of $R^+_{s,2}$ and $R^+_{s,3}$ are two alternative achievable secrecy rates given in [[11], Eq. (11)] and [[11], Eq. (12)], respectively. It is demonstrated that the secrecy rate calculated based on (4.7) with TDGN provides a tighter secrecy capacity lower bound compared to the other available lower bounds. Besides peak power constraint, VLC also has illumination and dimming requirements which further put an average constraint on the transmitted optical signal. Motivated by this, Wang et al. established the upper and lower bounds on secrecy capacity for the VLC SISO wiretap system accounting for both average and peak constraints [49].

In many studies exploring VLC wiretap systems, the channel modeling focuses exclusively on the LoS component. However, in some practical scenarios, NLoS components are comparable to LoS one and are non-negligible, e.g., when receivers are close to reflectors. Therefore, it is valuable to consider the secrecy performance of SISO wiretap systems in VLC with a complete channel model by taking multipath reflection-induced NLoS components into account. Considering both LoS and NLoS components, the channel impulse response (CIR) can be achieved through the iterative site-based method proposed in [13]. The combined CIR can be expressed as

$$h^{c}(t; \mathcal{S}, \mathcal{D}) = \sum_{l=0}^{\infty} h^{(l)}(t; \mathcal{S}, \mathcal{D}), \tag{4.9}$$

where $\mathcal{S}$ and $\mathcal{D}$ refer to the transmitter and the receiver, t is the elapsed time, l denotes the number of reflections the light undergoes, and $h^{(l)}$ denotes the l-bounce CIR. Note that when $l = 0$, the impulse response $h^{(0)}(t; \mathcal{S}, \mathcal{D})$ converges to the LoS channel gain given in (4.3). To calculate the CIR when $l > 0$, we can partition the whole reflector, e.g., a wall, into numerous elementary reflectors. The cumulative CIR for this scenario is obtained by summing the contributions from each individual small reflector. The approximated l-bounce response can be written as [13]

$$h^{(l)}(t; \mathcal{S}, \mathcal{D}) = \sum_{j=1}^{N_r} \rho_j h^{(l-1)}(t; \mathcal{S}, \delta_j) \otimes h^{(0)}(t; \delta_j, \mathcal{D}), \tag{4.10}$$

where N_r denotes the number of elementary reflectors, and δ_j refers to the ith element with a reflectivity of ρ_j. Note that with the increase of l, $h^{(l)}(t; \mathcal{S}, \mathcal{D})$ decreases and in the practical simulation the contribution of CIRs with $l > 5$ in (4.9) becomes negligible and can be safely ignored [13]. The combined CIR for Bob (Eve) can be achieved by substituting $\mathcal{S}$ and $\mathcal{D}$ by Alice and Bob (Eve), respectively. By convolving the $h^{c}(t; \mathcal{S}, \mathcal{D})$ with the pulse shaping filter during transmission, followed by the matched filter and subsequent downsampling process at the receiver, the overall discrete-time CIR, denoted by $h_{B}^{c,d}[k]$ and $h_{E}^{c,d}[k]$ for Bob and Eve, respectively, where k is the discrete index, can be achieved [52]. The discrete signal received by Bob and Eve considering the combined signal components can be expressed as

$$Y_{B}^{c}[k] = \alpha_c h_{B}^{c,d}[0] X[k] + \sum_{i=1}^{\infty} \alpha_c h_{B}^{c,d}[i] X[k-i] + n_B[k], \tag{4.11}$$

$$Y_{E}^{c}[k] = \alpha_c h_{E}^{c,d}[0] X[k] + \sum_{i=1}^{\infty} \alpha_c h_{E}^{c,d}[i] X[k-i] + n_E[k], \tag{4.12}$$

where $X[k]$ denotes the transmitted symbol sequence and $n_B[k]$ and $n_E[k]$ are the AWGN for Bob and Eve, respectively. The second term in (4.11) refer to the inter-symbol-interference (ISI) induced by the multipath reflections which can be combined with the AWGN term resulting in a total noise term, i.e., $n_{B}^{c}[k] \triangleq \sum_{i=1}^{\infty} \alpha_c h_{B}^{c,d}[i] X[k-i] + n_B[k]$. Similarly, at the Eve side, one has $n_{E}^{c}[k] \triangleq \sum_{i=1}^{\infty} \alpha_c h_{E}^{c,d}[i] X[k-i] + n_E[k]$. Considering the amplitude constraint on

$X[k]$, a lower bound on the secrecy capacity is given by [52]

$$R_s^c = \frac{1}{2}\log_2\left[\frac{12\sigma_E^2 + \alpha_c^2 A_c^2 \sum_{k=0}^{\infty}\left(h_E^{c,d}[k]\right)^2}{12\sigma_B^2 + \alpha_c^2 A_c^2 \sum_{k=0}^{\infty}\left(h_B^{c,d}[k]\right)^2} \times \frac{12\sigma_B^2 + \frac{6}{\pi e}\alpha_c^2 A_c^2\left(h_B^{c,d}[0]\right)^2 + \alpha_c^2 A_c^2 \sum_{k=0}^{\infty}\left(h_B^{c,d}[k]\right)^2}{12\sigma_E^2 + \alpha_c^2 A_c^2 \sum_{k=0}^{\infty}\left(h_E^{c,d}[k]\right)^2}\right], \tag{4.13}$$

where the continuous uniform distributed input is considered. In the above discussion, symbol-by-symbol transmission is considered. The capacity lower bound when block (i.e., multiple symbols) transmission is employed is also derived in [52]. For block transmission, precoding schemes can be adopted to mitigate the ISI and, hence, improve the secrecy performance. The precoding matrix can be optimized through brute-force search method. Yang et al. also proposed a low-complexity amplitude scaling scheme for precoding which can provide a suboptimal solution whose performance is close to the optimal one [52].

When multiple EDs exist in the coverage area of the LED, the collusion of EDs might significantly influence the secrecy performance. Cho et al. examined the SISO VLC channel considering the existence of colluding EDs with random positions. The binomial point process (BPP) and Poisson point process (PPP) were employed to model the randomness of the single LR and multiple ED positions, respectively. By assuming a continuous pulse amplitude modulation (PAM) scheme and the maximum-ratio combining (MRC) of the colluding EDs, the closed-form SOP of the considered wiretap system was derived.

4.3 PLS for MISO VLC

In this section, the PLS for VLC systems with multiple LEDs or APs at the transmitter and single PD at each receiver (both LRs and EDs) as shown in Figure 4.4 is investigated. For the three-node model discussed in Section 4.2, the signal received by LR and ED when only the LoS channel is considered is given by

$$Y_B = \alpha_c \mathbf{h}_B^T \mathbf{X} + n_B, \quad Y_E = \alpha_c \mathbf{h}_E^T \mathbf{X} + n_E, \tag{4.14}$$

where α_c denotes the transmitter electro-optical conversion factor, $\mathbf{h}_E$ represents the channel characteristic vector of the ED, while $\mathbf{h}_B$ represents that of the LR, both having a size of $M \times 1$, M refers to the number of LEDs, vector $\mathbf{X}$ which also possesses dimensions $M \times 1$ denotes the transmitted signal, and n_B and n_E refer to the AWGN noise with variance σ_B^2 and σ_E^2, respectively.

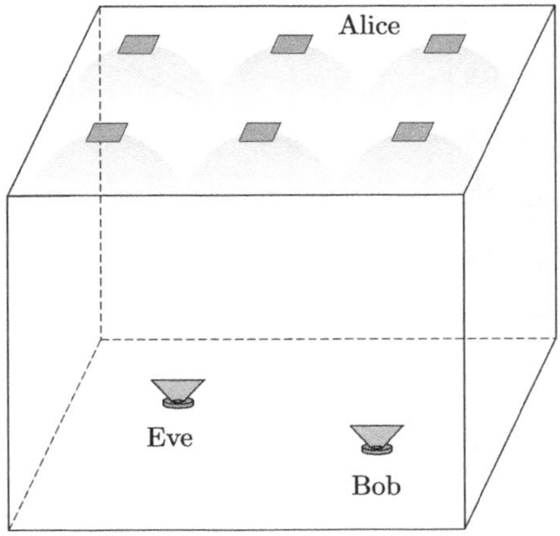

Figure 4.4 The indoor MISO VLC wiretap system.

4.3.1 MISO VLC Secrecy Analysis

Compared to SISO systems, the extra spatial DoFs at Alice can be exploited in MISO wiretap systems by means of beamforming to provide enhanced secrecy performance. The objective of beamforming is to sufficiently reduce or eliminate the information detected by the unintended user. Denoting the beamforming vector as $\mathbf{w}$, the transmitted vector can be expressed as $\mathbf{X} = \mathbf{w}u$, with u representing the transmitted confidential signal. For LR and ED, the received signals can be rewritten as

$$Y_{\mathrm{B}} = \alpha_{\mathrm{c}}\mathbf{h}_{\mathrm{B}}^{T}\mathbf{w}u + n_{\mathrm{B}}, \quad Y_{\mathrm{E}} = \alpha_{\mathrm{c}}\mathbf{h}_{\mathrm{E}}^{T}\mathbf{w}u + n_{\mathrm{E}}, \tag{4.15}$$

respectively. Assuming a uniform distributed input signal and $\sigma_{\mathrm{B}}^2 = \sigma_{\mathrm{E}}^2 = \sigma^2$, the expression for the achievable secrecy rate is given by [35]

$$R_{\mathrm{s}} = \max_{\mathbf{w}} \left[\frac{1}{2} \log \frac{6A_{\mathrm{c}}^2\alpha_{\mathrm{c}}^2\mathbf{w}^T\mathbf{h}_{\mathrm{B}}\mathbf{h}_{\mathrm{B}}^T\mathbf{w} + 3\pi e\sigma^2}{\pi e A_{\mathrm{c}}^2\alpha_{\mathrm{c}}^2\mathbf{w}^T\mathbf{h}_{\mathrm{E}}\mathbf{h}_{\mathrm{E}}^T\mathbf{w} + 3\pi e\sigma^2} \right]^{+}, \tag{4.16}$$

where $[y]^+ = \max\{y, 0\}$. Since the optimization problem seeking the best beamformer $\mathbf{w}$ to maximize the secrecy rate while adhering to amplitude constraints is non-convex and mathematically intractable, numerical algorithms must be employed for its solution. A simple but suboptimal design is the null steering which is also known as the zero-forcing (ZF) beamforming. With such a beamforming technique, the design ensures that the beamforming aligns with the nullspace of the ED, i.e., $\mathbf{h}_{\mathrm{E}}^T\mathbf{w}_{\mathrm{ZF}} = 0$. Note that Eve's CSI is required to design the ZF beamformer. As a consequence, what Eve receives is pure noise. The optimization of $\mathbf{w}_{\mathrm{ZF}}$ maximizing the secrecy rate is a linear problem that is able to be addressed with low computational complexity [35]. In this situation,

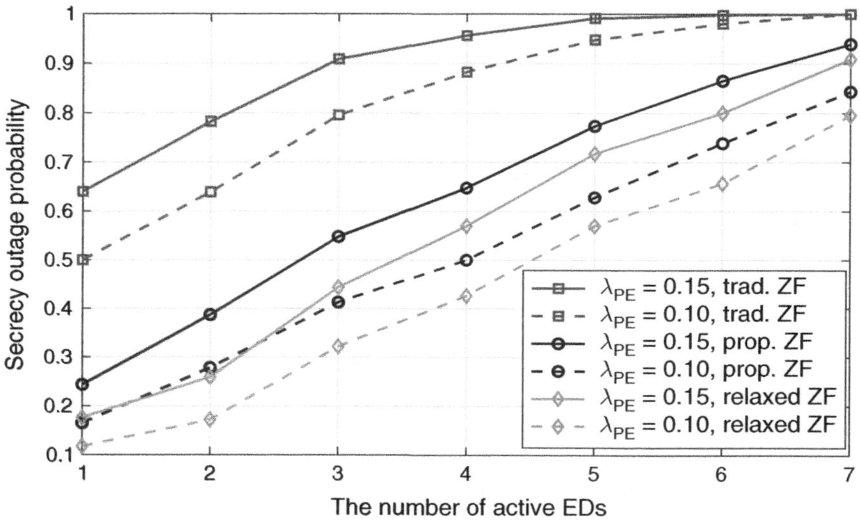

Figure 4.5 The SOP versus the number of active EDs for various ZF beamforming schemes. Source: Adapted from [18].

the achievable secrecy rate (4.16) can be rewritten as

$$R_{\mathrm{S}}^{\mathrm{ZF}} = \frac{1}{2}\log\left(1 + \frac{2A_{\mathrm{c}}^{2}\alpha_{\mathrm{c}}^{2}\mathbf{w}_{\mathrm{ZF}}^{T}\mathbf{h}_{\mathrm{B}}\mathbf{h}_{\mathrm{B}}^{T}\mathbf{w}_{\mathrm{ZF}}}{\pi e\sigma^{2}}\right). \tag{4.17}$$

However, null steering does not necessarily result in the highest secrecy rate and, hence, is not the optimal beamforming design [34]. More importantly, ZF beamforming purely designed to suppress the signal received by an active ED whose CSI is known at Alice might result in some unintentional areas where the received SNR is even higher than that of LR. If unknown passive EDs reside in these areas, huge secrecy degradation can be introduced. To this end, a novel ZF beamforming scheme was proposed in [18] taking into account both active and passive EDs. In addition, since the capability of suppressing the passive EDs can be improved by slightly relaxing the strict ZF condition, Cho et al. further proposed an alternative scheme called relaxed ZF beamforming to cope with the passive EDs more effectively [18]. Figure 4.5 presents the SOP versus the number of active EDs for different ZF beamforming techniques under various densities of passive EDs λ_{PE}. It is shown that with the increase of the number of active EDs and λ_{PE}, SOP performance is degraded, as expected. In addition, both the proposed novel ZF and relaxed ZF beamforming approaches present reduced

SOPs compared to the conventional ZF scheme. The placement of transmitters for ZF-based VLC MISO wiretap system is discussed in [32], which concludes that by employing the optimal transmitter placement an SNR gain of up to 6 dB can be achieved compared to the traditional uniform placement.

Apart from the ZF beamforming, Mostafa and Lampe [36] proposed to transform the non-convex beamforming optimization problem into a solvable quasiconvex line search problem. The beamforming design for VLC wiretap systems with discrete input signals was investigated in [6]. The joint optimization of the discrete input distribution and beamforming vector is non-convex and non-linear, which should be solved by approaches like generalized reduced gradient [23] and sequential quadratic programming methods [12]. It was demonstrated that the secrecy rate achieved by discrete input is generally superior to those achieved by continuous ones.

4.3.2 Secrecy Improvement in MISO VLC

Artificial noise (AN), also known as jamming, is the common scheme employed in VLC PLS to effectively enhance the secrecy performance, especially for systems without Eve's CSI at Alice [9, 17]. In practical VLC systems, it should be noted that Alice can be aware of Bob's CSI through, for example, a feedback link. However, as Eve is a passive or malicious user, its CSI might not be known to Alice. Therefore, a reasonable assumption is that the position of ED is distributed randomly within the cellular coverage, and Alice is only aware of its statistical information. The jamming signal can introduce strong interference to the EDs prohibiting their information reception whereas have little effect on the LR. When AN is adopted, the AP's transmitted signal can be split into two components. One is the information-bearing (IB) signal intended for Bob and another is the jamming signal, i.e., $\mathbf{X} = \mathbf{w}_x x + \mathbf{w}_J s_J$, where $\mathbf{w}_x$ and $\mathbf{w}_J$ with size $M \times 1$ denote the beamforming vectors for IB and jamming signals, respectively, and x and s_J refer to the random IB and jamming symbols, respectively. The combined signal $\mathbf{X}$ should be subject to peak and/or amplitude constraints, if applicable. By applying the considered signal, the received signal for Bob and Eve given in (4.14) can be expressed as

$$Y_B = \alpha_c \mathbf{h}_B^T \mathbf{w}_x x + \alpha_c \mathbf{h}_B^T \mathbf{w}_J s_J + n_B, \quad Y_E = \alpha_c \mathbf{h}_E^T \mathbf{w}_x x + \alpha_c \mathbf{h}_E^T \mathbf{w}_J s_J + n_E. \quad (4.18)$$

If $\mathbf{w}_J$ is designed in the nullspace of $\mathbf{h}_B^T$, the second term in Y_B can be removed which means that the jamming signal will have no impact on Bob's reception but only introduce extra interference to Eve's reception [34]. In this context, the achievable secrecy rate given any discrete input signals can be expressed as

$\max(R_S^{AN}, 0)$ with [9]

$$R_S^{AN}(\mathbf{h}_E) = \frac{1}{2}\log\left[\frac{1+2\gamma_B\sigma_X^2}{1+\gamma_B\sigma_X^2}\right] + \frac{1}{2}\log\left[\frac{1+2\gamma_{E,J}\sigma_X^2}{1+\gamma_{E,J}\sigma_X^2}\right] - \log\left[\sum_{i=1}^{K_X}\sum_{j=1}^{K_X} p_i p_j e^{d_{i,j}}\right]$$
$$- \frac{1}{2}\log\left[1+\gamma_{E,x}^2\sigma_X^2+\gamma_{E,J}^2\sigma_J^2\right] - \log\left[\sum_{i=1}^{K_J}\sum_{j=1}^{K_J} q_i q_j e^{d_{i,j}^J}\right], \tag{4.19}$$

with

$$d_{i,j} = \frac{-\gamma_B}{4\gamma_B\sigma_X^2+2}\left[\gamma_B\sigma_X^2(x_i-x_j)^2 - 2x_ix_j\right], \tag{4.20}$$

$$d_{i,j}^J = \frac{-\gamma_{E,J}}{4\gamma_{E,J}\sigma_J^2+2}\left[\gamma_{E,J}\sigma_J^2(s_{J,i}-s_{J,j})^2 - 2s_{J,i}s_{J,j}\right], \tag{4.21}$$

where $\gamma_B = \alpha_c^2(\mathbf{h}_B^T\mathbf{w}_x)^2/\sigma_B^2$, $\gamma_{E,J} = \alpha_c^2(\mathbf{h}_E^T\mathbf{w}_J)^2/\sigma_E^2$, $\gamma_{E,x} = \alpha_c^2(\mathbf{h}_E^T\mathbf{w}_x)^2/\sigma_E^2$, σ_X^2 and σ_J^2 represent the variance of x and s_J, respectively, x_i and $s_{J,i}$ refer to the ith mass points of the signals with probability p_i and q_i, respectively. Since the instantaneous $\mathbf{h}_E$ is unknown at Alice, the average secrecy rate $\mathbb{E}_{\mathbf{h}_E}[R_S^{AN}(\mathbf{h}_E)]$ could be selected as the performance metric. The optimal beamforming vectors $\mathbf{w}_x$ and $\mathbf{w}_J$ which maximize $\mathbb{E}_{\mathbf{h}_E}[R_S^{AN}(\mathbf{h}_E)]$ subject to the amplitude constraints can be solved numerically. Arfaoui et al. [9] further proposed a suboptimal and low-complexity approach to address this optimization problem. Figure 4.6 presents the relationship between the average secrecy capacity, average achievable secrecy rates, and A_c^2. The capacity result is achieved through numerical simulation. The secrecy rate achieved by brute-force search method is the optimal beamforming solution but with high complexity. The results demonstrate that the proposed low-complexity scheme provides marginal performance degradation compared to that of the optimal solution, and both schemes can result in a secrecy rate close to the capacity. In addition, a decent secrecy enhancement provided by AN can be observed by comparing the result of the proposed scheme with that without AN.

Another approach to optimize the beamforming vectors for the channel given in (4.18) is directly maximizing the SINR of the Bob subject to the average constraint of Eve's SINR [17]. The SINR of Bob can be expressed as

$$\gamma_B = \frac{\vartheta^2 I_{DC}^2\alpha_c^2\mathbf{w}_x^T\mathbf{h}_B\mathbf{h}_B^T\mathbf{w}_x}{\vartheta^2 I_{DC}^2\alpha_c^2\mathbf{w}_J^T\mathbf{h}_B\mathbf{h}_B^T\mathbf{w}_J + \sigma_B^2}, \tag{4.22}$$

where ϑ is the modulation index and I_{DC} refers to the average bias current of LED. The average SINR of Eve is given by

$$\overline{\gamma}_E = \mathbb{E}_{\mathbf{h}_E}\left[\frac{\vartheta^2 I_{DC}^2\alpha_c^2\mathbf{w}_x^T\mathbf{h}_E\mathbf{h}_E^T\mathbf{w}_x}{\vartheta^2 I_{DC}^2\alpha_c^2\mathbf{w}_J^T\mathbf{h}_E\mathbf{h}_E^T\mathbf{w}_J + \sigma_E^2}\right]. \tag{4.23}$$

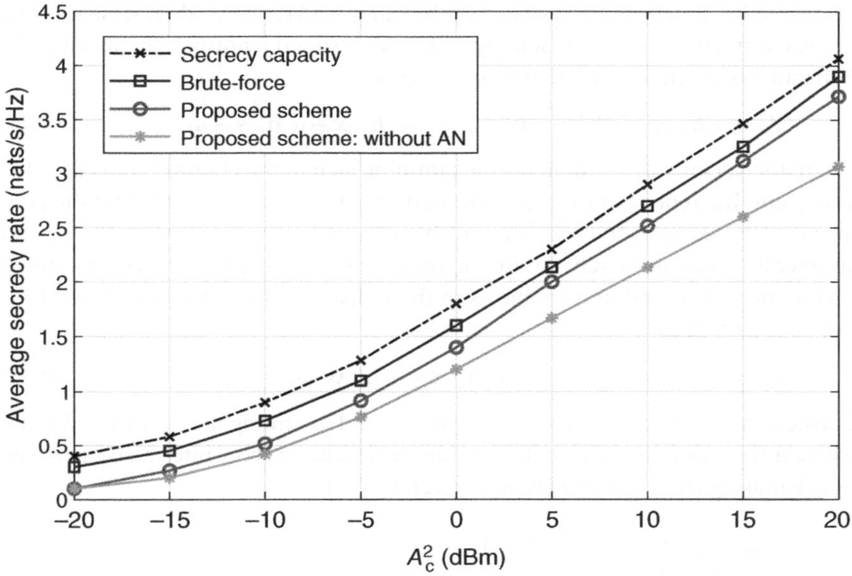

Figure 4.6 The average secrecy capacity and achievable secrecy rates versus the A_c^2. Source: Adapted from [9].

This SINR is only relevant to the statistical information of Eve rather than its instantaneous CSI. The optimal beamforming vectors $\mathbf{w}_X$ and $\mathbf{w}_J$ that maximize γ_B subject to the constraint that $\overline{\gamma}_E$ is less than a target value can be achieved by using the CCP-based approach [17]. The joint optimization of $\mathbf{w}_X$ and $\mathbf{w}_J$ for the MISO systems in the presence of multiple active EDs was investigated in [46]. Since the active and passive EDs might coexist in the coverage area, it would be crucial to consider the secrecy performance of such systems. To this end, an AN-based beamforming scheme maximizing the LR's SINR with the constraints of entirely suppressing the active EDs' SINR and restricting the passive EDs' average SINR was proposed in [19].

Besides the design of all the APs with the capability of transmitting both IB and jamming signals, a simpler AN-based beamforming design is letting some of the APs purely act as the IB signal transmitter and some just act as jammers [16, 33]. The beamforming matrix is designed to maximize the jamming signals at the EDs, while strategically aligning them within the null space of the LRs' channel. Adopting jamming techniques can effectively decrease the information received by the EDs, leading to elevated achievable secrecy rates across legitimate channels. Let us consider a MISO VLC wiretap system with one LR (Bob) and ED (Eve) and a single AP (Alice) is used to transmit IB signal, most likely the AP nearest to Bob. The jammer consists of the other N_J APs which do not have access to the confidential messages sent by Alice. With signal jamming, the received by Bob and Eve given in (4.14) can be modified as

$$Y_B = \alpha_c h_{AB} x + \alpha_c \mathbf{h}_{JB}^T \mathbf{s} + n_B, \quad Y_E = \alpha_c h_{AE} x + \alpha_c \mathbf{h}_{JE}^T \mathbf{s} + n_E, \tag{4.24}$$

where the vector $\mathbf{s} \in \mathbb{R}^{N_J}$ denotes the jamming signal, the channel gains denoted as h_{AB} and $\mathbf{h}_{JB}$ represent the signal strengths from Alice to Bob and from the jammer to Bob, respectively, and h_{AE} and $\mathbf{h}_{JE}$ channel gains from Alice to Eve and jammer to Eve, respectively. Assuming that $\mathbf{s} = \mathbf{w}s_J$, where $\mathbf{w}$ is the beamforming vector and s_J is a random variable, is in the nullspace of $\mathbf{h}_{JB}^T$, the signals in (4.24) can be rewritten as

$$Y_B = \alpha_c h_{AB} x + n_B, \quad Y_E = \alpha_c h_{AE} x + \alpha_c \mathbf{h}_{JE}^T \mathbf{w} s_J + n_E. \tag{4.25}$$

Considering an amplitude constraint for APs and assuming that both x and s_J are with continuous uniform distribution, the achievable secrecy rate can be achieved by calculating $\max(R_s, 0)$ where R_s is given by [33]

$$\begin{aligned} R_s = \frac{1}{2}\log\left(1 + \frac{2\alpha_c^2 h_{AB}^2 A_c^2}{\pi e \sigma^2}\right) \\ - \min\left(\log\frac{h_{AE}}{|\mathbf{h}_{JE}^T\mathbf{w}|} + \frac{|\mathbf{h}_{JE}^T\mathbf{w}|}{h_{AE}}\log\sqrt{e}, \frac{h_{AE}}{|\mathbf{h}_{JE}^T\mathbf{w}|}\log\sqrt{e}\right), \end{aligned} \tag{4.26}$$

where $\sigma_B^2 = \sigma_E^2 = \sigma^2$ is adopted. The beamformer $\mathbf{w}$ can be further optimized to maximize R_s. Upon Alice and the jammer having ED's CSI, the maximization of R_s is equivalent to the maximization of $\mathbf{h}_{JE}^T\mathbf{w}$. However, knowing Eve's CSI at transmitters is sometimes not practical, since Eve is usually a passive attacker. When Eve's CSI is not known but the transmitters know the bounded area where Eve is located, the optimization issue for the beamformer can be expressed as a max–min problem in which the secrecy rate in the worst-case scenario of Eve's position is maximized [33].

4.4 PLS for Multiuser VLC

4.4.1 Precoding Designs

In this section, the PLS of the VLC systems with multiple LRs (also known as multiuser (MU) VLC systems), as illustrated in Figure 4.7, is discussed. Note that in this section, the users refer to legitimate users. In such systems, the non-cooperative LRs can also be treated as potential EDs when they receive confidential messages intended for other LRs. Therefore, it is valuable to investigate the secrecy performance even in the absence of EDs [8, 42]. In this context, assume

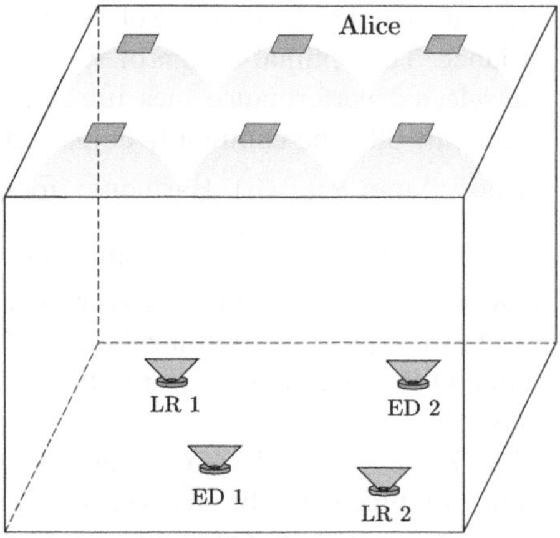

Figure 4.7 The indoor multiuser VLC wiretap system.

that there are N_T LEDs at the transmitter to broadcast information to K LRs. Denoting the vector of the transmitted symbols as $\mathbf{d} = [d_1\ d_2\ \dots\ d_K]^T$ and the precoding vector at the nth LED as $\mathbf{w}_n = [w_{n,1}\ w_{n,2}\ \cdots\ w_{n,K}]^T$, the transmitted signal at the nth LED is given by $x_n = \mathbf{w}_n^T\mathbf{d} + I_{\mathrm{DC},n}$ where $I_{\mathrm{DC},n}$ is the additive DC bias. The received signal at the kth LR can be expressed as

$$Y_{\mathrm{B},k} = \alpha_\mathrm{c}\mathbf{h}_{\mathrm{B},k}^T\mathbf{X} + n_{\mathrm{B},k}, \tag{4.27}$$

where $\mathbf{h}_{\mathrm{B},k} = [h_{k,1}\ h_{k,2}\ \cdots\ h_{k,N_\mathrm{T}}]^T$ is the channel gain vector for the kth receiver, $\mathbf{X} = [x_1\ x_2\ \dots\ x_{N_\mathrm{T}}]^T$ is the transmitted optical power vector, and $n_{\mathrm{B},k}$ is the AWGN noise term. After removing the DC bias and some mathematical manipulations, this received signal can be written by

$$Y_{\mathrm{B},k} = \alpha_\mathrm{c}\mathbf{h}_{\mathrm{B},k}^T\mathbf{W}\mathbf{d} + n_{\mathrm{B},k}, \tag{4.28}$$

where the matrix $\mathbf{W}$ with size $N_\mathrm{T} \times K$ is defined as $\mathbf{W} = [\tilde{\mathbf{w}}_1\ \tilde{\mathbf{w}}_2\ \dots\ \tilde{\mathbf{w}}_K]$ with the vector $\tilde{\mathbf{w}}_k = [w_{1,k}\ w_{2,k}\ \cdots\ w_{N_\mathrm{T},k}]^T$. Since only the kth symbol in $\mathbf{d}$ are the desired information for the kth LR, the received signal (4.28) can be rewritten as

$$Y_{\mathrm{B},k} = \alpha_\mathrm{c}\mathbf{h}_{\mathrm{B},k}^T\tilde{\mathbf{w}}_k d_k + \alpha_\mathrm{c}\mathbf{h}_{\mathrm{B},k}^T \sum_{i=1,i\neq k}^{K} \tilde{\mathbf{w}}_i d_i + n_{\mathrm{B},k}, \tag{4.29}$$

where the first term represents the intended signal for the kth the LR, while the second term accounts for multi-user interference (MUI). The receiver SINR can be expressed as

$$\gamma_k = \frac{\alpha_\mathrm{c}^2\left(\mathbf{h}_{\mathrm{B},k}^T\tilde{\mathbf{w}}_k\right)^2\sigma_k^2}{\sum_{i=1,i\neq k}^{K}\alpha_\mathrm{c}^2\left(\mathbf{h}_{\mathrm{B},k}^T\tilde{\mathbf{w}}_i\right)^2\sigma_i^2 + \sigma_\mathrm{B}^2}, \tag{4.30}$$

where σ_i^2 denotes the variance of the signal for the ith user and σ_{B}^2 is AWGN noise variance. The optimal design of $\tilde{\mathbf{w}}_k$, or equivalently the matrix $\mathbf{W}$, varies with the selected performance measure. Denote $R_{\mathrm{s},k}$ as the achievable secrecy rate of the kth LR. The commonly employed approaches include [30]: (i) Max–Min fairness: $\min_{1\leq k\leq K} R_{\mathrm{s},k}$; (ii) Harmonic mean: $K\left(\sum_{k=1}^{K} R_{\mathrm{s},k}^{-1}\right)^{-1}$; (iii) Proportional fairness: $\left(\prod_{k=1}^{K} R_{\mathrm{s},k}\right)^{1/K}$; (iv) Sum rate: $\sum_{k=1}^{K} R_{\mathrm{s},k}$. These measures are with an increasing order of achievable secrecy sum rate whereas a decreasing order of user fairness [11]. The optimization of $\mathbf{W}$ can be formulated as the maximization of one of the measures subject to the potential amplitude and/or average power constraints.

The ZF precoding design for the considered system is investigated in [42]. To realize ZF precoding, $\mathbf{h}_{\mathrm{B},k}^T\tilde{\mathbf{w}}_i = 0$, $\forall i \neq k$ should be satisfied. With ZF precoding, the MUI can be eliminated and for each LR no signals intended for other LRs can be detected as $Y_{\mathrm{B},k} = \alpha_{\mathrm{c}}\mathbf{h}_{\mathrm{B},k}^T\tilde{\mathbf{w}}_k d_k + n_{\mathrm{B},k}$. Since the analytical expression for the capacity of such a channel in the presence of non-negativity and amplitude constraints is mathematically intractable, the capacity lower bounds can be adopted as the achievable secrecy rate [27]. Pham et al. investigated the optimal ZF precoding designs when either max-min fairness or the sum rate is adopted as the performance measure [11]. When max-min fairness is considered, the optimization is a linear problem which can be efficiently solved using standard optimization tools; whereas, when maximum sum rate is considered, the problem is non-convex and a suboptimal solution based on CCP is proposed which provides local optimality.

Since ZF type of precoding is in general not optimal, following this line, numerous research initiatives have been committed to the optimal design of the precoding. In [37], the precoding of MU MISO VLC with two LRs was investigated. The objective of this work was finding the precoders which achieve the boundary points of the secrecy rate region. The weighted sum rate was adopted as the performance measure and the problem was formulated as the maximization of this sum rate considering the constraints on both total power and power per-antenna, along with an amplitude constraint. The proposed non-convex problem can be solved by transforming it into a simpler equivalent problem which can be solved iteratively through a subgradient search. Arfaoui et al. further investigated the optimal precoding designs for a more general MU MISO VLC broadcast systems with an arbitrary number of users under all the aforementioned four performance measures [8]. More specifically, considering that the users, excluding the kth user, are collectively regarded as an effective ED for the confidential link between Alice and kth LR, and assuming that the signals for different users have the same variance σ_u^2 (i.e., $\sigma_k^2 = \sigma_u^2$), the achievable secrecy rate can be calculated as max $\left(R_{s,k}, 0\right)$ with [8]

$$R_{s,k}(p_u, \mathbf{W}) = \tag{4.31}$$

$$\frac{1}{2}\log\left[\frac{1+\zeta_1\alpha_c^2\sum_{l=1}^{K}\left(\mathbf{h}_{B,k}^T\tilde{\mathbf{w}}_l\right)^2}{1+\zeta_2\alpha_c^2\sum_{l=1,l\neq k}^{K}\left(\mathbf{h}_{B,k}^T\tilde{\mathbf{w}}_l\right)^2}\right] - \frac{1}{2}\log\left[1+\zeta_2\alpha_c^2\sum_{i=1,i\neq k}^{K}\left(\mathbf{h}_{B,i}^T\tilde{\mathbf{w}}_k\right)^2\right],$$

where p_u denotes the probability density function of d_k, $\zeta_1 = \exp(2h_u)/2\pi e\sigma_B^2$, $\zeta_2 = \sigma_u^2/\sigma_B^2$, and h_u represents the differential entropy of the variable d_k. Note that it is assumed that p_u is continuous with support $[-A_c, A_c]$. The optimal $\mathbf{W}$ can be achieved by substituting $R_{s,k}(p_u, \mathbf{W})$ into one of the above-mentioned performance measures and calculating the matrix which results in the maximal performance measure subject to the amplitude constraint. It is demonstrated the proposed precoding scheme outperforms the ZF precoding scheme with regard to the secrecy performance. However, this comes with the trade-off of increased computational complexity. The precoding designs for MU VLC systems from the energy efficiency point-of-view were investigated in [22]. All the works mentioned above assume that there are no external EDs and LRs are considered as potential EDs. Some other works also investigate the case in the presence of external EDs [41]. In this case, not only the confidentiality among LRs but also that between LRs and EDs should be addressed.

Besides ZF precoding, the AN-aided precoding scheme can also be employed in MU MISO VLC systems to improve secrecy performance. AN is designed to maintain the SINRs of the LRs while at the same time jam the EDs' channel to reduce their SINRs. When AN-aided system is considered, the expression for the signal emitted by the nth LED is given by $x_n = \mathbf{w}_n^T\mathbf{d} + \rho_n\mathbf{G}_n^T\mathbf{z} + I_{DC,n}$, where $\mathbf{G}_n^T$ with a size of $1 \times (N_T - K)$ is the nth row of the precoding matrix $\mathbf{G}$, and $\mathbf{z} = [z_1\ z_2\ \dots\ z_{N_T-K}]^T$ is the AN noise vector. When the ED is passive, i.e., without EDs' CSI at Alice, the AN precoder can be designed in the null-space of $h_{B,k}^T$, i.e., $\mathbf{h}_{B,k}^T\mathbf{G} = \mathbf{0}$. Then, the signal received by the kth LR is still given by (4.29) which is without AN term, but the signal received by the ED turns to [43]

$$Y_{E,k} = \alpha_c\mathbf{h}_{E,k}^T\mathbf{W}\mathbf{d} + \alpha_c\mathbf{h}_{E,k}^T(\boldsymbol{\rho}\circ\mathbf{G})\mathbf{z} + n_{E,k}, \tag{4.32}$$

where $\boldsymbol{\rho} = [\rho_1\ \rho_2\ \dots\ \rho_{N_T}]^T$. The second term in (4.32) denotes the AN term. On the other hand, when the ED is active with its CSI available at Alice, steering AN precoder into the null space of the users' channel is not necessary. One can in effect jointly design the user's precoder $\mathbf{W}$ and AN's precoder optimally. Denote the AN signal vector as $\tilde{\mathbf{z}} = [\tilde{z}_1\ \tilde{z}_2\ \dots\ \tilde{z}_{N_\mathrm{T}}]^T$ and AN's precoding matrix as $\tilde{\mathbf{G}}$ with size $N_\mathrm{T} \times N_\mathrm{T}$. The expressions for the signals received by LRs and EDs are

$$Y_{\mathrm{B},k} = \alpha_\mathrm{c}\mathbf{h}_{\mathrm{B},k}^T\tilde{\mathbf{w}}_k d_k + \alpha_\mathrm{c}\mathbf{h}_{\mathrm{B},k}^T \sum_{i=1,i\neq k}^{K} \tilde{\mathbf{w}}_i d_i + \alpha_\mathrm{c}\mathbf{h}_{\mathrm{B},k}^T\tilde{\mathbf{G}}\tilde{\mathbf{z}} + n_{\mathrm{B},k}, \tag{4.33}$$

$$Y_{\mathrm{E},k} = \alpha_\mathrm{c}\mathbf{h}_{\mathrm{E},k}^T\mathbf{W}\mathbf{d} + \alpha_\mathrm{c}\mathbf{h}_{\mathrm{E},k}^T\tilde{\mathbf{G}}\tilde{\mathbf{z}} + n_{\mathrm{E},k}. \tag{4.34}$$

The AN term is denoted by the third term on the right-hand side of $Y_{\mathrm{B},k}$. Assuming there is a single ED present in the system, Pham et al. investigated the optimal design of both users' and AN's precoders when the ED was either passive or active [43]. For both cases, the max–min problem of the users' SINR was formulated subject to the amplitude constraint. For the system with active ED, an additional constraint imposed on the ED's SINR was employed to limit ED's SINR below a threshold.

4.4.2 PLS for NOMA-Based VLC

Because of its intrinsic broadcast characteristics, VLC has the potential to facilitate massive connectivity. The commonly used orthogonal multiple access (OMA) schemes include space division multiple access (SDMA), time division multiple access (TDMA), orthogonal frequency division multiple access (OFDMA), and code division multiple access (CDMA). With these schemes, different users are allocated orthogonal wireless resources and no mutual interference exists. However, because of the limited wireless resource, systems with only OMA schemes might not be able to support massive connectivities. To this end, non-orthogonal multiple access (NOMA) is proposed to overcome this issue. In NOMA, users are usually multiplexed based on the power allocation. It enables several users to utilize the same frequency and time resources through the utilization of superposition coding and successive interference cancellation (SIC) at the transmitter and receivers, respectively. Applying NOMA allows each user to leverage the entire spectrum resource consistently, leading to a substantial improvement in the total data transmission rate when compared to its OMA counterpart.

Several research endeavors have been dedicated to the PLS analysis of VLC systems with NOMA [21, 55]. Let us first consider a downlink MU SISO VLC system with a single ED and K LRs. The transmitted signal at the LED generated using superposition coding can be expressed as

$$X = \sum_{i=1}^{K} \sqrt{P_\mathrm{tot}\, p_i}\, s_i + I_\mathrm{DC}, \tag{4.35}$$

where P_tot denotes the overall electrical power carried by the modulated signals, p_i represents the power allocation coefficient for the ith user, s_i is the modulated symbol for the ith user, and I_{DC} is the DC bias. After propagating through free space and received by the PD at the receiver, the signal received by the kth LR, following the elimination of the DC component, is expressed as

$$Y_{B,k} = \alpha_c h_{B,k} \sum_{i=1}^{K} \sqrt{P_{tot}} p_i s_i + n_{B,k}, \tag{4.36}$$

where $h_{B,k}$ denotes the channel gain for the kth user and $n_{B,k}$ is the AWGN noise with variance σ_B^2. Without sacrificing generality, let us assume that the channel gains satisfy $0 < h_{B,1} \leq h_{B,2} \leq \cdots \leq h_{B,K}$. Based on the NOMA principle, the SIC should be applied to all the users except the first one. Therefore, user k is capable of decoding the signal for jth user with $j < k$ and then removing it from the received signal. However, when decoding its own data, the signal from the ith user with $i > k$ should be treated as interference. Hence, the SINR for the kth user can be formulated as

$$\gamma_{B,k} = \frac{\alpha_c^2 h_{B,k}^2 p_k^2 P_{tot}}{\alpha_c^2 h_{B,k}^2 P_{tot} \sum_{i=k+1}^{K} p_i^2 + \sigma_B^2}. \tag{4.37}$$

The secrecy performance of a special case of only two users and one ED was investigated in [55]. In this scenario, the SINR of the two users is given by $\gamma_{B,1} = \vartheta_1 p_1^2/(\vartheta_1 p_2^2 + 1)$ and $\gamma_{B,2} = \vartheta_2 p_2^2$ where $\vartheta_i = \alpha_c^2 h_{B,i}^2 P_{tot}/\sigma_B^2$. With the derived SINRs, the achievable data rates, denoted by $C(\gamma_{B,1})$ and $C(\gamma_{B,2})$, can be calculated by using various capacity lower bounds available in the literature [14]. In terms of the ED, one can assume the worst-case scenario that ED knows the orders of decoding and the coefficients used for power allocation. This assumption is commonly assumed in NOMA-based RF systems to provide a worst-case secrecy performance benchmark. With this assumption, the SINR of ED can be given by $\gamma_E = \vartheta_E p_2^2$ where $\vartheta_E = \alpha_c^2 h_E^2 P_{tot}/\sigma_E^2$ and the achievable rate can be expressed as $C(\gamma_E)$ [55]. The secrecy rates for the two LRs can be written as $R_{s,1} = [C(\gamma_{B,1}) - C(\gamma_E)]^+$ and $R_{s,2} = [C(\gamma_{B,1}) - C(\gamma_E)]^+$, respectively. Considering that ED is randomly located and Alice lacks knowledge of the instantaneous CSI of the ED, instead she has information only about its statistics, the SOP of both LRs, i.e., $\Pr\left\{R_{s,i} \leq C_{th}\right\}$, can be derived as presented in [55]. The authors further extended the considered system to that with multiple EDs whose locations obey the homogeneous PPP and derived the corresponding SOPs for both LRs via the stochastic geometry theory. It was concluded that higher LED transmission power and lower ED density render enhanced SOP performance. In addition, the secrecy performance of VLC attocells employing NOMA influenced by the user mobility was investigated [54]. The optimization problems are formulated to determine the optimal resource allocation strategy which consists of the AP allocation in the presence of user mobility and also the transmitted power allocation of each AP to its associated users.

Besides the discussed SISO system, the secrecy performance for MISO VLC employing NOMA has also been investigated, highlighting the superior performance of the MISO system compared to the SISO counterpart [21]. Due to the extra spatial DoFs provided by MISO, the beamforming design can be optimized to improve the confidentiality performance. Specifically, consider a MISO VLC system with N_{T} LEDs, K LRs, and a single ED. The signal vector at the transmitter side is given by

$$\mathbf{X} = \sum_{k=1}^{K} \mathbf{w}_k s_k + \mathbf{b}, \tag{4.38}$$

where s_k denotes the signal for the kth LR, $\mathbf{b} = [I_{\mathrm{DC}}\ I_{\mathrm{DC}}\ \cdots\ I_{\mathrm{DC}}]^T$ with length N_{T} is the bias vector, and $\mathbf{w}_k$ is the beamforming vector for s_k which should be designed to satisfy the potential constraints, e.g., the transmission electrical power of LED, the peak optical power constraint, and practical illumination constraints by the dimming [21]. The signals received by the LRs and ED are given by

$$Y_{\mathrm{B},k} = \alpha_{\mathrm{c}} \mathbf{h}_{\mathrm{B},k}^T \sum_{j=1}^{K} \mathbf{w}_j s_j + \alpha_{\mathrm{c}} \mathbf{h}_{\mathrm{B},k}^T \mathbf{b} + n_{\mathrm{B},k}, \quad 1 \le k \le K, \tag{4.39}$$

$$Y_{\mathrm{E}} = \alpha_{\mathrm{c}} \mathbf{h}_{\mathrm{E}}^T \sum_{j=1}^{K} \mathbf{w}_j s_j + \alpha_{\mathrm{c}} \mathbf{h}_{\mathrm{E}}^T \mathbf{b} + n_{\mathrm{E}}, \tag{4.40}$$

respectively. Again assuming that the norm of the channel gain vector increases with the index of the LR, for the kth LR, the signal $\{s\}_{i=1}^{k-1}$ can be decoded and the corresponding interference can be removed. If the kth user is used to decode the ith signal, the signal after SIC can be formulated as $Y_{\mathrm{B},k,i}^{\mathrm{SIC}} = \alpha_{\mathrm{c}} \mathbf{h}_{\mathrm{B},k}^T \sum_{j=i}^{K} \mathbf{w}_j s_j + \mathbf{h}_{\mathrm{B},k}^T \mathbf{b} + n_{\mathrm{B},k}$ with $i \le k$. The secrecy capacity for the signal s_i is given by [31]

$$R_{\mathrm{s},i} = \max_{p_i(s_i)} \left[\min_{1 \le k \le K} I(s_i; Y_{\mathrm{B},k,i}^{\mathrm{SIC}}) - I(s_i; Y_{\mathrm{E}}) \right]^+, \tag{4.41}$$

where $p_i(s_i)$ denotes the probability density function of s_i. However, the closed-form expression of $R_{\mathrm{s},i}$ is mathematically intractable. To this end, Du et al. derived both upper and lower bounds of secrecy capacity, upon which the optimal beamforming vectors $\{\mathbf{w}_k\}_{k=1}^{K}$ designed for distinct objectives, i.e., total transmitted power minimization and minimum secrecy rate maximization, were developed [21].

4.5 PLS for VLC with Emerging Technologies

Machine learning (ML) and intelligent reflecting surfaces (IRS) as two of the emerging technologies that have been vastly studied for VLC to improve the PLS

[4, 15, 48, 51]. In this section, we first present the ML-based techniques in VLC for PLS. Then, we discuss how IRS can improve the PLS.

ML has emerged as an effective solution to deal with complex problems in many applications. It has been widely used in communication systems for resource allocations, signal detection, channel prediction as well as PLS. Methods, such as Q-learning based on reinforcement learning (RL) [50], offer an efficient approach for beamforming selection in the MISO VLC systems, as the dynamic anti-eavesdropping communication system's beamforming policy can be represented through a Markov decision process (MDP). However, the Q-learning-based methods fail to effectively handle discrete and low-dimensional action spaces. Deep reinforcement learning can achieve the optimal beamforming policy against the ED to remarkably reduce the LR's error rate and enhance both the secrecy rate and utility of the MISO VLC systems with anti-eavesdropping feature [51].

In case of the MISO VLC wiretap channel, one has the constraint on average power as $|\mathbf{x}| \preceq \alpha I\mathbf{1}$, where $\preceq$ denotes the entry-wise inequality, meaning that the amplitude of every element in vector $\mathbf{x}$ does not exceed αI. Here, I is the current bias and $\alpha \in [0, 1]$ is the modulation index. The achievable secrecy rate for the MISO VLC wiretap channel with beamforming, represented by R_s, can be calculated as [35]:

$$R_s = \frac{1}{2}\log\frac{6\alpha^2 I^2 \mathbf{w}^T \mathbf{h}_B \mathbf{h}_B^T \mathbf{w} + 3\pi e\sigma^2}{\pi e\alpha^2 I^2 \mathbf{w}^T \mathbf{h}_E \mathbf{h}_E^T \mathbf{w} + 3\pi e\sigma^2}. \quad (4.42)$$

The beamforming vector $\mathbf{w}^*$ that achieves the maximum secrecy rate R_s in the above equation is theoretically derived by the solution of the following problem:

$$\mathbf{w}^* = \arg\max \frac{1}{2}\log\frac{6\alpha^2 I^2 \mathbf{w}^T \mathbf{h}_B \mathbf{h}_B^T \mathbf{w} + 3\pi e\sigma^2}{\pi e\alpha^2 I^2 \mathbf{w}^T \mathbf{h}_E \mathbf{h}_E^T \mathbf{w} + 3\pi e\sigma^2}, \quad (4.43)$$

$$\text{s.t.} \ |\mathbf{w}| \preceq \mathbf{1}. \quad (4.44)$$

Solving this problem presents a challenge as it is a problem that is not convex and falls within the category of NP-hard problems [35].

In a dynamic VLC system, the transmitter determines its beamforming strategy based on the acquired Q-values of the present system state, directing the transmission of the data signal to the LR Bob. The current state includes the current CSI of Bob, the previous BER, and the secrecy rate of the system. The transmitter's action, i.e., its beamforming strategy, influences the next system state, hence impacting the future actions in the learning process and the future rewards. Hence, the succeeding system state solely relies on the state and action of the present time slot and is unrelated to preceding states or actions [50].

The performance of the deep reinforcement learning (DRL)-based PLS has been compared with the ZF method, RL-based, and the fixed friendly jamming (FFJ) method [33] through simulations with respect to different SNR values as

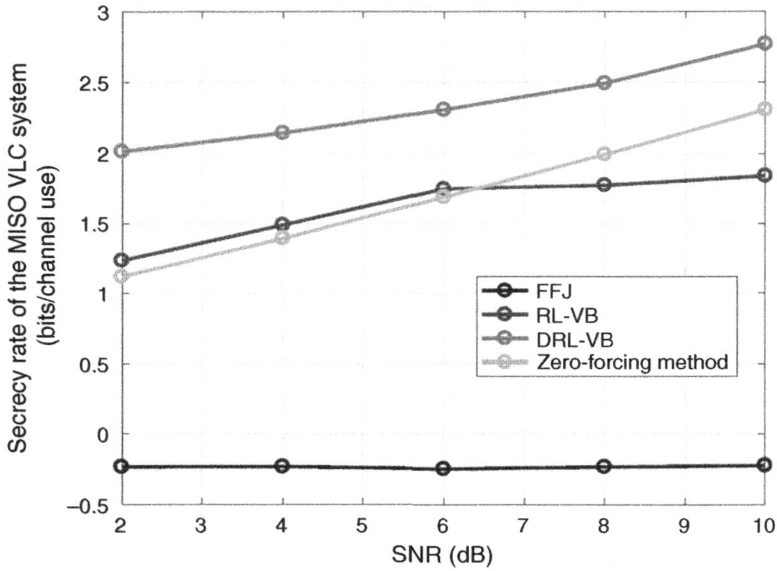

Figure 4.8 Secrecy rate of the MISO VLC beamforming system. Source: Adapted from [50].

shown in Figure 4.8. The LR's secrecy rate exhibits an upward trend with the increase in SNR, as observed in the presented average performance calculated across 5000 time slots. The results confirm that the DRL outperforms both RL and ZF by a secrecy rate of more than 1 bits per channel use and 0.5 bits per channel use, respectively.

Most of the studies in VLC have considered only the LoS channel gain due to the fact that exploring the secrecy rate when both LoS and NLoS channel gains are included is more time-consuming and complicated. However, ML-aided algorithms can help to evaluate the VLC system performance in the presence of NLoS channel gains. For instance, in [15], a deep neural network model that characterizes the spatial response of the VLC channel is introduced. The model considers both specular and diffusive reflections and is built upon a limited set of channel response samples generated from the ray tracing model. Utilizing the suggested model enables the derivation of upper and lower bounds for VLC secrecy capacity, which facilitates an assessment of VLC communication confidentiality considering various factors such as transceiver and eavesdropper positions. It is demonstrated that because of different reflection types, the vulnerability of the VLC system varies across locations, with heightened susceptibility observed in the presence of strong reflections.

Nowadays, RISs have become a popular solution for addressing situations where a barrier obstructs the LoS between a transmitter and receiver, disturbing the transmission of signals. The advantages of using IRS for VLC mainly include improved coverage and energy efficiency, enhanced data rates and PLS, flexibility, and also low cost [1]. In particular, enhancing the security of the physical layer can be achieved by the IRS through the establishment of a more intricate wireless communication environment, which makes it more difficult for attackers to intercept or eavesdrop on transmissions. By reflecting and scattering the signals, IRS can also create multiple paths for the signals to reach the receiver, increasing the diversity of the received signals and making it more challenging for an attacker to interfere with the communication. Additionally, IRS can dynamically adjust the phase and amplitude of the reflected signals to control the channel conditions, contributing to an increased level of security in communication.

For VLC systems, incorporating IRS can also add a PLS advantage by exploiting the directional nature of light. By controlling the direction of the reflected light, IRS can create highly directional channels, which makes it more difficult for an attacker to intercept the signals from an off-angle direction. This provides an extra layer of security compared to traditional wireless communication systems that rely on omnidirectional antennas. An IRS consists of many passive or active reflecting elements that can be intelligently directed. This allows the light rays to be reflected toward a specific user, as shown in Figure 4.9 [4]. Therefore, managing the light propagation can be employed to improve the overall performance of the system. An in-depth examination of the advantages and disadvantages of the IRS in the context of VLC is discussed in [5].

The enhancement in security at the physical layer through the use of IRS may differ based on the particular implementation and deployment situation. Configuring numerous parameters is essential when addressing PLS with IRS. These parameters encompass IRS reflectivity, the quantity of IRS units, placement of IRS units, and orientation of IRS units [48]. The influence of IRS reflectivity and the quantity of IRS units on the secrecy rate is demonstrated in Figure 4.10. It is

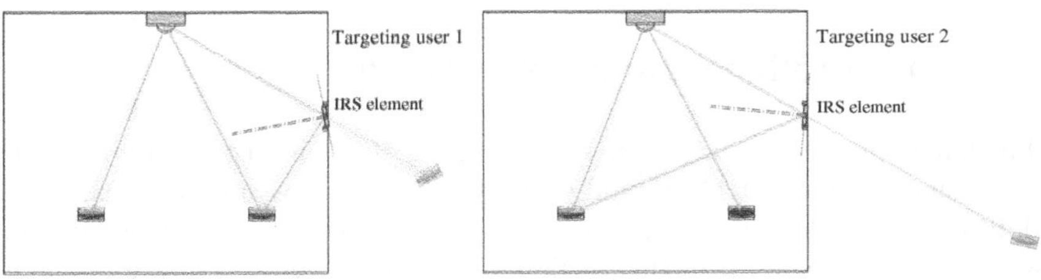

Figure 4.9 Configuration of the IRS: each reflecting element can direct the reflected light in a specific direction, allowing it to serve individual users sequentially. Source: [4] /with permission from IEEE.

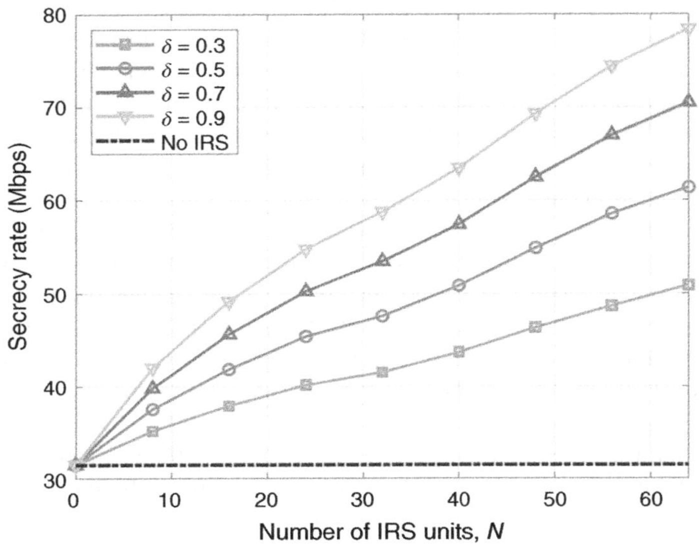

Figure 4.10 The relationship between the secrecy rate and the quantity of IRS units across various levels of reflectivity denoted by δ. Source: Adapted from[48].

shown that, for a transmitted power of 10 dBW, the overall secrecy rate is positively affected by the number of IRS units N. Specifically, when δ is set to 0.5, increasing N from 0 to 64 results in a twofold increase in the secrecy rate. The findings also indicate that elevating the reflectivity parameter (δ) would contribute to an increased secrecy rate. For example, achieving a secrecy rate of 40 Mbps is possible when δ is 0.7 with 8 IRS units. However, as the value of δ decreases to 0.3, the required IRS unit number would increase to 24. These results imply that optimizing the overall secrecy rate should involve a thoughtful selection of the quantity of IRS units and reflectivity values.

4.6 Open Challenges and Future Works

This chapter presents a study of the PLS for indoor VLC. Some commonly adopted VLC systems, e.g., SISO, MISO, and multiuser VLC systems, are investigated. In addition, the PLS for VLC systems with emerging technologies, such as ML and IRS, are also introduced. In this section, some challenges and future research trends of the VLC PLS are presented.

Considering the perspective of information theory, it is essential to dedicate research endeavors to explore the secrecy capacity of VLC and determine the optimal probability distribution for achieving this capacity. In the existing literature, solely the analytical upper and lower limits have been formulated, while the closed-form expression for secrecy capacity remains unavailable, constituting an ongoing challenge. Despite the derivations of the secrecy capacity, it is also valuable to come up with novel bounds with tighter gaps to the capacity and propose some numerical methods with low computational complexity for the capacity calculation.

In most of the VLC studies, the receivers are assumed to be stationary and vertically upward and without any blockage. However, this is not the case in many real-world situations where user mobility, and random device orientation and blockage are present. In particular, recent studies illustrated that the polar angle of the device orientation can be modeled by either Laplace for static users and Gaussian distribution for mobile users [47]. It is concluded that the device orientation can strongly influence communication performance. Some models have also been proposed to model user mobility in indoor scenarios, e.g., the random waypoint model. The link blockage in VLC can be introduced by moving objects such as mobile users in indoor environments which can cause a temporary loss of connection. To simplify the blockage model, the blockers are normally modeled as cylindrical objects. In the literature, there is still a lack of studies investigating the PLS for VLC with a comprehensive model considering practical factors like user mobility, random blockage and device orientation. Analyzing the secrecy performance with such a complete channel is crucial which can reflect indoor VLC systems in practical scenarios more accurately.

Since the received optical power in VLC systems is mainly determined by the transceiver geometry, EDs can easily enhance their eavesdropping capability through the modifications of receivers, for example, employing larger detectors or lenses and smartly adjusting their orientations toward the APs. The currently available secrecy improvements techniques such as ZF precoding and AN-based precoding could be failed in the presence of the advanced EDs [20]. Therefore, more robust PLS techniques designed considering the possible presence of intelligent EDs are desired.

Most of the studies in the context of PLS for VLC consider either SISO or MISO systems, whereas few works have investigated the PLS for MIMO VLC systems [10]. The extra spatial degree of freedom (DoFs) provided by MIMO at the receiver side should have the potential to enhance the secrecy performance which deserve to be further exploited. Besides the PLS for VLC systems investigated in the chapter, the investigations of PLS for other OWC-related systems, e.g., FSO systems [44], hybrid VLC/RF systems [26], FSO-VLC cooperative systems [45], and RF-FSO cooperative systems [40], are also of great importance and deserve more research efforts.

Future work should also focus on developing algorithms that jointly consider both the PLS and VLC aspects of these systems. This involves optimizing multiple factors, including the quantity and positioning of IRS units, the modulation and coding techniques used, and the power levels of the transmitted signals. Balancing the trade-off between PLS and VLC performance is a challenge, as increasing the PLS often reduces the VLC performance and vice versa. This requires the development of algorithms that can jointly optimize both the PLS and VLC aspects of these systems. In addition, the implementation and experimentation of PLS algorithms in real-world VLC systems is another area of future work. This demands the development of hardware prototypes and the creation of testbeds to assess how effectively these algorithms perform in real-world scenarios. The degradation of the received signal quality can significantly reduce the overall PLS of the VLC system, making it more vulnerable to eavesdropping attacks. This highlights the importance of developing algorithms that can effectively mitigate possible signal impairments and ensure the secure operation of VLC systems.

References

1 S. Aboagye, A. R. Ndjiongue, T. M. N. Ngatched, O. A. Dobre, and H. V. Poor. RIS-assisted visible light communication systems: A tutorial. *IEEE Communications Surveys & Tutorials*, 25(1):251–288, 2022. doi: 10.1109/COMST.2022.3225859.

2 H. Abumarshoud, M. D. Soltani, M. Safari, and H. Haas. Secrecy capacity of LiFi systems. In *Emerging Imaging and Sensing Technologies for Security and Defence V; and Advanced Manufacturing Technologies for Micro- and Nanosystems in Security and Defence III*, volume 11540, page 115400Y. International Society for Optics and Photonics, SPIE, 2020.

3 H. Abumarshoud, M. D. Soltani, M. Safari, and H. Haas. Realistic secrecy performance analysis for LiFi systems. *IEEE Access*, 9:120675–120688, 2021. doi: 10.1109/ACCESS.2021.3108727.

4 H. Abumarshoud, C. Chen, I. Tavakkolnia, H. Haas, and M. A. Imran. Intelligent reflecting surfaces for enhanced physical layer security in NOMA VLC systems. *arXiv preprint arXiv:2211.09456*, 2022.

5 H. Abumarshoud, L. Mohjazi, O. A. Dobre, M. Di Renzo, M. A. Imran, and H. Haas. LiFi through reconfigurable intelligent surfaces: A new frontier for 6G? *IEEE Vehicular Technology Magazine*, 17(1):37–46, 2022. doi: 10.1109/MVT.2021.3121647.

6 M. A. Arfaoui, Z. Rezki, A. Ghrayeb, and M. S. Alouini. Discrete input signaling for MISO visible light communication channels. In *2017 IEEE Wireless Communications and Networking Conference (WCNC)*, pages 1–6, 2017. doi: 10.1109/WCNC.2017.7925912.

7 M. A. Arfaoui, A. Ghrayeb, and C. Assi. Secrecy rate closed-form expressions for the SISO VLC wiretap channel with discrete input signaling. *IEEE Communications Letters*, 22(7):1382–1385, 2018. doi: 10.1109/LCOMM.2018.2829479.

8 M. A. Arfaoui, A. Ghrayeb, and C. M. Assi. Secrecy performance of multi-user MISO VLC broadcast channels with confidential messages. *IEEE Transactions on Wireless Communications*, 17(11):7789–7800, 2018. doi: 10.1109/TWC.2018.2871055.

9 M. A. Arfaoui, H. Zaid, Z. Rezki, A. Ghrayeb, A. Chaaban, and M.-S. Alouini. Artificial noise-based beamforming for the MISO VLC wiretap channel. *IEEE Transactions on Communications*, 67(4):2866–2879, 2019. doi: 10.1109/TCOMM.2018.2889649.

10 M. A. Arfaoui, A. Ghrayeb, and C. M. Assi. Secrecy performance of the MIMO VLC wiretap channel with randomly located eavesdropper. *IEEE Transactions on Wireless Communications*, 19(1):265–278, 2020. doi: 10.1109/TWC.2019.2944144.

11 M. A. Arfaoui, M. D. Soltani, I. Tavakkolnia, A. Ghrayeb, M. Safari, C. M. Assi, and H. Haas. Physical layer security for visible light communication systems: A survey. *IEEE Communications Surveys & Tutorials*, 22(3):1887–1908, 2020. doi: 10.1109/COMST.2020.2988615.

12 P. T. Boggs and J. W. Tolle. Sequential quadratic programming. *Acta Numerica*, 4:1–51, 1995.

13 J. B. Carruthers and P. Kannan. Iterative site-based modeling for wireless infrared channels. *IEEE Transactions on Antennas and Propagation*, 50(5):759–765, 2002. doi: 10.1109/TAP.2002.1011244.

14 A. Chaaban, J.-M. Morvan, and M.-S. Alouini. Free-space optical communications: Capacity bounds, approximations, and a new sphere-packing perspective. *IEEE Transactions on Communications*, 64(3):1176–1191, 2016. doi: 10.1109/TCOMM.2016.2524569.

15 J. Chen and T. Shu. Statistical modeling and analysis on the confidentiality of indoor VLC systems. *IEEE Transactions on Wireless Communications*, 19(7):4744–4757, 2020. doi: 10.1109/TWC.2020.2986964.

16 S. Cho, G. Chen, and J. P. Coon. Securing visible light communications with spatial jamming. In *ICC 2019 - 2019 IEEE International Conference on Communications (ICC)*, pages 1–6, 2019. doi: 10.1109/ICC.2019.8761165.

17 S. Cho, G. Chen, and J. P. Coon. Enhancement of physical layer security with simultaneous beamforming and jamming for visible light

communication systems. *IEEE Transactions on Information Forensics and Security*, 14(10):2633–2648, 2019. doi: 10.1109/TIFS.2019.2904440.

18 S. Cho, G. Chen, and J. P. Coon. Zero-forcing beamforming for active and passive eavesdropper mitigation in visible light communication systems. *IEEE Transactions on Information Forensics and Security*, 16:1495–1505, 2021. doi: 10.1109/TIFS.2020.3036806.

19 S. Cho, G. Chen, and J. P. Coon. Cooperative beamforming and jamming for secure VLC system in the presence of active and passive eavesdroppers. *IEEE Transactions on Green Communications and Networking*, 5(4):1988–1998, 2021. doi: 10.1109/TGCN.2021.3092146.

20 S. Cho, G. Chen, J. P. Coon, and P. Xiao. Challenges in physical layer security for visible light communication systems. *Network*, 2(1):53–65, 2022.

21 C. Du, F. Zhang, S. Ma, Y. Tang, H. Li, H. Wang, and S. Li. Secure transmission for downlink NOMA visible light communication networks. *IEEE Access*, 7:65332–65341, 2019. doi: 10.1109/ACCESS.2019.2916548.

22 S. T. Duong, T. V. Pham, C. T. Nguyen, and A. T. Pham. Energy-efficient precoding designs for multi-user visible light communication systems with confidential messages. *IEEE Transactions on Green Communications and Networking*, 5(4):1974–1987, 2021. doi: 10.1109/TGCN.2021.3091829.

23 G. A. Gabriele and K. M. Ragsdell. The generalized reduced gradient method: A reliable tool for optimal design. *ASME Journal of Engineering for Industry*, 99(2):394–400, 1977.

24 H. Haas, L. Yin, Y. Wang, and C. Chen. What is LiFi? *Journal of Lightwave Technology*, 34(6):1533–1544, 2016. doi: 10.1109/JLT.2015.2510021.

25 J. M. Kahn and J. R. Barry. Wireless infrared communications. *Proceedings of the IEEE*, 85(2):265–298, 1997. doi: 10.1109/5.554222.

26 A. Kumar, P. Garg, and A. Gupta. PLS analysis in an indoor heterogeneous VLC/RF network based on known and unknown CSI. *IEEE Systems Journal*, 15(1):68–76, 2021. doi: 10.1109/JSYST.2020.2964033.

27 A. Lapidoth, S. M. Moser, and M. A. Wigger. On the capacity of free-space optical intensity channels. *IEEE Transactions on Information Theory*, 55(10):4449–4461, 2009. doi: 10.1109/TIT.2009.2027522.

28 Q. C. Li, H. Niu, A. T. Papathanassiou, and G. Wu. 5G network capacity: Key elements and technologies. *IEEE Vehicular Technology Magazine*, 9(1):71–78, 2014.

29 Y. Liu, H.-H. Chen, and L. Wang. Physical layer security for next generation wireless networks: Theories, technologies, and challenges. *IEEE Communications Surveys & Tutorials*, 19(1):347–376, 2017.

30 Z.-Q. Luo and S. Zhang. Dynamic spectrum management: Complexity and duality. *IEEE Journal of Selected Topics in Signal Processing*, 2(1):57–73, 2008. doi: 10.1109/JSTSP.2007.914876.

31 S. Ma, Z.-L. Dong, H. Li, Z. Lu, and S. Li. Optimal and robust secure beamformer for indoor MISO visible light communication. *Journal of Lightwave Technology*, 34(21):4988–4998, 2016. doi: 10.1109/JLT.2016.2605000.

32 M. Mohammadi, S. M. S. Sadough, and Z. Ghassemlooy. Physical layer security enhancement in VLC using zero-forcing beamforming and optimized LED placement. In *2022 13th International Symposium on Communication Systems, Networks and Digital Signal Processing (CSNDSP)*, pages 452–456, 2022. doi: 10.1109/CSNDSP54353.2022.9907923.

33 A. Mostafa and L. Lampe. Securing visible light communications via friendly jamming. In *2014 IEEE Globecom Workshops (GC Wkshps)*, pages 524–529, 2014. doi: 10.1109/GLOCOMW.2014.7063485.

34 A. Mostafa and L. Lampe. Physical-layer security for indoor visible light communications. In *2014 IEEE International Conference on Communications (ICC)*, pages 3342–3347, 2014. doi: 10.1109/ICC.2014.6883837.

35 A. Mostafa and L. Lampe. Physical-layer security for MISO visible light communication channels. *IEEE Journal on Selected Areas in Communications*, 33(9):1806–1818, 2015. doi: 10.1109/JSAC.2015.2432513.

36 A. Mostafa and L. Lampe. Optimal and robust beamforming for secure transmission in MISO visible-light communication links. *IEEE Transactions on Signal Processing*, 64(24):6501–6516, 2016. doi: 10.1109/TSP.2016.2603964.

37 A. Mostafa and L. Lampe. On linear precoding for the two-user MISO broadcast channel with confidential messages and per-antenna constraints. *IEEE Transactions on Signal Processing*, 65(22):6053–6068, 2017. doi: 10.1109/TSP.2017.2745454.

38 M. Obeed, A. M. Salhab, M.-S. Alouini, and S. A. Zummo. On optimizing VLC networks for downlink multi-user transmission: A survey. *IEEE Communications Surveys & Tutorials*, 21(3):2947–2976, 2019. doi: 10.1109/COMST.2019.2906225.

39 O. Ozel, E. Ekrem, and S. Ulukus. Gaussian wiretap channel with amplitude and variance constraints. *IEEE Transactions on Information Theory*, 61(10):5553–5563, 2015. doi: 10.1109/TIT.2015.2459705.

40 D. R. Pattanayak, V. K. Dwivedi, V. Karwal, I. S. Ansari, H. Lei, and M.-S. Alouini. On the physical layer security of a decode and forward based mixed FSO/RF co-operative system. *IEEE Wireless Communications Letters*, 9(7):1031–1035, 2020. doi: 10.1109/LWC.2020.2979442.

41 T. V. Pham and A. T. Pham. Secrecy sum-rate of multi-user MISO visible light communication systems with confidential messages. *Optik*, 151:65–76, 2017.

42 T. V. Pham, H. Le-Minh, and A. T. Pham. Multi-user visible light communication broadcast channels with zero-forcing precoding. *IEEE Transactions on Communications*, 65(6):2509–2521, 2017. doi: 10.1109/TCOMM.2017.2691710.

43 T. V. Pham, T. Hayashi, and A. T. Pham. Artificial-noise-aided precoding design for multi-user visible light communication channels. *IEEE Access*, 7:3767–3777, 2019. doi: 10.1109/ACCESS.2018.2889119.

44 M. J. Saber and S. M. S. Sadough. On secure free-space optical communications over Málaga turbulence channels. *IEEE Wireless Communications Letters*, 6(2):274–277, 2017. doi: 10.1109/LWC.2017.2671872.

45 V. N. Saxena, J. Gupta, and V. K. Dwivedi. Secured end-to-end FSO-VLC based IoT network with randomly positioned VLC: Known and unknown CSI. *IEEE Internet of Things Journal*, 10(2):1347–1357, 2022.

46 H. Shen, Y. Deng, W. Xu, and C. Zhao. Secrecy-oriented transmitter optimization for visible light communication systems. *IEEE Photonics Journal*, 8(5):1–14, 2016. doi: 10.1109/JPHOT.2016.2598684.

47 M. D. Soltani, A. A. Purwita, Z. Zeng, H. Haas, and M. Safari. Modeling the random orientation of mobile devices: Measurement, analysis and LiFi use case. *IEEE Transactions on Communications*, 67(3):2157–2172, 2019. doi: 10.1109/TCOMM.2018.2882213.

48 S. Sun, F. Yang, J. Song, and Z. Han. Optimization on multiuser physical layer security of intelligent reflecting surface-aided VLC. *IEEE Wireless Communications Letters*, 11(7):1344–1348, 2022. doi: 10.1109/LWC.2022.3167108.

49 J.-Y. Wang, C. Liu, J.-B. Wang, Y. Wu, M. Lin, and J. Cheng. Physical-layer security for indoor visible light communications: Secrecy capacity analysis. *IEEE Transactions on Communications*, 66(12):6423–6436, 2018. doi: 10.1109/TCOMM.2018.2859943.

50 L. Xiao, D. Jiang, D. Xu, H. Zhu, Y. Zhang, and H. V. Poor. Two-dimensional antijamming mobile communication based on reinforcement learning. *IEEE Transactions on Vehicular Technology*, 67(10):9499–9512, 2018. doi: 10.1109/TVT.2018.2856854.

51 L. Xiao, G. Sheng, S. Liu, H. Dai, M. Peng, and J. Song. Deep reinforcement learning-enabled secure visible light communication against eavesdropping. *IEEE Transactions on Communications*, 67(10):6994–7005, 2019. doi: 10.1109/TCOMM.2019.2930247.

52 F. Yang, J. Wang, and Y. Dong. Physical-layer security for indoor VLC wiretap systems under multipath reflections. *IEEE Transactions on Wireless Communications*, 21(12):11179–11192, 2022. doi: 10.1109/TWC.2022.3190362.

53 L. Yin and H. Haas. Physical-layer security in multiuser visible light communication networks. *IEEE Journal on Selected Areas in Communications*, 36(1):162–174, 2018. doi: 10.1109/JSAC.2017.2774429.

54 X. Zhao and J. Sun. Physical-layer security for mobile users in NOMA-enabled visible light communication networks. *IEEE Access*, 8:205411–205423, 2020. doi: 10.1109/ACCESS.2020.3037180.

55 X. Zhao, H. Chen, and J. Sun. On physical-layer security in multiuser visible light communication systems with non-orthogonal multiple access. *IEEE Access*, 6:34004–34017, 2018. doi: 10.1109/ACCESS.2018.2847744.

56 X. Zhou, L. Song, and Y. Zhang. *Physical Layer Security in Wireless Communications*. CRC Press, 2013.

5

The Impact of Secrecy on Stable Throughput and Delay

Parthajit Mohapatra[1] *and Nikolaos Pappas*[2]

[1]*Department of Electrical Engineering, Indian Institute of Technology Tirupati, Tirupati, India*
[2]*Department of Computer and Information Science, Linköping University, Linköping, Sweden*

5.1 Introduction

The field of wireless communication currently stands at the forefront of a new era characterized by the transition to the new generation of wireless communications, namely, the rise of 5G. With the recent emergence of novel communication scenarios such as machine-to-machine (M2M) communications and Internet-of-Things (IoT) applications, Internet traffic is experiencing exponential growth. Future networks must accommodate diverse Quality of Service (QoS) demands, spanning from low throughput and moderate latency (M2M) to exceedingly high throughput and minimal latency (high-definition real-time video). Hence, characterizing the performance limits is essential for designing 5G and beyond communication networks efficiently. Besides these, data security has been another concern as most devices will be connected to the Internet through a wireless medium. Thus, meeting the need for high throughput, reliability, and low latency along with data security poses new challenges for future wireless networks and the need to develop new techniques and performance metrics.

Physical-layer secrecy is a potential scheme to ensure secure communication over wireless channels. However, most works in physical-layer secrecy ignore the underlying traffic model. In general, information theoretic-inspired metrics such as secrecy rate and secrecy capacity region do not take the underlying traffic model, and it is assumed that the user always has data to send. On the other hand, network theory provides the mathematical framework or tools to capture communication's underlying traffic model and latency aspects. One such relevant metric is the *stability region* introduced in [20]. The stability region in network theory

Physical-Layer Security for 6G, First Edition. Edited by Parthajit Mohapatra, Nikolaos Pappas, Arsenia Chorti, and Stefano Tomasin.

is a counterpart of the capacity region in multiuser information theory. Multiple users must be served simultaneously with different security requirements in many emerging communication scenarios. To take account of the bursty nature of sources and secrecy requirement jointly, a 2-user broadcast channel (BC) [3] with secrecy constraint is considered. The model captures a heterogeneous scenario where secrecy needs to be ensured for one user, and there is no secrecy requirement for the other user. It is assumed that the intended receiver of the confidential message possesses the full-duplex capability, allowing it to transmit a jamming or interfering signal to disrupt another receiver's decoding of its data packets. Exploring the impact of secrecy requirements on the stability region is limited in the existing literature. Furthermore, the stability of the queues may be affected by both secrecy requirements and the self-interference resulting from the receiver's full-duplex operation. Studying this model can provide valuable insights into system performance.

5.1.1 Related Works

The BC is an important communication model studied under different settings in the existing literature. Superposition coding facilitates the concurrent transmission of messages to multiple users. Extensive analysis of the capacity region of BC has been carried out both with and without secrecy constraints [3–6, 14]. Despite these efforts, the secrecy capacity region of the general BC remains an area for improvement. The achievable rate region obtained in [4] is the most comprehensive known rate region for a general discrete memoryless 2-user BC. The work in [6] considers the problem of capacity characterization for 2-user Gaussian BC under fading.

The main philosophy of physical-layer secrecy lies in exploiting the randomness of wireless channels, such as noise and fading. The result of the wiretap channel for the first time shows that noise in communication system can be helpful for secure communication [25]. This pioneering work demonstrated the feasibility of secure communication between legitimate users without necessitating a secret key exchange. Building upon this, the findings of the wiretap channel were extended to the BC in [4], where the objective is to keep one of the messages secret from unintended receivers. Moreover, user cooperation's impact on BC's secrecy capacity was investigated in [5]. The challenge of secure broadcasting over fading channels has also been addressed in [7].

The works discussed above assume that users always have data to transmit. One of the common assumptions involved in the works related to information theory is that users always have data to send [4, 5, 13, 25]. However, data arrivals at the transmitter occur randomly in many real-world scenarios, such as IoT and M2M applications. The notion of stability region has been used to capture the random data arrival at the users. The existing literature studies stability regions for multiuser scenarios such as broadcast and interference channels. The stability region of the two-user BC without secrecy constraints has been investigated in [17, 19]. Similarly, stability regions have been explored for various communication models, including the two-user interference channel [18]. The main challenge in deriving stability regions lies in the coupling among the queues. The technique of stochastic dominant systems, proposed in [20], addresses this challenge. However, the stochastic dominance technique does not extend for more than two users and more discussion in this regard can be found in [9].

The stability region has been examined across various user scenarios incorporating secrecy constraints [11, 12, 23]. In [23], stability conditions are derived for a slotted ALOHA network with secrecy constraints, wherein users not transmitting are regarded as potential eavesdroppers. For a 2-user system, optimal transmission probabilities to maximize the system throughout under secrecy constraints. The work in [11, 12] considers a K-user BC where the message intended for a particular user needs to be kept secret from other users. A queue-length-based rate scheduling algorithm is proposed for the K-user BC with secrecy constraint under fading.

5.2 System Model

This chapter explores a scenario involving a 2-user BC, where one of the receivers can jam. The transmitter is tasked with catering to two users, each with distinct secrecy needs (see Figure 5.1). The transmitter has two queues S_1 (Queue 1) and S_2 (Queue 2) to store the packet corresponding to each receiver. The packet from queue S_i is intended for receiver i (D_i) ($i \in \{1, 2\}$). The packet from S_2 has no confidentiality requirements, whereas the packet from S_1 needs to be kept confidential

Figure 5.1 2-User BC with secrecy constraint with D_1 having jamming ability.

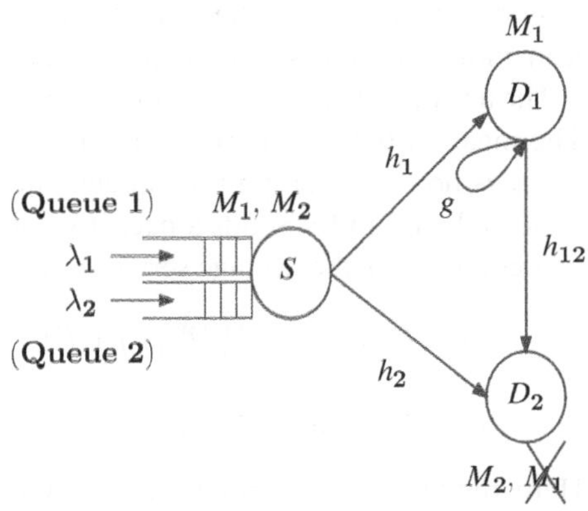

from D_2. While there are no secrecy constraints for packets from S_2, those from the S_1 must be kept secret from D_2. To improve the secrecy performance of the system, receiver D_1 possesses the full-duplex capability, enabling it to receive packets from the transmitter while simultaneously sending an interfering signal to create additional interference at the other receiver. The system model outlined herein captures scenarios where heterogeneous secrecy requirements arise based on the type of data being sent to multiple users from a single transmitter.

The signal received at the receiver is given by the following relation:

$$\begin{aligned} y_1^{(t)} &= h_1^{(t)} x^{(t)} + g x_J^{(t)} + z_1^{(t)}, \\ y_2^{(t)} &= h_2^{(t)} x^{(t)} + h_{12}^{(t)} x_J^{(t)} + z_2^{(t)}, \end{aligned} \tag{5.1}$$

where $z_i^{(t)} \sim \mathcal{CN}(0, 1)$. Note that $h_i^{(t)}$ and $h_{ij}^{(t)}$ denote the channel gain between various nodes as shown in Figure 5.1. The signal sent by the transmitter depends on whether the queue is empty or not, as described below:

1. If both queues have packets, the transmitter sends superposition of the packets, i.e., $x^{(t)} = x_1^{(t)} + x_2^{(t)}$, where $x_i^{(t)}$ is the packet intended for receiver i.
2. When only S_i has packet to send, then $x^{(t)} = x_i^{(t)}$ is sent by the transmitter.

The total power budget of the transmitter is denoted as P, which is the sum of power allocations P_i for sending messages from each queue, where $i \in \{1, 2\}$. A fixed power level is utilized for packet transmission. Receiver 1 sends an interfering signal with power P_J. Channels between nodes experience independent fading, modeled as Rayleigh fading. Self-interference resulting from the receiver's full-duplex operation is represented as a scalar quantity, with g indicating the degree of self-interference cancelation [8, 16, 21]. Perfect self-interference cancelation at the receiver corresponds to $g = 0$.

The event $\mathcal{D}_{i/\mathcal{T}}$ denotes that D_i is able to decode the packet sent from S_i, given a set of non-empty queues $\mathcal{T}$. Conversely, $\mathcal{D}_{i/\mathcal{T}}^{s}$ indicates that D_i can decode its intended packet while the other users $j \in \mathcal{T} \backslash i$ are unable to decode the packet transmitted from S_i, satisfying the secrecy requirement for D_i.

The arrival of packets at S_i is modeled as an independent and stationary process. The mean of the arrival rate is assumed to be λ_i packets per time slot $i \in \{1, 2\}$. There is no packet drop in the queue due to the waiting time. The length of S_i is denoted by Q_i and S_i sends a packet whenever they have a packet. The following expression gives the average service rate for the queues:

$$\mu_1 = \mathcal{P}(Q_2 > 0)\, \mathcal{P}\left(\mathcal{D}_{1/1,2}^{s}\right) + \mathcal{P}(Q_2 = 0)\, \mathcal{P}\left(\mathcal{D}_{1/1}^{s}\right), \tag{5.2}$$

$$\mu_2 = \mathcal{P}(Q_1 > 0)\, \mathcal{P}(\mathcal{D}_{2/1,2}) + \mathcal{P}(Q_1 = 0)\, \mathcal{P}(\mathcal{D}_{2/2}). \tag{5.3}$$

Here, the term $\mathcal{P}(\mathcal{A})$ denotes the probability of occurrence of the event $\mathcal{A}$.

5.3 Stability Region for the General Case

The following definition of queue stability is used [24].

Definition 5.1 Let $Q_i^{(t)}$ represent the length of S_i at the beginning of the time-slot t. A queue is deemed stable if $\lim_{t\to\infty} \mathcal{P}(Q_i^{(t)} < x) = F(x)$ and $\lim_{x\to\infty} F(x) = 1$. Furthermore, the queue is classified as sub-stable if the following condition is met:

$$\lim_{x\to\infty} \liminf_{t\to\infty} \mathcal{P}(Q_i^{(t)} < x) = 1. \tag{5.4}$$

If a queue is stable, then it is also sub-stable. A queue is said to be unstable if the queue is not sub-stable.

For multiuser scenarios, the notion of stability region becomes a meaningful metric under the random arrival of data at the users. The definition of stability region is based on the stability of queues, and it is the set of all possible arrival rates at the queues for which these queues remain stable. The average service rate of one user is influenced by whether the other queue has any packet to send or not. The average service rates of the first and second queues are given by (5.2) and (5.3), respectively. The average service rate of a queue depends on the status of another queue. Due to the coupling between the queues, characterizing the stability region is a nontrivial problem. The stochastic dominance technique [20] can be employed to decouple the queues, which in turn allows obtaining the stability region.

5.3.1 First Dominant System

In this case, when the first queue is empty ($Q_1 = 0$), the source sends a dummy packet designated for the first queue. In contrast, the operational behavior of the second queue mirrors that of the original system. Consequently, in this dominant setup, S_1 maintains a non-empty status, thereby establishing a constant service rate for S_2, denoted as $\mu_2 = \mathcal{P}(\mathcal{D}_{2|1,2})$. The stability of S_2 depends upon the condition $\lambda_2 < \mu_2$, specifically $\lambda_2 < \mathcal{P}(\mathcal{D}_{2|1,2})$. Using Loynes's criteria, the stability condition for S_2 is obtained. The probability that the second queue being empty is given by:

$$\mathcal{P}(Q_2 = 0) = 1 - \frac{\lambda_2}{\mathcal{P}(\mathcal{D}_{2|1,2})}. \tag{5.5}$$

Using (5.5) in (5.2), the service rate for S_1 can be expressed as given below:

$$\mu_1 = \mathcal{P}(\mathcal{D}^s_{1/1}) - \frac{\mathcal{P}(\mathcal{D}^s_{1/1}) - \mathcal{P}(\mathcal{D}^s_{1/1,2})}{\mathcal{P}(\mathcal{D}_{2/1,2})}\lambda_2. \tag{5.6}$$

Note that S_1 is stable if the following condition is satisfied: $\lambda_1 < \mu_1$. The stability region $\mathcal{R}_1$ for this case is given by

$$\mathcal{R}_1 = \left\{(\lambda_1, \lambda_2) : \frac{\lambda_1}{\mathcal{P}(\mathcal{D}^s_{1/1})} + \frac{\mathcal{P}(\mathcal{D}^s_{1/1}) - \mathcal{P}(\mathcal{D}^s_{1/1,2})}{\mathcal{P}(\mathcal{D}^s_{1/1})\mathcal{P}(\mathcal{D}_{2/1,2})}\lambda_2 < 1, \lambda_2 < \mathcal{P}(\mathcal{D}_{2/1,2})\right\}. \tag{5.7}$$

5.3.2 Second Dominant System

In this case, when $Q_2 = 0$, the source initiates the transmission of a dummy packet designated for S_2. In contrast, the operational behavior of the first queue remains consistent with that of the original system. Hence, S_2 maintains a non-empty state. Following a similar process as in the first dominant system, the stability region $\mathcal{R}_2$ is given by the following expression:

$$\mathcal{R}_2 = \left\{(\lambda_1, \lambda_2) : \frac{\lambda_2}{\mathcal{P}(\mathcal{D}_{2/2})} + \frac{\mathcal{P}(\mathcal{D}_{2/2}) - \mathcal{P}(\mathcal{D}_{2/1,2})}{\mathcal{P}(\mathcal{D}_{2/2})\,\mathcal{P}(\mathcal{D}^s_{1/1,2})}\lambda_1 < 1, \lambda_1 < \mathcal{P}(\mathcal{D}^s_{1/1,2})\right\}. \tag{5.8}$$

Finally, the stability region is given by

$$\mathcal{R} = \mathcal{R}_1 \cup \mathcal{R}_2, \tag{5.9}$$

where $\mathcal{R}_1$ and $\mathcal{R}_2$ correspond to the stability regions of dominant systems as described above.

A pictorial representation for the stability region is shown in Figure 5.2.

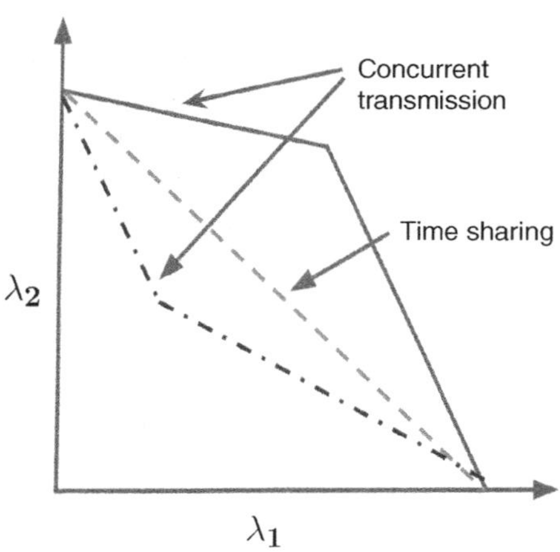

Figure 5.2 The stability region for 2-user BC.

Remarks:

1. Stability region obtained using (5.7) and (5.8) is general, and it does not necessitate the assumption of any particular decoding scheme at the receivers. It is noteworthy that the stability criteria for the queues are both necessary and sufficient for ensuring the stability of the original system, as established in [19, 20].
2. The indistinguishability argument used in [20] also applies to the system model considered in this paper. Leveraging the construction of the dominant system, it is observed that the queue sizes within the dominant system consistently surpass those of the original system, assuming identical initialization values and arrival rates in both systems. If, for a specified $\lambda_2 < \mu_2$, the first queue remains stable at a particular λ_1 in the dominant system, stability also ensues in the corresponding queue within the original system. Conversely, suppose the first queue saturates at a certain λ_1 within the dominant system. In that case, the behavior aligns with that of the original system, given the elimination of dummy packet transmission near boundary points. Consequently, the original and dominant systems exhibit identical characteristics at the boundary points. This argument is a pivotal strategy for resolving the challenge of interacting queues in the existing literature.
3. The stability region may exhibit either convex or non-convex characteristics (refer to Figure 5.2). When the stability region is non-convex, a straightforward time-sharing strategy will result in an expanded stability region compared to concurrent transmission methods.

5.4 Stability Region Analysis: Receivers with Different Decoding Abilities

The stability region obtained in the previous section depends on the probability of successful decoding at the receiver with and without secrecy constraints. The probability of successful decoding depends on factors such as underlying channel conditions, power budget available at the nodes and degree of self-interference cancelation. Users may possess diverse decoding capabilities in multiuser scenarios, consequently influencing overall system performance. To investigate the impact of divergent receiver decoding capabilities, secrecy constraints, and self-interference, we examine the following cases:

1. *Users with limited decoding capability*: Treating interference as noise (TIN) is used to decode the packet at both receivers.
2. *Users with varied decoding capabilities*: We consider the following scenarios to address heterogeneous decoding abilities among devices:

(a) The intended receiver of a confidential message employs successive decoding (SD), while the other receiver uses TIN. In SD, decoding occurs in two stages. Initially, the receiver decodes the unintended user's packet by treating its transmission as noise. Subsequently, it decodes its packet after mitigating the interference.
(b) The intended receiver of the confidential message decodes the packet using TIN, while the other receiver performs SD.

The following subsections present the probability of success for the first and third scenarios. Details regarding the second scenario can be found in [15].

5.4.1 Receivers with Limited Decoding Abilities

In this case, both the receivers employ TIN. It is an attractive scheme under channel uncertainty and in terms of the complexity of the receiver. A signal-to-noise ratio (SNR) or signal-to-interference-noise ratio (SINR) based decoding approach is used to capture the reliability aspect of communication. It can take account of the distance between the nodes and path-loss exponent (α). The distance between the transmitter and the receiver i is denoted by d_i $(i = 1, 2)$, and the distance between the receivers is d_3. When the SNR/SINR at receiver i is above a threshold termed decoding threshold ($\gamma_i, i \in \{1, 2\}$), the packet can be decoded successfully.

5.4.1.1 When Only the Second Queue Is Non-empty

There is no requirement for secrecy for the packet sent from S_2 and the corresponding event for successful decoding of packet at D_2 is given by the following expression:

$$\mathcal{D}_{2/2} = \left\{P_2|h_2|^2 d_2^{-\alpha} \geq \gamma_2\right\}. \tag{5.10}$$

In this case, D_1 does not send any interfering signal intended for D_2 as $Q_1 = 0$. Receiver 1 can come to know in advance whether the first queue is sending any packet or not through the exchange of control signals with the transmitter. For the considered model, the probability of successful decoding is given by the following expression [15]:

$$\mathcal{P}(\mathcal{D}_{2/2}) = \exp\left(-\frac{\gamma_2 d_2^{\alpha}}{P_2}\right). \tag{5.11}$$

5.4.1.2 When Only the First Queue Is Non-empty

In this case, S_2 does not send a packet intended for D_2 as $Q_2 = 0$. Receiver D_2 eavesdrops to decode the packet designated for D_1, as S_1 has a confidential packet intended for D_1. Hence, it is required to consider the reliability aspect at D_1 and the secrecy aspect at D_2. The following event captures these jointly:

$$\mathcal{D}^s_{1/1} = \left\{ \frac{P_1|h_1|^2 d_1^{-\alpha}}{1+P_J g^2} \geq \gamma_1, \frac{P_1|h_2|^2 d_2^{-\alpha}}{1+P_J|h_{12}|^2 d_3^{-\alpha}} \leq \gamma_1 \right\}. \tag{5.12}$$

Note that the information-theoretic metrics such as rate, secrecy capacity, and ergodic secrecy rate require the block length to be arbitrarily large. To overcome this problem, the secrecy criteria is defined in terms of SNR/SINR as in [2, 10, 22]. If the received SNR/SINR is less than the decoding threshold, the receiver cannot decode the packet, and the packet can be assumed to be secure. Determining the decoding threshold is a difficult problem and depends on many factors such as use-case scenarios, rate, encoding, and decoding mechanism at the receiver [22].

Using the fact that channels between the various nodes are independent of each other in (5.12), the probability of successful decoding for receiver 1 is given by the following expression:

$$\begin{aligned}\mathcal{P}(\mathcal{D}^s_{1/1}) &= \mathcal{P}\left\{ \frac{P_1|h_1|^2 d_1^{-\alpha}}{1+P_J g^2} \geq \gamma_1 \right\} \mathcal{P}\left\{ \frac{P_1|h_2|^2 d_2^{-\alpha}}{1+P_J|h_{12}|^2 d_3^{-\alpha}} \leq \gamma_1 \right\}, \\ &= \exp\left(-\frac{\gamma_1 d_1^{\alpha}}{P_1}(1+P_J g^2)\right)\left\{1-\exp\left(-\frac{\gamma_1 d_2^{\alpha}}{P_1}\right)\left[1+\gamma_1\frac{P_J}{P_1}\left(\frac{d_2}{d_3}\right)^{\alpha}\right]^{-1}\right\}.\end{aligned} \tag{5.13}$$

In the above expression, the probability of the first term corresponds to the reliability aspect, whereas the probability of the second term corresponds to the secrecy aspect. In this case, D_1 sends the interfering signal intended for D_2, which can help to improve its secrecy performance. However, due to the simultaneous transmission and reception at D_1, interference results at D_1. The impact of self-interference in (5.13) depends on P_J and g.

5.4.1.3 When Both the Queues Are Non-empty

As $Q_1 \neq 0$ and $Q_2 \neq 0$, transmitter sends a superposition of both the packets. The event which captures the successful decoding of packet at D_1 with secrecy constraint at D_2 is captured by the following:

$$\mathcal{D}^s_{1/1,2} = \left\{ \frac{P_1|h_1|^2 d_1^{-\alpha}}{1+P_2|h_1|^2 d_1^{-\alpha}+P_J g^2} \geq \gamma_1, \frac{P_1|h_2|^2 d_2^{-\alpha}}{1+P_2|h_2|^2 d_2^{-\alpha}+P_J|h_{12}|^2 d_3^{-\alpha}} < \gamma_1 \right\}. \tag{5.14}$$

The packet sent from the S_2 remains as noise along with the residual self-interference due to the full-duplex operation of D_1. The success probability for this event simplifies to the following expression:

$$\mathcal{P}(\mathcal{D}^s_{1/1,2}) = \mathcal{P}\left\{\frac{P_1|h_1|^2d_1^{-\alpha}}{1+P_2|h_1|^2d_1^{-\alpha}+P_Jg^2} \geq \gamma_1, \right.$$

$$\left.\frac{P_1|h_2|^2d_2^{-\alpha}}{1+P_2|h_2|^2d_2^{-\alpha}+P_J|h_{12}|^2d_3^{-\alpha}} < \gamma_1\right\},$$

$$= \exp\left(-\frac{\gamma_1 d_1^\alpha}{P_1-\gamma_1P_2}(1+P_Jg^2)\right)\left\{1-\exp\left(-\frac{\gamma_1 d_2^\alpha}{P_1-\gamma_1P_2}\right)\right.$$

$$\left.\left[1+\gamma_1\frac{P_J}{P_1-\gamma_1P_2}\left(\frac{d_2}{d_3}\right)^\alpha\right]^{-1}\right\}. \tag{5.15}$$

From (5.15), one can see that the $\mathcal{P}(\mathcal{D}^s_{1/1,2}) > 0$ provided the following condition is satisfied

$$\frac{P_1}{P_2} > \gamma_1. \tag{5.16}$$

The event which captures the successful decoding of packet at D_2 is given by the following expression:

$$\mathcal{D}_{2/1,2} = \left\{\frac{P_2|h_2|^2d_2^{-\alpha}}{1+P_1|h_2|^2d_2^{-\alpha}+P_J|h_{12}|^2d_3^{-\alpha}} \geq \gamma_2\right\}. \tag{5.17}$$

In this case, the packet sent from S_1 remains as noise along with the interfering signal. The success probability corresponding to this event is

$$\mathcal{P}\left(\mathcal{D}_{2/1,2}\right) = \exp\left(-\frac{\gamma_2 d_2^\alpha}{P_2-\gamma_2P_1}\right)\left[1+\gamma_2\frac{P_J}{P_2-\gamma_2P_1}\left(\frac{d_2}{d_3}\right)^\alpha\right]^{-1}. \tag{5.18}$$

To ensure $\mathcal{P}\left(\mathcal{D}_{2/1,2}\right) > 0$, following condition is required to be satisfied:

$$\frac{P_2}{P_1} > \gamma_2. \tag{5.19}$$

Hence, for $\mathcal{P}(\mathcal{D}_{2/1,2}) > 0$ and $\mathcal{P}(\mathcal{D}^s_{1/1,2}) > 0$, the conditions in (5.16) and (5.19) are required to be satisfied.

Using (5.13), (5.15), and (5.18), the stability region $\mathcal{R}_1$ in (5.7) can be obtained. Using (5.11), (5.15), and (5.18), the stability region $\mathcal{R}_2$ in (5.8) can be obtained. Finally, the stability region of the 2-user BC with secrecy constraint is obtained by taking the union of $\mathcal{R}_1$ and $\mathcal{R}_2$.

Now, consider the scenario where the receivers possess varying decoding capabilities. When ($Q_1 \neq 0, Q_2 = 0$) and ($Q_1 = 0, Q_2 \neq 0$), the probabilities of successful decoding for D_1 and D_2 are determined by (5.12) and (5.11), respectively. However, the expressions for the probability of successful decoding differ across all cases when S_1 and S_2 have packets to send. Therefore, in the subsequent analysis, we focus on the situation where both queues are non-empty ($Q_1 \neq 0$ and $Q_2 \neq 0$).

5.4.2 Receiver 1 with Limited Decoding Ability and Receiver 2 Uses SD

To determine the probability of successful decoding at D_1, it is required to consider the decoding ability of D_2 due to the confidentiality of the packet sent from S_1. The packet from S_1 is considered secure from D_2, if D_2 cannot decode the packet of S_1 using SD or TIN. The interfering signal sent by D_1 remains as noise in both stages of SD. The interfering signal sent by D_1 introduces additional noise at D_2, potentially aiding D_1 in achieving increased secrecy throughput. Nevertheless, the interfering signal sent by D_1 induces self-interference at D_1, with the impact of interference depending upon the efficiency of self-interference cancelation technique used at D_1. The probability of successful decoding at D_2 needs to account for: (a) D_2 uses SD to decode its intended packet, and (b) if it fails in SD, then uses TIN to decode the packet. The details of the derivation of probability of successful decoding with secrecy constraint (for receiver 1) and without secrecy constraint (for receiver 2) can be found in [15].

5.5 Impact of Secrecy on Delay Performance

With the introduction of 5G technology, there is a need to develop various communication services to support low-latency applications. It is required to provide guarantees on throughput, secrecy, and latency. A standard metric used to measure latency in communication is delay. For many future wireless networks, security and delay are required to be considered jointly for the performance analysis of the networks. At the same time, the users' decoding ability can impact the users' performance significantly. The considered system model is analyzed under various decoding schemes to explore the impact of security and delay constraints on performance. In this case, it is assumed that S_2 is saturated. The saturated queue captures the case of a heavy traffic scenario, where $\mathcal{P}(Q_2 \neq 0) \approx 1$.

5.5.1 Delay Analysis for User with Confidential Data

The delay is determined by the following two factors:

1. Queueing delay ($\bar{D}_T$)
2. Transmission delay ($D_{\bar{Q}_1}$)

The average transmission delay of a packet from S_1 when the source transmits it to the legitimate receiver is given by

$$\bar{D}_T = \frac{1}{\mu_1}, \tag{5.20}$$

where μ_1 is the service probability for the packet from S_1. The following expression gives the average queue size $\bar{Q_1}$, provided that the queue is stable [1]

$$\bar{Q_1} = \frac{\lambda_1}{\mu_1 - \lambda_1}\left(1 - \lambda_1\right). \tag{5.21}$$

Using Little's law, the average waiting time of a packet in S_1 is given by

$$D_{\bar{Q_1}} = \frac{\bar{Q_1}}{\lambda_1} = \frac{1 - \lambda_1}{\mu_1 - \lambda_1}. \tag{5.22}$$

Using (5.20) and (5.22), average delay is given by

$$\bar{D}_1 = \bar{D}_T + D_{\bar{Q_1}} = \frac{1}{\mu_1} + \frac{1 - \lambda_1}{\mu_1 - \lambda_1}. \tag{5.23}$$

Using the results derived in Section 5.4 for service probability (μ_1), the delay performance for the BC with secrecy constraint under various decoding schemes can be obtained. More details related to the delay analysis for the BC with one of the queues saturated can be found in [1].

5.6 Results and Discussion

This subsection presents the stability region for different decoding schemes at receivers. In all figures, (DC_1, DC_2) represents that the decoding scheme at receiver i is DC_i, which is decoding by either TIN or SD.

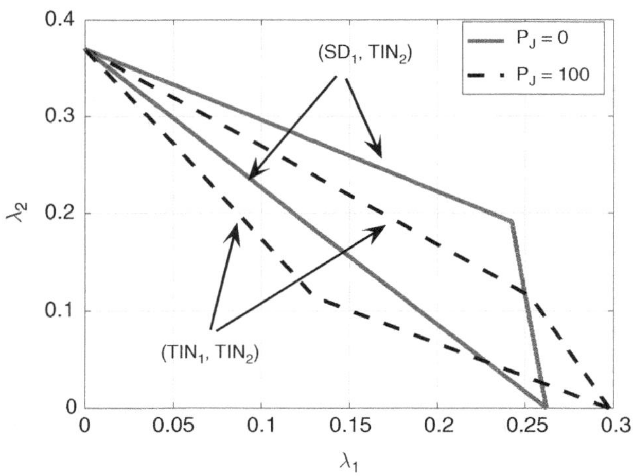

Figure 5.3 Impact of full-duplex ability of receiver 1 on the stability region: $P_1 = P_2 = 100$, $\gamma_1 = 0.5$, $\gamma_2 = 0.4$, $d = 10$, $d_1 = d$, $d_2 = d_3 = 1.1d$, and $\alpha = 2.3$. [15].

5.6.1 Stability Region with Secrecy Constraint

This section explores the impact of interfering signal sent from D_1 on the stability region. In Figure 5.3, the stability regions are plotted using the results derived in Section 5.4.1 for this decoding $(\text{TIN}_1, \text{TIN}_2)$ scenario and using the result in [15] for the decoding $(\text{SD}_1, \text{TIN}_2)$ scenario. The impact of interfering signal power on the stability region is also explored when the degree of self-interference cancelation is $\beta \triangleq -10 \log g^2 = 60\,\text{dB}$. It can be noticed that with an increase in the power of the interfering signal, there is also an increase in the maximum achievable throughput for D_1. Nonetheless, this occurs within a confined region where the maximum achievable throughput for D_2 remains limited.

We now explore the scenario where D_1 cannot perform SD, whereas D_2 can perform SD. Consequently, the packet from S_2 and any residual interfering signal remain as noise at D_1. The SINR thresholds for decoding packets intended to D_1 and D_2 are $\gamma_1 = 0.2$ and $\gamma_2 = 0.25$, respectively. In Figure 5.4, the stability region obtained in Section 5.4.2 is plotted for $P_1 = P_2 = P_J = 100$, and $\beta = 60\,\text{dB}$. The plot indicates the feasibility of ensuring a nonzero stable throughput for S_1, even when D_2 can perform SD. With increased power of the interfering signal, D_1 can accommodate larger arrival rates. However, the arrival rate for D_2 diminishes with the increasing power of the interfering signal, given the additional interference it incurs.

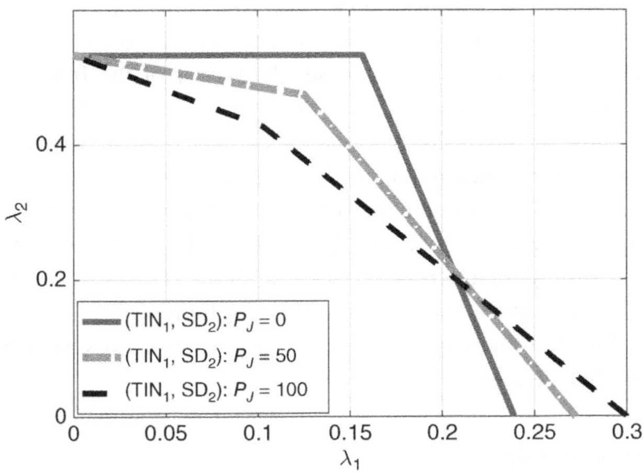

Figure 5.4 The stability region for $(\text{TIN}_1, \text{SD}_2)$: $P_1 = P_2 = 100$, $\alpha = 2.8$, $d = 10$, $c = 0.8$, $c' = 1.1$, $d_1 = d$, $d_2 = cd$, $d_3 = c'd$, $\gamma_1 = 0.2$, $\gamma_2 = 0.25$, and $\beta = 60\,\text{dB}$ [15]/with permission from IEEE.

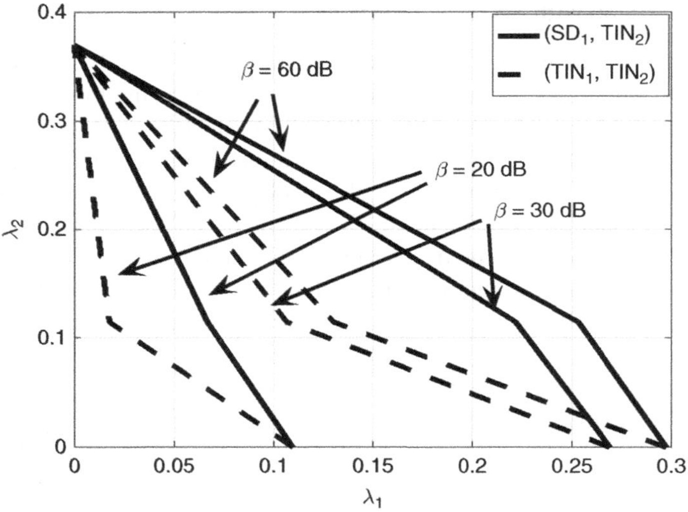

Figure 5.5 Impact of degree of self-interference cancelation on the stability region: $P_1 = P_2 = P_J = 100$, $\gamma_1 = 0.5$, $\gamma_2 = 0.4$, $d = 10$, $d_1 = d$, $d_2 = d_3 = 1.1d$, and $\alpha = 2.3$ [15].

5.6.2 Impact of Imperfect Self-interference Cancelation on the Stability Region

This subsection explores the impact of imperfect self-interference cancelation on the stability region. It is assumed that D_2 has limited decoding ability and, hence, cannot perform SD. In Figure 5.5, the stability regions are plotted for different degrees of self-interference cancelation. In this case, the power of the interfering signal is kept constant $P_J = 100$. The stability regions for $(\text{TIN}_1, \text{TIN}_2)$ and $(\text{SD}_1, \text{TIN}_2)$ are obtained using the results in Sections 5.4.1 and [15], respectively. It can be noticed that the stability region enlarges with an increase in the value of β. Note that a higher β value signifies a more effective self-interference cancelation technique at the full-duplex receiver. Consequently, if the self-interference technique is inefficient, the advantages gained in service rate due to D_1 transmitting an interfering signal to D_2 may diminish for both decoding schemes.

The numerical results for the other case $(\text{TIN}_1, \text{SD}_2)$ can be obtained similarly as obtained in the case of $(\text{SD}_1, \text{TIN}_2)$.

5.6.3 Impact of Secrecy on Delay

In Figure 5.6, both receivers are assumed to have limited decoding ability, i.e., $(\text{TIN}_1, \text{TIN}_2)$. Note that S_2 is assumed to be saturated. In Figure 5.6, the average packet delay is plotted against the power of the interfering signal at D_1 (P_J).

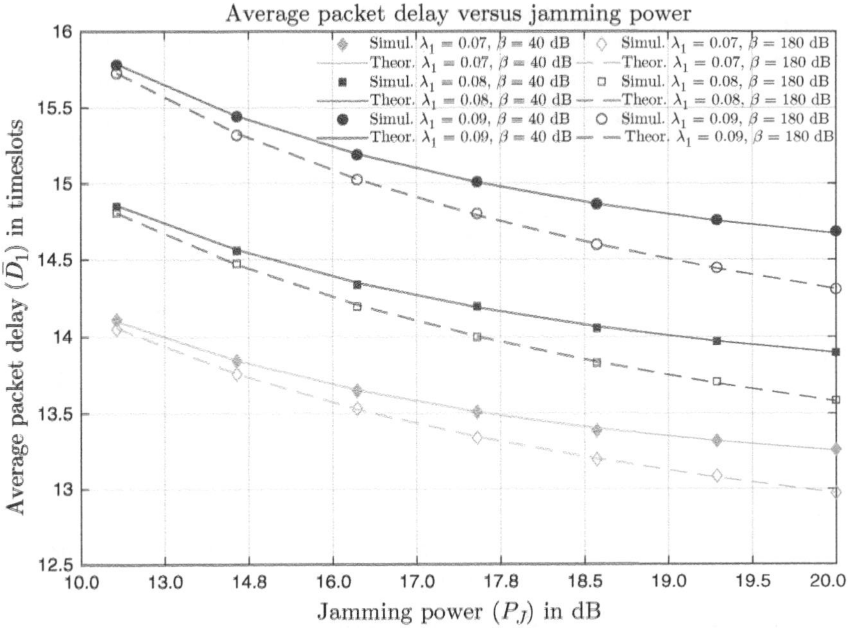

Figure 5.6 Impact of jamming power on average packet delay at the legitimate receiver when both receivers treat interference as noise. Source: [1]/Springer Nature/ CC BY 4.0.

The different values of λ_1 and β are also considered to explore the impact of arrival rate and the degree of self-interference cancelation on the average delay. The average packet delay decreases as the power of the interfering signal increases since the service probability increases as P_J increases. It is essential to highlight that the simulation results corroborate the analytical expressions. Furthermore, we observe that the average packet delay rises with the increase in the arrival probability at S_1. This is attributable to the prolonged waiting time of packets at queue S_1. The average packet delay also escalates when β is small, indicating imperfect self-interference cancelation techniques, leading to packet retransmissions.

In Figure 5.7, D_1 performs SD and D_2 treats interference as noise. The average packet delay is plotted against P_J for different values of λ_1 and β. One can observe that the average packet delay decreases as P_J decreases, and it increases as λ_1 increases. A comparison between Figures 5.6 and 5.7 reveals that employing SD for receiver D_1 results in a reduced average packet delay compared to the scenario where both receivers use TIN under the same λ_1 and P_J.

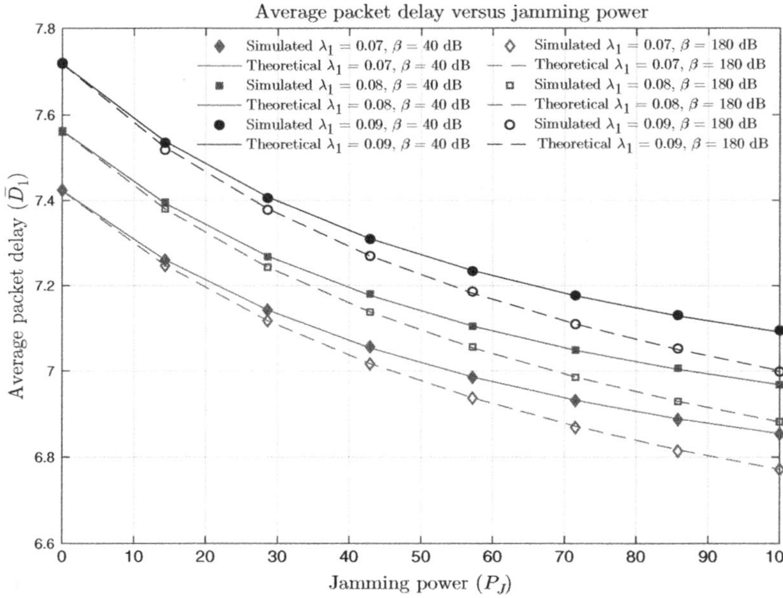

Figure 5.7 Impact of power of the interfering signal on average packet delay: D_1 uses SD and D_2 uses TIN [1]/Springer Nature/CC BY 4.0.

5.7 Conclusion

This chapter shows how secrecy considerations influence the stability region and the average delay within a two-user BC. Initially, the stability region of the BC under secrecy constraints is derived in a general context. Subsequently, the stability region is analyzed in relation to various decoding schemes implemented at the receivers. The investigation encompasses the imposition of secrecy constraints at the receiver and the concurrent effects of imperfect self-interference cancelation on the stability region. When receiver 2 can perform SD, the significance of sending an interfering signal to improve the secrecy performance for receiver 1 becomes more pronounced.

References

1 A. Arvanitaki, N. Pappas, P. Mohapatra, and N. Carlsson. Delay performance of a two-user broadcast channel with security constraints. *SN Computer Science*, 1(1):53, 2019. doi: 10.1007/s42979-019-0055-3.

2 C. Capar, D. Goeckel, B. Liu, and D. Towsley. Secret communication in large wireless networks without eavesdropper location information. In

Proceedings IEEE INFOCOM, pages 1152–1160, Mar 2012. doi: 10.1109/INFCOM.2012.6195474.

3 T. M. Cover. An achievable rate region for the broadcast channel. *IEEE Transactions on Information Theory*, 21(4):399–404, Jul 1975. ISSN 0018-9448. doi: 10.1109/TIT.1975.1055418.

4 I. Csiszar and J. Korner. Broadcast channels with confidential messages. *IEEE Transactions on Information Theory*, 24(3):339–348, May 1978. ISSN 0018-9448. doi: 10.1109/TIT.1978.1055892.

5 E. Ekrem and S. Ulukus. Secrecy in cooperative relay broadcast channels. *IEEE Transactions on Information Theory*, 57(1):137–155, Jan 2011. ISSN 0018-9448. doi: 10.1109/TIT.2010.2090215.

6 A. Jafarian and S. Vishwanath. The two-user Gaussian fading broadcast channel. In *Proceedings of IEEE International Symposium on Information Theory*, pages 2964–2968, Jul 2011. doi: 10.1109/ISIT.2011.6034121.

7 A. Khisti, A. Tchamkerten, and G. W. Wornell. Secure broadcasting over fading channels. *IEEE Transactions on Information Theory*, 54(6):2453–2469, Jun 2008. ISSN 0018-9448. doi: 10.1109/TIT.2008.921861.

8 D. Kim, H. Ju, S. Park, and D. Hong. Effects of channel estimation error on full-duplex two-way networks. *IEEE Transactions on Vehicular Technology*, 62(9):4666–4672, Nov 2013. ISSN 0018-9545. doi: 10.1109/TVT.2013.2265407.

9 S. Kompella and A. Ephremides. Stable throughput regions in wireless networks. *Foundations and Trends in Networking*, 7(4):235–338, 2014. ISSN 1554-057X. doi: 10.1561/1300000039.

10 J. Lee, A. Conti, A. Rabbachin, and M. Z. Win. Distributed network secrecy. *IEEE Journal on Selected Areas in Communications*, 31(9):1889–1900, Sept 2013. ISSN 0733-8716. doi: 10.1109/JSAC.2013.130920.

11 Y. Liang, H. V. Poor, and L. Ying. Wireless broadcast networks: Reliability, security, and stability. In *Information Theory and Applications Workshop*, pages 249–255, Jan 2008. doi: 10.1109/ITA.2008.4601057.

12 Y. Liang, H. V. Poor, and L. Ying. Secure communications over wireless broadcast networks: Stability and utility maximization. *IEEE Transactions on Information Forensics and Security*, 6(3):682–692, Sept 2011. ISSN 1556-6013. doi: 10.1109/TIFS.2011.2158311.

13 R. Liu, I. Maric, P. Spasojević, and R. D. Yates. Discrete memoryless interference and broadcast channels with confidential messages: Secrecy rate regions. *IEEE Transactions on Information Theory*, 54(6):2493–2507, Jun 2008. ISSN 0018-9448. doi: 10.1109/TIT.2008.921879.

14 K. Marton. A coding theorem for the discrete memoryless broadcast channel. *IEEE Transactions on Information Theory*, 25(3):306–311, May 1979. ISSN 0018-9448. doi: 10.1109/TIT.1979.1056046.

15 P. Mohapatra, N. Pappas, J. Lee, T. Q. S. Quek, and V. Angelakis. Secure communications for the two-user broadcast channel with random traffic. *IEEE*

Transactions on Information Forensics and Security, 13(9):2294–2309, 2018. doi: 10.1109/TIFS.2018.2818076.

16 D. W. K. Ng, E. S. Lo, and R. Schober. Dynamic resource allocation in MIMO-OFDMA systems with full-duplex and hybrid relaying. *IEEE Transactions on Communications*, 60(5):1291–1304, May 2012. ISSN 0090-6778. doi: 10.1109/TCOMM.2012.031712.110233.

17 N. Pappas and M. Kountouris. The stability region of the two-user broadcast channel. In *Proceedings of IEEE International Conference on Communications (ICC)*, May 2016.

18 N. Pappas, M. Kountouris, and A. Ephremides. The stability region of the two-user interference channel. In *Proceedings of IEEE Information Theory Workshop (ITW)*, pages 1–5, Sept 2013. doi: 10.1109/ITW.2013.6691293.

19 N. Pappas, M. Kountouris, A. Ephremides, and V. Angelakis. Stable throughput region of the two-user broadcast channel. *CoRR*, abs/1604.08418, 2016.

20 R. R. Rao and A. Ephremides. On the stability of interacting queues in a multiple-access system. *IEEE Transactions on Information Theory*, 34(5):918–930, Sept 1988. ISSN 0018-9448. doi: 10.1109/18.21216.

21 T. Riihonen, S. Werner, and R. Wichman. Mitigation of loopback self-interference in full-duplex MIMO relays. *IEEE Transactions on Signal Processing*, 59(12):5983–5993, Dec 2011. ISSN 1053-587X. doi: 10.1109/TSP.2011.2164910.

22 A. K. Sadek, K. J. R. Liu, and A. Ephremides. Cognitive multiple access via cooperation: Protocol design and performance analysis. *IEEE Transactions on Information Theory*, 53(10):3677–3696, Oct 2007. ISSN 0018-9448. doi: 10.1109/TIT.2007.904784.

23 Y. Sarikaya and O. Ercetin. On physically secure and stable slotted aloha system. In *Proceedings of the 47th Annual Allerton Conference on Communication, Control, and Computing*, pages 1090–1096, Sept 2009. doi: 10.1109/ALLERTON.2009.5394871.

24 W. Szpankowski. Stability conditions for some distributed systems: Buffered random access systems. *Advances in Applied Probability*, 26(2):498–515, 1994. doi: 10.2307/1427448.

25 A. D. Wyner. The wire-tap channel. *The Bell System Technical Journal*, 54(8):1334–1387, Oct 1975.

6

Physical-Layer Secrecy for Ultrareliable Low-Latency Communication

Parthajit Mohapatra[1] *and Nikolaos Pappas*[2]

[1] *Department of Electrical Engineering, Indian Institute of Technology Tirupati, Tirupati, India*
[2] *Department of Computer and Information Science, Linköping University, Linköping, Sweden*

6.1 Introduction

Given the emergence of novel communication scenarios facilitated by the advent of 5G technology, there is a pressing need to develop secure communication methods that ensure both reliability and minimal latency. In cyber-physical systems and the Internet of Things (IoT), timely dissemination of status updates is of utmost importance. The concept of *Age of Information* (AoI), introduced in the literature, addresses this need by quantifying the freshness of information [1]. However, existing research in physical-layer security often overlooks latency requirements, assuming backlogged users and large packet sizes. The obtained results rely on the assumption of large packet sizes to leverage the law of large numbers. Consequently, the interplay between reliability, latency, and secrecy still needs to be explored in the literature, specifically when users are constrained to transmit short packets. As wireless communication progresses toward 6G, it becomes imperative to redesign secure communication techniques at the physical layer to accommodate these latency constraints, including AoI and delay. This chapter delves into this challenge, focusing primarily on the impact of secrecy on reliability and latency.

In machine-to-machine (M2M) communication and the IoT, sensor-generated traffic typically manifests as short packets. Existing information-theoretic findings often prove inapplicable in such scenarios due to their assumption of large packet sizes. When users are constrained to transmit short packets, errors become inevitable, leading to a delicate balance between rate, error, and block length (packet length). Recent advancements in finite block-length information theory have offered valuable insights into this trade-off in point-to-point and

Physical-Layer Security for 6G, First Edition. Edited by Parthajit Mohapatra, Nikolaos Pappas, Arsenia Chorti, and Stefano Tomasin.

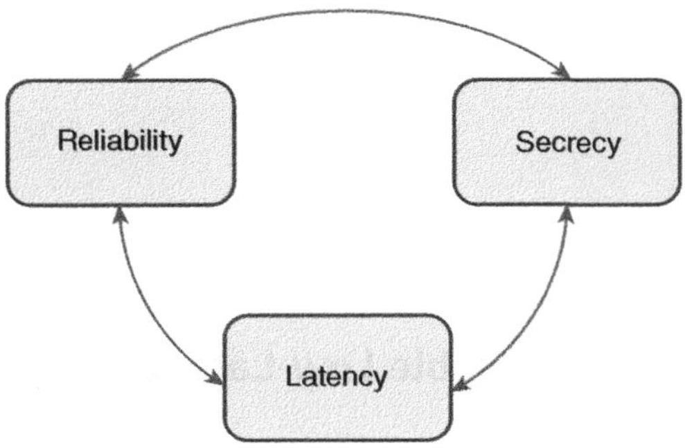

Figure 6.1 Interplay between reliability, secrecy and latency.

multiuser settings like multiple access channels. These developments serve as a foundational framework for addressing short-packet communication challenges. However, minimizing only packet length may not effectively reduce communication latency, as it fails to consider queuing delays or information freshness. Delay, encompassing transmission and queuing delays, and AoI, measuring information freshness, emerge as pertinent metrics in capturing latency and timing aspects of information flow [8, 15, 17, 27]. Consequently, there is a pressing need to establish a mathematical framework that comprehensively addresses the interaction between reliability, secrecy, and latency in communication. Figure 6.1 illustrates this high-level conceptual framework.

6.2 Background

6.2.1 Finite Block-Length Information Theory

Most existing works in information theory have the underlying assumption that packet length or block length (N) is large. When N is large, good channel codes with high probability exist, so the error in recovering the information from the received signal can be arbitrarily small. When N is small, one cannot use the law of large numbers, and, hence, the thermal noise and the distortion due to the channel cannot be averaged out. In addition, one cannot ignore the meta-data size when N is small. For short-packet communication, it is not possible to use the law of large numbers, and the size of the metadata cannot be ignored. Due to the recent development in finite block-length information theory, progress has been made in developing a short-packet communication theory. One of the standard measures used in a communication system is rate $R = \frac{K}{N}$, where K is the number of information bits that need to be encoded in a codeword of length N. The general objective of any communication system is to make R as large as possible, and block-length error rate (P_e) (BLER) is made as small as possible. Let $R^*(N, \epsilon)$ denote the maximal coding rate where a codeword of length N is used with BLER ϵ. The analysis of short-packet communication requires a fundamental understanding of $R^*(N, \epsilon)$, which captures the interplay between BLER, rate, and block length. Determining the exact value of $R^*(N, \epsilon)$ is a challenging problem even for simpler communication models such as point-to-point additive white Gaussian noise (AWGN) channel. In recent years, progress has been made in developing results for point-to-point and multiuser scenarios. In the following section, results related to the AWGN channel are presented.

6.2.1.1 Results for the AWGN Channel

For the AWGN channel with a single transmitter and receiver pair, the transmitter needs to choose the following rate to achieve an error probability of ϵ [16]

$$R = \log_2(1 + \text{SNR}) - \sqrt{\frac{V(\text{SNR})}{N}} Q^{-1}(\epsilon) + \mathcal{O}\left(\frac{\log N}{N}\right), \tag{6.1}$$

where $V(\text{SNR}) = \log_2^2(e)\left(1 - \frac{1}{(1+\text{SNR})^2}\right)$ and SNR is the signal-to-noise ratio for the AWGN channel.

6.2.1.2 Results for the AWGN Wiretap Channel

Wiretap channel is one of the fundamental models which is studied in information theory extensively under various assumptions. Under finite-block length regime, the following result is achievable [26]:

$$R \geq C_s - \sqrt{\frac{V_1}{n}} Q^{-1}(\epsilon) - \sqrt{\frac{V_2}{N}} Q^{-1}(\delta) + \mathcal{O}\left(\frac{\log N}{N}\right) \tag{6.2}$$

where the probability of block error at Bob is not larger than ϵ while the information leakage at the eavesdropper is bounded by δ.

Note that V_1 and V_2 correspond to the channel dispersion of the channel between Alice and Bob, and Alice and Eve, respectively. These terms are defined in the later part of this chapter. This equation is applicable when the legitimate receiver has better SNR compared to SNR at Eve.

6.2.1.3 Stability Criteria of a Queue

The stability region is defined as the set of all arrival rates for which all queues within the system remain stable. Loynes' theorem [11, 22] establishes the stability criterion for a queue, stating that when the average arrival rate (λ) is lower than the average service rate (μ), denoted by $\lambda < \mu$, assuming strict joint stationarity of the arrival and service processes. Conversely, when this condition is not met, the queue is deemed unstable, almost surely leading to an unbounded queue length.

6.2.1.4 Age of Information

The concept of AoI emerged in the early 2010s with the aim of quantifying the freshness of information possessed by a system regarding a remotely observed process. AoI is defined as the duration since the generation of the most recent successfully received message containing updated information about its source system. Serving as an end-to-end metric, AoI finds utility in characterizing latency within status-updating systems and associated applications. Further elaboration on AoI can be found in [8, 27].

6.2.2 Related Works

The finite block-length information theory framework provides the fundamental results on the interplay between reliability, rate, and block length. It will play an essential role in developing the theory for short-packet communication [4, 16, 25]. Most information-theoretic results are based on the law of large numbers, so these results do not apply to short-packet communication. For the finite-block length regime, the work in [16] obtains several lower and upper bounds on the maximal rate, which captures the relationship between rate, error, and block length. One of the fundamental parameters that play a central role in this characterization is channel dispersion, which measures the stochastic variability of the underlying channel. The work in [25] obtains bounds on the maximum achievable rate of the quasi-static fading channel, where the transmitter and the receiver are equipped with multiple antennas. Extending the developed results from point-to-point scenario to a multiuser scenario is a nontrivial problem in finite block-length information theory, and some of the results in this regard can be found in [6, 7, 28].

When it is required to ensure secure communication and reliability requirements in a finite block-length regime, one of the fundamental problems of interest is the interplay between error, rate, information leakage, and block length. The wiretap channel [24], which captures the problem of reliable and secure communication jointly, has been studied extensively in the existing literature under different settings. Significant progress has been made in developing theory and signal-processing techniques for physical-layer secrecy. Most of the existing works in physical-layer secrecy suffer from the following limitations: (i) developed results do not take the random nature of the sources, and (ii) the length of the codeword can be arbitrarily large. In recent years, some progress has been made in developing results for secure communication for the finite-block length regime. Bounds on the maximal secrecy rate are obtained for discrete memoryless wiretap channel for the finite block-length regime [26]. The recent progress made in finite block-length information theory has led to developing a theory for secure short-packet communication. The impact of fading on the average secrecy throughput is explored for the wiretap channel in [23] when the eavesdropper is equipped with multiple antennas. One standard metric used in physical-layer secrecy for fading scenario is the secrecy outage probability (SOP) or ergodic secrecy rate. For short-packet communication, there is a need to redefine or introduce new performance metrics that take into account the finite size of the codeword or packet. Some of the results in these directions can be found in [5, 29]. The problem of secure short-packet communication under various communication settings can be found in [2, 3, 19].

In many emerging communication scenarios, it is required to take account of the bursty nature of sources and latency aspects of communication. Using a short packet for communication can reduce the physical-layer latency, but it cannot take account of the queuing delay. Hence, results must be developed that consider packet length and queuing delay. The works considering random data arrival at the user under finite block-length regime can be found in [10, 18, 21]. The problem of information freshness is explored for downlink cellular networks using the finite block-length information theory [21]. The works above assume that keeping the message secure from unauthorized users is not required. The impact of secrecy on the stability region has been explored in [14]. The results of the study do not take into account the size of the packet. The impact of secrecy on reliability and latency aspects of communication under random arrival of data in the finite-block length regime has yet to be explored in the existing literature, and this chapter primarily focuses on this.

6.3 System Model

This chapter considers the problem of secure communication over a wiretap channel with a friendly jammer, where the channel between the different nodes undergoes Rayleigh fading. This chapter considers the problem of secure communication over a wiretap channel with a friendly jammer, where the channel between the different nodes undergoes Rayleigh fading (See Figure 6.2). The wiretap channel models a communication scenario where the legitimate transmitter needs to communicate securely with the legitimate receiver over a

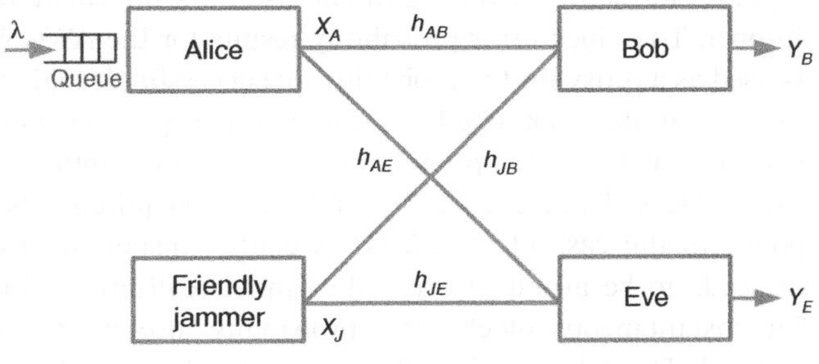

Figure 6.2 Rayleigh fading wiretap channel with a trusted jammer and queue at the transmitter to store the incoming packet.

noisy channel in the presence of an eavesdropper. All the nodes are assumed to have a single antenna. The input–output relation for the considered model is

$$\begin{aligned} Y_B &= h_{AB}X_A + h_{JB}X_J + Z_B, \\ \text{and } Y_E &= h_{AE}X_A + h_{JE}X_J + Z_E, \end{aligned} \tag{6.3}$$

where h_{UW} represents to the channel gain between the nodes U and W, where $U \in \{\text{Alice, Jammer}\}$ and $W \in \{\text{Bob, Eve}\}$, and noise is modeled as: $Z_W \sim \mathcal{CN}(0, 1)$. The transmitter and jammer have a power budget of P_A and P_J, respectively. The signal sent by the transmitter (friendly jammer) to the channel is given by X_A (X_J). The data at the transmitter arrives randomly, and the transmitter has a queue to store the incoming packets. It is assumed that the queue has infinite capacity and no packet is being dropped at the queue. The packet arrival process is modeled as independent and stationary with an arrival probability λ. The length of the queue is denoted by Q. The K information bits at the queue are encoded into a packet of length N, and the corresponding rate is defined as $R \triangleq \frac{K}{N}$. The transmitter sends the packet whenever its queue has a packet to send. The attacker and the friendly jammer know the queue status. The friendly jammer only sends the jamming signal when the transmitter sends data. The signal sent by the jammer is assumed to be drawn from the Gaussian distribution, i.e., $X_J \sim \mathcal{CN}(0, P_J)$. In the following sections, the impact of secrecy on stable throughput and latency is explored.

6.4 Impact of Secrecy on Stable Throughput

This section derives the results for the secure stable throughput for the considered system model. Stable throughput is a meaningful metric to consider where the data arrival at the source is random or bursty. The secure stable throughput must account for successful packet decoding by the legitimate user and information leakage by the eavesdropper. The block error probability results for the AWGN wiretap channel [26] is used as a proxy for the probability of successful decoding at the legitimate receiver. The finite block-length information theory framework is used to derive the results for the wiretap channel. It allows for capturing the interplay between rate, packet length, reliability, and secrecy requirements. As this result corresponds to the case of the AWGN scenario, instantaneous block-error probability needs to be averaged over all channel realizations for the various nodes. The instantaneous block error probability also needs to consider the jamming signal. The role of a friendly jammer has been explored in physical-layer secrecy under asymptotic settings for various communication scenarios. However, the role of a friendly jammer in short-packet communication needs to be explored to the best of the author's knowledge. The expression for instantaneous block error probability for the wiretap channel in the presence of a friendly jammer is given by the following expression:

$$\epsilon \approx Q\left(\sqrt{\frac{N}{V_{\gamma_{AB}}}}\left(\ln\frac{1+\gamma_{AB}}{1+\gamma_{AE}} - \sqrt{\frac{V_{\gamma_{AE}}}{N}}Q^{-1}(\delta) - \frac{K}{N}\ln 2\right)\right), \tag{6.4}$$

where

$$\gamma_{AB} = \frac{|h_{AB}|^2 P_A}{1+|h_{JB}|^2 P_J}, \text{ and } \gamma_{AE} = \frac{|h_{AE}|^2 P_A}{1+|h_{JE}|^2 P_J}. \tag{6.5}$$

For the above equation to hold, the following condition is required to be satisfied: $\gamma_{AB} > \gamma_{AE}$. The quantity $V_{\gamma_{AB}} = 1-(1+\gamma_{AB})^{-2}$ and $V_{\gamma_{AE}} = 1-(1+\gamma_{AE})^{-2}$ are the channel dispersion corresponding to the channel between Alice to Bob and Alice to Eve, respectively. The term $Q^{-1}(x)$ represents the inverse of the Gaussian Q function. The quantity δ corresponds to the information leakage at the eavesdropper, and the lower the value of δ, the lower the amount of information leakage at the eavesdropper. In general, the value of δ is chosen in the range of $(0, \frac{1}{2})$ [23]. Using the instantaneous block error probability in (6.4), the average successful decoding probability is obtained in the following theorem.

Theorem 6.1 The average service rate for the considered system model is given by the following equation:

$$\begin{aligned}\mu \approx{}& \frac{\phi_{JB}\,\phi_{AE}\,\phi_{JE}}{\phi_{AB}}\frac{P_A}{P_J}\exp\left(\frac{\phi_{JB}}{P_J}\right)\\ &\int_0^{\infty}\alpha_{AE}\left[\mathrm{Ei}\left(\frac{-\phi_{AB}}{P_A}\left(\beta_{AE}+\frac{1}{2\alpha_{AE}}\right)-\frac{\phi_{JB}}{P_J}\right)\right.\\ &\left.-\mathrm{Ei}\left(\frac{-\phi_{AB}}{P_A}\left(\beta_{AE}-\frac{1}{2\alpha_{AE}}\right)-\frac{\phi_{JB}}{P_J}\right)\right]\frac{\phi_{JE}\,P_A+\phi_{AE}\,y\,P_J+P_J\,P_A}{(\phi_{JE}\,P_A+\phi_{AE}\,y\,P_J)^2}\\ &\exp\left(-\frac{\phi_{AE}\,y}{P_A}\right)dy,\end{aligned} \tag{6.6}$$

where $\beta_{AE} \triangleq \exp\left(\sqrt{\frac{V_{\gamma_{AE}}}{N}}Q^{-1}(\delta)+\frac{K}{N}\ln 2\right)(1+\gamma_{AE})-1$, $\alpha_{AE} \triangleq \frac{d\epsilon_{\gamma_{AB}|\gamma_{AE}}(x)}{dx}\Big|_{x=\beta_{AE}} = -\sqrt{\frac{N}{2\pi\beta_{AE}(\beta_{AE}+2)}}$ and $\mathrm{Ei}(x)$ is the exponential integral function defined as follows, $\mathrm{Ei}(x) = \int_{-\infty}^{x}\frac{e^t}{t}\,dt$.

Proof: The probability of successful decoding is approximated by the following expression:

$$P_s \approx 1 - \bar{\epsilon} = 1 - \mathbb{E}_{\gamma_{AB},\gamma_{AE}}[\epsilon], \tag{6.7}$$

where ϵ corresponds to the instantaneous block-error probability as given in (6.4). Obtaining a closed-form expression for the probability of successful decoding is a nontrivial problem due to the following reasons:

1. The expectation should be computed with respect to γ_{AB} and γ_{AE}, which involve a ratio of exponential random variables.
2. The instantaneous block-error probability depends on γ_{AB}, γ_{AE}, and dispersion term through Q-function.

The distribution of γ_{AB} and γ_{AE} is given by the following expression:

$$F_{\gamma_{AB}}(x) = 1 - \frac{\exp\left(-\frac{\phi_{AB}\, x}{P_A}\right)}{1 + \frac{\phi_{AB}}{\phi_{JB}}\frac{P_J}{P_A}\, x},$$

$$\text{and } F_{\gamma_{AE}}(y) = 1 - \frac{\exp\left(-\frac{\phi_{AE}\, y}{P_A}\right)}{1 + \frac{\phi_{AE}}{\phi_{JE}}\frac{P_J}{P_A}\, y}. \tag{6.8}$$

Note that the random variables $X_{AB} \triangleq |h_{AB}|^2$ and $X_{JB} \triangleq |h_{JB}|^2$ are distributed as exponential distribution with parameters ϕ_{AB} and ϕ_{JB}, respectively, whose probability density functions are given by

$$f_{X_{AB}}(x_{AB}) = \phi_{AB} \exp(-\phi_{AB}\, x_{AB}), x_{AB} \geq 0, \tag{6.9}$$

$$\text{and } f_{X_{JB}}(x_{JB}) = \phi_{JB} \exp(-\phi_{JB}\, x_{JB}), x_{JB} \geq 0. \tag{6.10}$$

Using (6.7), the average probability of successful decoding can be expressed by the following:

$$\mathcal{P}(\mathcal{D}^{S}_{B|\{A,J\}}) = \int_0^\infty \mathcal{S}(y)\, f_{\gamma_{AE}}(y)\, dy, \tag{6.11}$$

$$\text{where } \mathcal{S}(y) \triangleq \int_y^\infty (1 - \epsilon_{\gamma_{AB}|\gamma_{AE}=y}(x)) f_{\gamma_{AB}}(x)\, dx. \tag{6.12}$$

One can notice that $\epsilon_{\gamma_{AB}|\gamma_{AE}=y}(x)$ corresponds to the block error decoding probability in (6.4) when γ_{AE} takes a realization of y. Evaluation of the integral in (6.11) is difficult because the inner integral (6.12) contains the Q-function, and the lower limit of the integral depends on y. It is overcome by considering the following:

1. Using linear approximation for the Q-function.
2. Lower limit of the integral $\mathcal{S}(y)$ in (6.12) is modified to 0.

Using the linearization technique used in [12] and [13], the term $\epsilon_{\gamma_{AB}|\gamma_{AE}}(x)$ in (6.12) is approximated as given below:

$$\epsilon_{\gamma_{AB}|\gamma_{AE}}(x) \approx \begin{cases} 1 & x < \beta_{AE} + \frac{1}{2\alpha_{AE}}, \\ \frac{1}{2} + \alpha_{AE}(x - \beta_{AE}) & \beta_{AE} + \frac{1}{2\alpha_{AE}} \leq x \\ & \leq \beta_{AE} - \frac{1}{2\alpha_{AE}}, \\ 0 & x > \beta_{AE} - \frac{1}{2\alpha_{AE}}. \end{cases} \tag{6.13}$$

Now, changing the lower limit of $S(y)$ to 0, (6.12) reduces to the following:

$$S(y) \approx 1 - \int_{y=0}^{\infty} \epsilon_{\gamma_{AB}|\gamma_{AE}=y}(x)\, f_{\gamma_{AB}}(x)\, dx. \tag{6.14}$$

Using the approximation of $\epsilon_{\gamma_{AB}|\gamma_{AE}}(x)$ in (6.13), the integral $S(y)$ is simplified to the following:

$$S(y) \approx \alpha_{AE} \frac{\phi_{JB}}{\phi_{AB}} \frac{P_A}{P_J} \exp\left(\frac{\phi_{JB}}{P_J}\right) \left[\mathrm{Ei}\left(\frac{-\phi_{AB}}{P_A}\left(\beta_{AE} + \frac{1}{2\alpha_{AE}}\right) - \frac{\phi_{JB}}{P_J}\right) \right.$$
$$\left. - \mathrm{Ei}\left(\frac{-\phi_{AB}}{P_A}\left(\beta_{AE} - \frac{1}{2\alpha_{AE}}\right) - \frac{\phi_{JB}}{P_J}\right)\right]. \tag{6.15}$$

By substituting $S(y)$ in (6.11), the result in the theorem is obtained. One can refer to [20] for more details of the proof. □

6.5 Impact of Secrecy on Latency

The derived average service rate in the previous section is used to determine the delay and average age of information (AAoI) of the system.

6.5.1 Delay Analysis

Delay is primarily composed of the following: (a) *transmission delay* (D_T) and (b) *queuing delay* (D_Q). The average transmission delay for a packet is inversely proportional to the average service rate, denoted as μ, and can be represented as:

$$D_T = \frac{1}{\mu}. \tag{6.16}$$

Determining the average queuing delay entails calculating the average waiting time of a packet in a queue. This depends on factors such as the average queue length and the arrival rate. When the queue is stable, the average packet length is defined as:

$$Q = \frac{\lambda(1 - \lambda)}{\mu - \lambda}. \tag{6.17}$$

Utilizing Little's theorem, the queuing delay is defined as $D_Q = \frac{Q}{\lambda}$, yielding:

$$D_Q = \frac{1-\lambda}{\mu-\lambda}. \tag{6.18}$$

By combining Eqs. (6.17) and (6.18), the average delay can be simplified as follows:

$$D = D_T + D_Q = \frac{1}{\mu} + \frac{1-\lambda}{\mu-\lambda}. \tag{6.19}$$

One can use the result of Theorem 6.1 for the service rate (μ) to explore the role of secrecy on delay.

6.5.2 AAoI Analysis

The concept of AAoI pertains to how up-to-date the information from a source is when received at the end device, which differs fundamentally from considering the delay. In conjunction with delay, this measure captures time-sensitive information within a system. The queuing model discussed in this chapter is the Geo/Geo/1 queue. In this context, the AAoI is given by the following equation:

$$\text{AAoI} = \frac{1}{\lambda} + \frac{1-\lambda}{\mu-\lambda} - \frac{\lambda}{\mu^2} + \frac{\lambda}{\mu}. \tag{6.20}$$

The derivation of the above expression is provided in [9]. By substituting the value of μ from (6.6) in (6.20), one can calculate the AAoI for the considered system model.

6.6 Results and Discussion

This section presents numerical and simulation results to investigate the interplay between block length (N), stable throughput, delay, and AAoI under secrecy requirements. The average channel gain between Alice and Bob (jammer and Bob) is set as $\phi_{AB} = 1$ ($\phi_{JB} = 1$). The average channel gain between Alice and Eve is set as $\phi_{AE} = 0.2$ ($\phi_{JE} = 0.2$).

In Figure 6.3, the impact of transmitting and jamming power is illustrated on the average service rate. The result stated in Theorem 6.1 is used to obtain these plots. Figure 6.3a shows that the average service rate does not increase linearly with the increase in the transmit power. When the jamming power is increased to $P_J = 20$ or $P_J = 30$, the average service rate increases compared to $P_J = 5$ only for higher values of P_A. To get more insights on the impact of jamming signal on the average service rate, consider Figure 6.3b. The average service rate increases as the value of jamming power increases at the friendly jammer as the jammer causes additional interference at the eavesdropper, which impacts the attacker's decoding

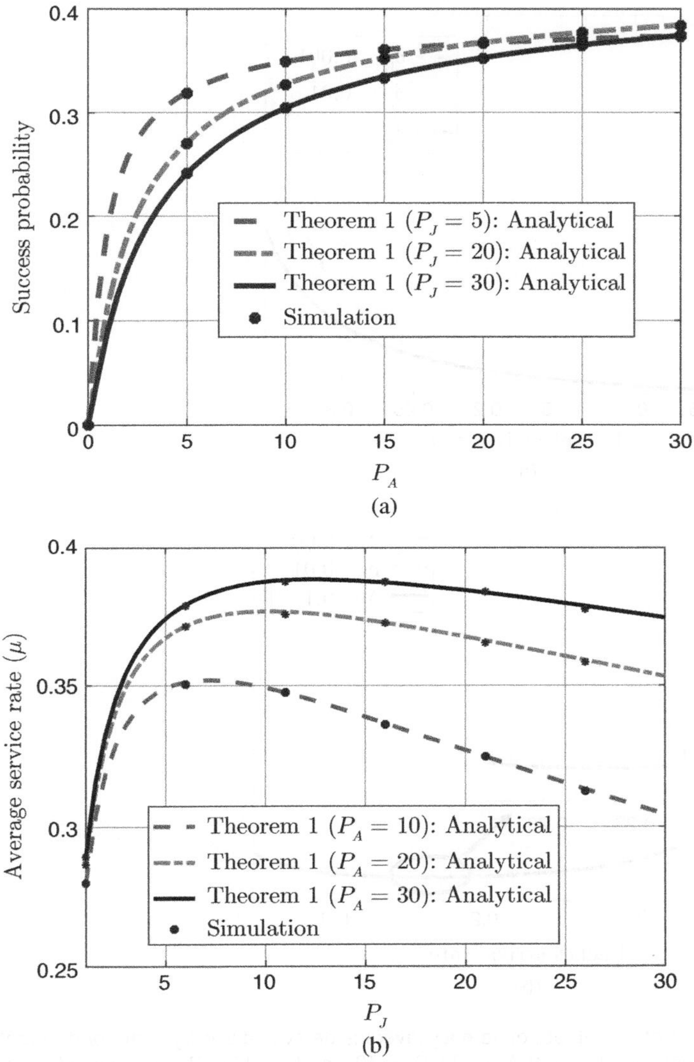

Figure 6.3 Impact of power at the transmitter and jammer on the average service rate: $(N, K) = (500, 100)$ [20]. (a) Impact of power at the transmitter on the average service rate. [20]/with permission from IEEE. (b) Impact of power at the jammer on the average service rate [20]/with permission from IEEE.

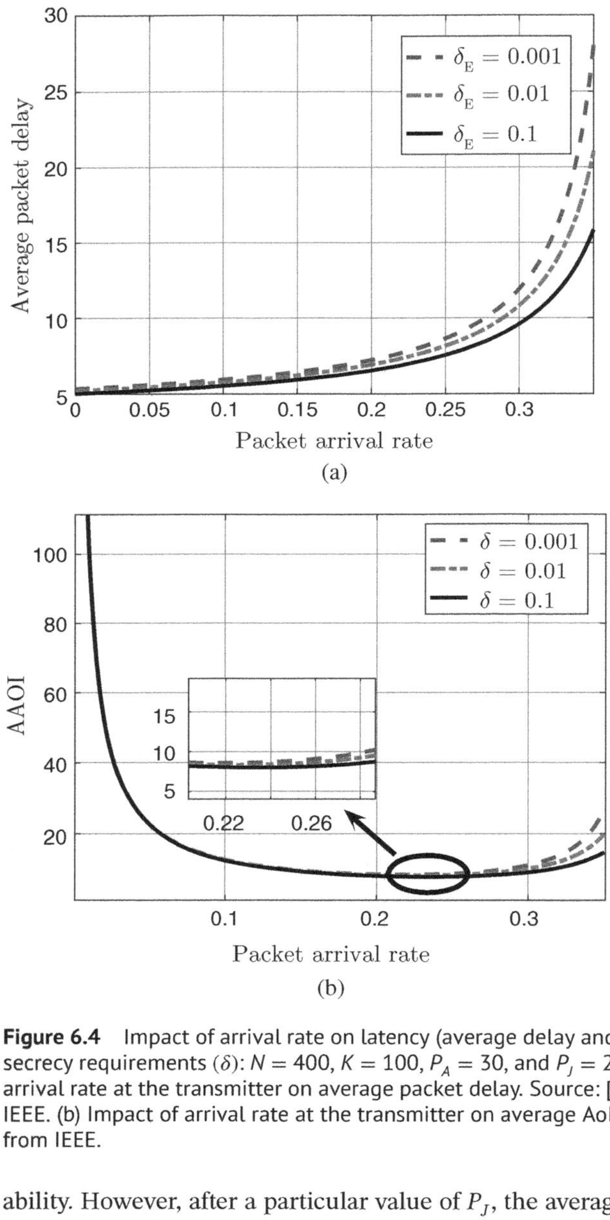

Figure 6.4 Impact of arrival rate on latency (average delay and average AoI) for different secrecy requirements (δ): $N = 400$, $K = 100$, $P_A = 30$, and $P_J = 20$ [20]. (a) Impact of arrival rate at the transmitter on average packet delay. Source: [20]/ with permission from IEEE. (b) Impact of arrival rate at the transmitter on average AoI [20]/ with permission from IEEE.

ability. However, after a particular value of P_J, the average service rate decreases with increased jamming power. It is because the gain obtained by jamming the eavesdropper diminishes as the error at the legitimate receiver increases due to the jamming signal. Hence, choosing the appropriate jamming power is necessary to benefit from the friendly jammer.

In Figure 6.4, the impact of packet arrival rate (λ) at the transmitter is explored on latency (average delay and AAoI). From Figure 6.4a, one can notice that an increase in the arrival rate results in an increase in the average delay. This is due to the queueing delay at the transmitter. It is interesting to note that AAoI decreases initially with an increase in arrival rate (see Figure 6.4b). When the arrival rate

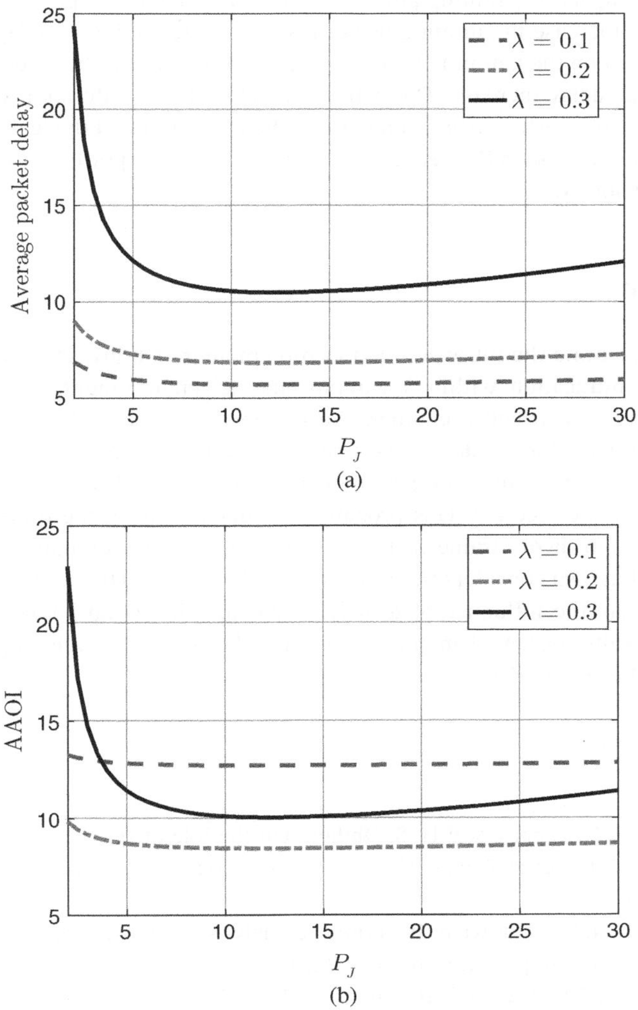

(a)

(b)

Figure 6.5 Impact of power budget of jammer on latency (average delay and average AoI) for different arrival rates at the transmitter: $N = 500$, $K = 100$, $P_A = 30$, and $\delta = 0.0001$ [20]. (a) Impact of jammer's power budget on average delay [20]. (b) Impact of jammer's power budget on average AoI [20]/ with permission from IEEE.

increases beyond a specific value, AAoI starts decreasing. Initially, AAoI was high as the source was not getting enough updates. The source gets enough updates as the arrival rate increases, reducing AAoI. However, when the arrival rate increases beyond a specific value, AAoI increases as the queueing delay increases. One can also see that it is possible to trade between information leakage and latency (AAoI and delay) for a given power budget at the transmitter and friendly jammer.

In Figure 6.5, the impact of jamming power is explored on latency aspects of communication. The increase in jamming power results in a reduction of delay and AAoI. However, when the jamming power increases beyond a specific value, the AAoI and average delay increase. The benefit obtained due to interference caused by the jammer diminishes with an increase in the interference at the legitimate receiver with an increase in P_J. Hence, one needs to choose the power of the friendly jammer judiciously.

6.7 Conclusion

In this chapter, we examine the intricate relationship among reliability, secrecy, and latency in communication within a wiretap channel environment, with the added complexity of a friendly jammer. We utilize the concept of stable throughput to account for the random data arrival at the users. However, the determination of stable throughput hinges upon the probability of successful decoding at the legitimate receiver. This probability is derived using the finite block-length information theory framework, which not only captures reliability and secrecy aspects but also considers the finite size of the packet. The findings enable us to investigate the influence of secrecy on average delay and latency. Furthermore, the results presented in this study pave the way for developing quality of security (QoSec) metrics.

References

1 M. A. Abd-Elmagid, N. Pappas, and H. S. Dhillon. On the role of age of information in the Internet of Things. *IEEE Communications Magazine*, 57(12):72–77, 2019.

2 N. Arı, N. Thomos, and L. Musavian. Performance analysis of short packet communications with multiple eavesdroppers. *IEEE Transactions on Communications*, 70(10):6778–6789, 2022. doi: 10.1109/TCOMM.2022.3198111.

3 J. Cao, J. Zhao, X. Zhu, Y. Jiang, and Z. Wei. Toward a green secure relay system for mission-critical IoT: Hybrid duplex relay selection and resource allocation in the finite block length regime. *IEEE Transactions*

on Green Communications and Networking, 5(4):1869–1879, 2021. doi: 10.1109/TGCN.2021.3093206.

4 G. Durisi, T. Koch, J. Östman, Y. Polyanskiy, and W. Yang. Short-packet communications over multiple-antenna Rayleigh-fading channels. *IEEE Transactions on Communications*, 64(2):618–629, 2015.

5 C. Feng, H.-M. Wang, and H. V. Poor. Reliable and secure short-packet communications. *IEEE Transactions on Wireless Communications*, 21(3):1913–1926, 2022. doi: 10.1109/TWC.2021.3108042.

6 Y. Gu, H. Chen, Y. Li, L. Song, and B. Vucetic. Short-packet two-way amplify-and-forward relaying. *IEEE Signal Processing Letters*, 25(2):263–267, 2017.

7 M. Haghifam, M. R. Mili, B. Makki, M. Nasiri-Kenari, and T. Svensson. Joint sum rate and error probability optimization: Finite blocklength analysis. *IEEE Wireless Communications Letters*, 6(6):726–729, 2017.

8 A. Kosta, N. Pappas, and V. Angelakis. Age of information: A new concept, metric, and tool. *Foundations and Trends® in Networking*, 12(3):162–259, 2017.

9 A. Kosta, N. Pappas, A. Ephremides, and V. Angelakis. Age of information performance of multiaccess strategies with packet management. *Journal of Communications and Networks*, 21(3):244–255, 2019.

10 C. K. Kourtellaris, C. Psomas, and I. Krikidis. Finite blocklength analysis of multiple access channels with/without cooperation. *IEEE Transactions on Communications*, 68(10):6317–6330, 2020.

11 R. M. Loynes. The stability of a queue with non-independent inter-arrival and service times. In *Mathematical Proceedings of the Cambridge Philosophical Society*, volume 58, pages 497–520. Cambridge University Press, 1962.

12 B. Makki, T. Svensson, and M. Zorzi. Finite block-length analysis of the incremental redundancy HARQ. *IEEE Wireless Communications Letters*, 3(5):529–532, 2014.

13 B. Makki, T. Svensson, and M. Zorzi. Wireless energy and information transmission using feedback: Infinite and finite block-length analysis. *IEEE Transactions on Communications*, 64(12):5304–5318, 2016.

14 P. Mohapatra, N. Pappas, J. Lee, T. Q. S. Quek, and V. Angelakis. Secure communications for the two-user broadcast channel with random traffic. *IEEE Transactions on Information Forensics and Security*, 13(9):2294–2309, 2018.

15 N. Pappas, M. A. Abd-Elmagid, H. S. Dhillon, B. Zhou, and W. Saad. *Age of Information: Foundations and Applications*. Cambridge University Press, 2023. doi: 10.1017/9781108943321.

16 Y. Polyanskiy, H. V. Poor, and S. Verdú. Channel coding rate in the finite blocklength regime. *IEEE Transactions on Information Theory*, 56(5): 2307–2359, 2010.

17 P. Popovski, F. Chiariotti, K. Huang, A. E. Kalør, M. Kountouris, N. Pappas, and B. Soret. A perspective on time toward wireless 6G. *Proceedings of the IEEE*, 110(8):1116–1146, 2022. doi: 10.1109/JPROC.2022.3190205.

18 S. Schiessl, M. Skoglund, and J. Gross. NOMA in the uplink: Delay analysis with imperfect CSI and finite-length coding. *IEEE Transactions on Wireless Communications*, 19(6):3879–3893, 2020.

19 U. Somalatha and P. Mohapatra. Short packet communication over 2-user non-orthogonal multiple access channel with confidential message. In *2023 National Conference on Communications (NCC)*, pages 1–6, 2023. doi: 10.1109/NCC56989.2023.10068105.

20 U. Somalatha, P. Mohapatra, and N. Pappas. Performance analysis of the wire-tap channel with a friendly jammer under finite blocklength. In *2022 IEEE Globecom Workshops (GC Wkshps)*, pages 239–244, 2022. doi: 10.1109/GCWkshps56602.2022.10008520.

21 H. Sung, M. Kim, S. Lee, and J. Lee. Age of information analysis for finite blocklength regime in downlink cellular networks. *IEEE Wireless Communications Letters*, 11(4):683–687, 2021.

22 W. Szpankowski. Stability conditions for some distributed systems: Buffered random access systems. *Advances in Applied Probability*, 26(2):498–515, 1994.

23 H.-M. Wang, Q. Yang, Z. Ding, and H. V. Poor. Secure short-packet communications for mission-critical IoT applications. *IEEE Transactions on Wireless Communications*, 18(5):2565–2578, 2019.

24 A. D. Wyner. The wire-tap channel. *The Bell System Technical Journal*, 54(8):1355–1387, 1975. doi: 10.1002/j.1538-7305.1975.tb02040.x.

25 W. Yang, G. Durisi, T. Koch, and Y. Polyanskiy. Quasi-static multiple-antenna fading channels at finite blocklength. *IEEE Transactions on Information Theory*, 60(7):4232–4265, 2014.

26 W. Yang, R. F. Schaefer, and H. V. Poor. Wiretap channels: Nonasymptotic fundamental limits. *IEEE Transactions on Information Theory*, 65(7):4069–4093, 2019.

27 R. D. Yates, Y. Sun, D. Richard Brown, S. K. Kaul, E. Modiano, and S. Ulukus. Age of Information: An introduction and survey. *IEEE Journal on Selected Areas in Communications*, 39(5):1183–1210, 2021. doi: 10.1109/JSAC.2021.3065072.

28 Y. Yu, H. Chen, Y. Li, Z. Ding, and B. Vucetic. On the performance of non-orthogonal multiple access in short-packet communications. *IEEE Communications Letters*, 22(3):590–593, 2017.

29 Y. Zhu, X. Yuan, Y. Hu, R. F. Schaefer, and A. Schmeink. Trade reliability for security: Leakage-failure probability minimization for machine-type communications in URLLC. *IEEE Journal on Selected Areas in Communications*, 41(7):2123–2137, 2023. doi: 10.1109/JSAC.2023.3280960.

Part III

Integration of Physical-Layer Security with 6G Communication

7

Security Challenges and Solutions for Rate-Splitting Multiple Access

Abdelhamid Salem[1,2] and Christos Masouros[1]

[1] *Department of Electronic and Electrical Engineering, University College London, London, UK*
[2] *Department of Electronic and Electrical Engineering, University of Benghazi, Benghazi, Libya*

7.1 Introduction

RSMA has been identified as a powerful scheme for boosting the performance of wireless communication networks [3, 5–7, 10]. With RSMA, the transmitters apply RS technique to transmit the messages, and the receivers use successive interference cancellation (SIC) to detect their signals [9]. Specifically, with RS, the messages are divided into a common/shared part and a private/confidential part, which are superimposed and transmitted as a single signal.[1] All receivers first decode the shared part, and then using SIC the intended receiver can decode its own private stream. Figure 7.1 shows the general architecture of RSMA. For clarity, RSMA strategy can be described as follows. The base station (BS) will transmit uncoded messages $Q_1, Q_2, \dots, Q_K$ in the same time and frequency to users, $1, 2, \dots, K$. Typically, the message to a user, Q_k, is divided into two parts, the shared (Q_{k0}) and the private (Q_{k1}). The shared portions of the users can be packed into a single shared message, $Q_0 = (Q_{10}, \dots, Q_{K0})$. Then, both the shared and K private signals are coded as streams of symbols. The shared stream is precoded in a multi-cast fashion, and the confidential streams are mapped to the antenna via appropriate multiuser precoders.

Every user decodes both streams contained in the received signal, the shared and its own specific, confidential message. This is often performed by, initially, considering the private messages as noise and decoding the shared message. After this, the SIC technique can be implemented to remove the shared stream, and the confidential stream is decoded. By efficiently allocating the transmit power between the confidential and shared messages, RSMA can improve the sum

1 Hereafter, common and shared, private and confidential are used interchangeably.

Physical-Layer Security for 6G, First Edition. Edited by Parthajit Mohapatra, Nikolaos Pappas, Arsenia Chorti, and Stefano Tomasin.

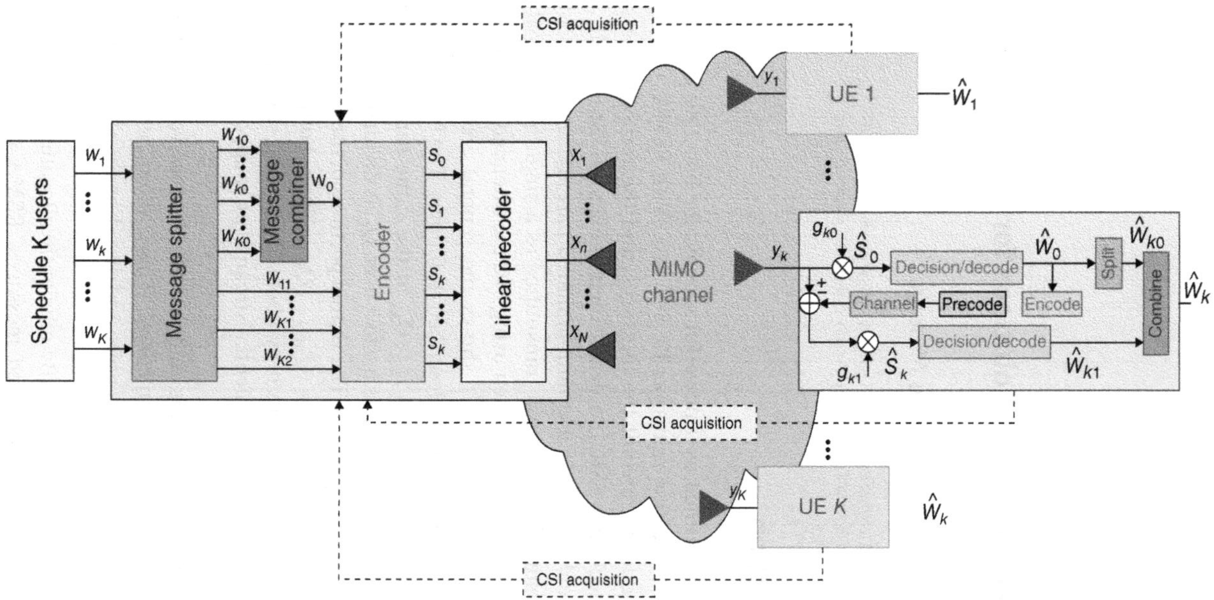

Figure 7.1 Overall architecture of RSMA scheme. Source: Adapted from [9].

rate achieved by wireless communication systems. Various RSMA architectures are now available for enhancing the performance of multiple users systems [3, 5–7, 10]. By adjusting the power and contents of the confidential and shared messages, RSMA can automatically reduce to conventional multiple access technique. However, the privacy/security of RSMA communication systems remains a most challenging and important problem. The challenge here is not just from external eavesdroppers, but also, and this is a real challenge, from internal eavesdroppers. Specifically, in RSMA parts of the users' signals are in the shared message, this makes them susceptible to eavesdropping by internal users/eavesdroppers. This raises unique security challenges, as splitting the signals can both improve the sum rate and deteriorate the system's security. Research on physical-layer security of RSMA systems is still in its early stages. Therefore, the scope of this chapter is to consider security issues relevant to RSMA and answer the question: How to provide secure RSMA transmission in a system including passive internal eavesdroppers?

Next Section 7.2 overviews security issues in RSMA systems. Section 7.3 responds to the challenge of how much of the split signal should be revealed, and Section 7.4 discusses secure beamforming design for RSMA transmission.

7.2 Security Issues in RSMA

Typical physical-layer security approaches can be implemented to provide secure RSMA transmission against any external eavesdropping attack. In this regard, the additional usage of the shared message as a jamming signal for any external eavesdroppers has been shown to provide secure RSMA communication [8]. This dual use of the shared message has also been employed to enhance the secrecy and fairness of wireless communication networks [2]. Additionally, RSMA scheme was implemented to transmit signals to legitimate users and jamming to illegitimate users in cognitive radio systems [4]. However, the real security challenge in RSMA is caused by the internal eavesdropper, where a legitimate user in the system can also be an eavesdropper. In such cases, splitting the messages into shared and confidential components increases the susceptibility to internal eavesdropping. Indeed, encoding an entire message into the shared stream cannot satisfy any realistic secrecy constraints because any user would be able to detect the complete message of any other user in the system. Thus, the revealing of the users' signals in RSMA poses unique security challenges. On the one hand, increasing the portion allocated to the shared signal might lead to an increase in the sum rate, but at the expense of the security performance. It is important to emphasize that the classical physical-layer security approaches cannot be straightforwardly applied to RSMA systems, since the passive eavesdroppers

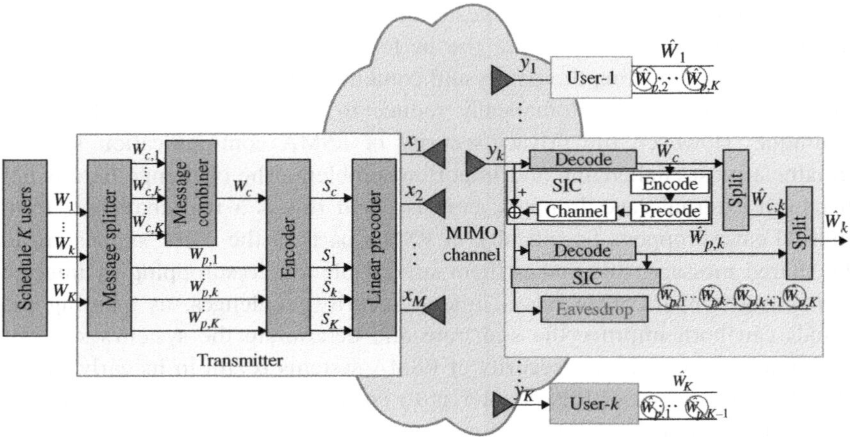

Figure 7.2 Internal eavesdropping attack in RSMA systems. Source: [14]/with permission from IEEE.

are within the system. This secrecy issue is further illustrated in Figure 7.2. Each receiver/user first decodes the shared message, and then using a first layer SIC detects its personal, confidential message, after that, another SIC layer can be used to eavesdrop any other private message in the system.

Therefore, unique and new physical-layer security techniques are needed here to provide secure RSMA transmission against any internal eavesdropping attack. In this context, two efficient security schemes have recently been shown to provide secure RSMA transmission. One depends on the relative allocation of the power to optimize the achievable total rate conditional upon specific secrecy rates. The other scheme is based on the design of an optimal RSMA precoder which can maximize the sum rate whilst attaining the required secrecy rates.

The following Sections 7.3 and 7.4 consider both schemes in more detail.

7.3 How Much of the Split Signal Should Be Revealed?

To answer this question, let us study a conventional multi-user multiple-input single-output (MU-MISO) RSMA communication system, in which N_t-antennas BS uses RSMA technique to communicate with K users each with a single antenna. It is assumed that the communication channels independently follow identical Gaussian distribution, the main channel matrix being $\mathbf{G} \in \mathbb{C}^{K \times N}$. Based on the symmetric channels' properties in time division duplex (TDD) mode, we assume similar channel responses for both up-link and down-link. By applying channel estimator schemes, one can describe the channel, G, using the estimated channel, $\hat{G}$, and the error, $\tilde{G}$, as, $G = \hat{G} - \tilde{G}$, where $\hat{G} \sim \mathcal{CN}\left(0, \hat{\boldsymbol{D}}\right)$, $\tilde{\mathbf{G}} \sim \mathcal{CN}\left(0, \tilde{\mathbf{D}}\right)$, the diagonal matrices $\tilde{\mathbf{D}}$ and $\hat{\boldsymbol{D}}$ present the variances of $\tilde{\mathbf{G}}$ and $\hat{G}$, respectively, with $\left[\tilde{\mathbf{D}}\right]_{kk} = \sigma^2_{\tilde{\mathbf{g}}_k}$ and $\left[\hat{\boldsymbol{D}}\right]_{kk} = \sigma^2_{\hat{\mathbf{g}}_k}$.

As explained above with RSMA each message, $x_{t,k}$, is divided into shared part, $x_{c,k}$, and confidential part, x_k, i.e., $x_{t,k} = \left\{x_{c,k}, x_k\right\}$. Then, the shared parts are grouped into one shared message, x_c. Consequently, the $K+1$ data symbols may be expressed in vector form as $\mathbf{x} = \left[x_c, x_1, \ldots, x_K\right]^T$, with $\mathcal{E}\left\{\mathbf{x}\mathbf{x}^H\right\} = I$. The data symbols are then linearly precoded using a precoding matrix $\mathbf{W} = \left[\mathbf{w}_c, \mathbf{w}_1^p, \ldots \mathbf{w}_K^p\right]$ where $\mathbf{w}_c \in \mathbb{C}^{N_t}$ and $\mathbf{w}_k^p \in \mathbb{C}^{N_t}$ are the precoders of the shared message and kth confidential message, respectively. Then, it is possible to express the transmitted signal, s, as

$$\mathbf{s} = \mathbf{W}\mathbf{x} = \sqrt{P_c}\mathbf{w}_c x_c + \sum_{k=1}^{K} \sqrt{P_p}\mathbf{w}_i^p x_i, \tag{7.1}$$

where $P_p = \frac{tP}{K}$, is the power of the confidential part, with P is the total power, t is the fractional factor, $0 < t \leq 1$, while $P_c = (1-t)P$ is the power allocated to the shared part.

Accordingly, user k, can receive the following signal:

$$y_k = \sqrt{P_c}\mathbf{g}_k\mathbf{w}_c x_c + \sum_{i=1}^{K} \sqrt{P_p}\mathbf{g}_k\mathbf{w}_i^p x_i + n_k, \tag{7.2}$$

where $\mathbf{g}_k$ denotes user k channel vector, and n_k is the additive white Gaussian noise (AWGN), $n_k \sim \mathcal{CN}\left(0, \sigma_k^2\right)$. Then, SIC is used to completely remove the shared message, thus, the confidential signal can be presented as,

$$y_k^p = \sum_{i=1}^{K} \sqrt{P_p}\mathbf{g}_k\mathbf{w}_i^p x_i + n_k, \tag{7.3}$$

RSMA sum-rate, r, is evaluated by,

$$r = r_c + \sum_{k=1}^{K} r_k^p, \tag{7.4}$$

where r_c represents the rate of the shared part and r_k^p is the achievable rate of the kth confidential signal. Any user-k in the system is assigned a portion C_k from r_c for its common part. Therefore, the portions of the shared rates at the users should satisfy $\sum_{k \in K} C_k = r_c$. To ensure the successful decoding of the shared message, and its removal by all users, the shared message would have to be sent at an ergodic rate less than, $\min_i \left(\mathbb{E}\left\{C_k\right\}\right)_{k=1}^{K}$. Thus, the ergodic sum rate is determined by,

$$\mathbb{E}\left\{r\right\} = \min_i \left(\mathbb{E}\left\{C_k\right\}\right)_{k=1}^{K} + \sum_{k=1}^{K} \mathbb{E}\left\{r_k^{\mathrm{p}}\right\}. \tag{7.5}$$

Here, we define secrecy by the ability of a user in the system to detect any other user's message in the system, e.g., the eavesdropper/user i exploits the common part and the leakage due to the partial knowledge of the channel state information (CSI), to eavesdrop the message intended for user k. Accordingly, we define the ergodic secrecy rate as

$$r_{\mathrm{s},k} = \mathbb{E}\left\{\left[\left(r_{\mathrm{c}} + r_k^{\mathrm{p}}\right) - \max\left\{\left(r_{\mathrm{c}} + r_{j\to k}^{\mathrm{p}}\right), j \in K, j \neq k\right\}\right]^{+}\right\}, \tag{7.6}$$

where $[r_s]^+ = \max(0, r_s)$, and the achievable rate at which user j decodes the signal assigned for user k is $r_{j\to k}^{\mathrm{p}}$. It can be seen from (7.4) and (7.6) that by suitably tuning the power between the confidential and shared parts, it is possible to achieve a flexible trade-off between the system rates and the security.

7.3.1 Ergodic Rates

Here, we implement maximum ratio transmission (MRT) precoding for the shared signal, while zero-forcing (ZF) scheme is used for the confidential signals, thus, one can write the precoding vectors for the shared and confidential messages, respectively, as:

$$\mathbf{w}_{\mathrm{c}} = \frac{\sum_{i=1}^{K} \hat{\mathbf{g}}_i^{\mathrm{H}}}{\left\|\sum_{i=1}^{K} \hat{\mathbf{g}}_i^{\mathrm{H}}\right\|}, \quad \text{and} \quad \mathbf{w}_k^{\mathrm{p}} = \frac{\mathbf{f}_k^{\mathrm{p}}}{\left\|\mathbf{f}_k^{\mathrm{p}}\right\|}, \tag{7.7}$$

where $\mathbf{f}_k^{\mathrm{p}}$ is the kth vector in $\mathbf{F}^{\mathrm{p}} = \hat{\mathbf{G}}^{\mathrm{H}}\left(\hat{\mathbf{G}}\hat{\mathbf{G}}^{\mathrm{H}}\right)^{-1}$. Now (7.2) is represented as

$$y_k = \sqrt{P_{\mathrm{c}}}\hat{\mathbf{g}}_k\mathbf{w}_{\mathrm{c}}x_{\mathrm{c}} + \sqrt{P_{\mathrm{c}}}\tilde{\mathbf{g}}_k\mathbf{w}_{\mathrm{c}}x_{\mathrm{c}} + \beta_{\mathrm{p}_k}\sqrt{P_{\mathrm{p}}}x_k^{\mathrm{p}} + \sum_{i=1}^{K}\sqrt{P_{\mathrm{p}}}\tilde{\mathbf{g}}_k\mathbf{w}_i^{\mathrm{p}}x_i + n_k, \tag{7.8}$$

where $\beta_{\mathrm{p}_k} = \frac{1}{\left\|\mathbf{f}_k^{\mathrm{p}}\right\|}$. It is now possible to present the signal to interference and noise ratio (SINR) of the shared part for the kth user as

$$\gamma_k^{\mathrm{c}} = \frac{P_{\mathrm{c}}\left|\frac{\hat{\mathbf{g}}_k\sum_{i=1}^{K}\hat{\mathbf{g}}_i^{\mathrm{H}}}{\left\|\sum_{i=1}^{K}\hat{\mathbf{g}}_i^{\mathrm{H}}\right\|}\right|^2 + P_{\mathrm{c}}\sigma_{\tilde{\mathbf{g}}_k}^2}{\frac{P_{\mathrm{p}}}{\left[\left(\hat{\mathbf{G}}\hat{\mathbf{G}}^{\mathrm{H}}\right)^{-1}\right]_{k,k}} + P_{\mathrm{p}}\sigma_{\tilde{\mathbf{g}}_k}^2 + \sigma_k^2}, \tag{7.9}$$

With an ergodic rate as

$$\mathbb{E}\left[C_k\right] = \mathbb{E}\left[\log_2\left(1+\frac{P_{\mathrm{c}}\left|\frac{\hat{\mathbf{g}}_k\sum_{i=1}^{K}\hat{\mathbf{g}}_i^{\mathrm{H}}}{\left\|\sum_{i=1}^{K}\hat{\mathbf{g}}_i^{\mathrm{H}}\right\|}\right|^2+P_{\mathrm{c}}\sigma_{\tilde{\mathbf{g}}_k}^2}{\frac{P_{\mathrm{p}}}{\left[\left(\hat{\mathbf{G}}\hat{\mathbf{G}}^{\mathrm{H}}\right)^{-1}\right]_{k,k}}+P_{\mathrm{p}}\sigma_{\tilde{\mathbf{g}}_k}^2+\sigma_k^2}\right)\right]. \tag{7.10}$$

We obtain the ergodic rate of the shared message in terms of the transmission powers of the confidential and shared signals P_{c}, P_{p}, respectively, as [12]

$$C_k = \frac{1}{\ln(2)}\sum_{k=1}^{K}\mathrm{T}_i\frac{1}{o_i}\left(1-\left(\left(1+\frac{P_{\mathrm{c}}\theta_k\,o_i}{\beta}\right)^{-K}\right)e^{-o_iP_{\mathrm{c}}\sigma_{\tilde{\mathbf{g}}_k}^2}\right)$$
$$\times\left(\left(1+\frac{P_{\mathrm{p}}\Psi_k\,o_i}{\beta}\right)^{-1+K-N}\right), \tag{7.11}$$

where $\beta = KP_{\mathrm{p}}\sigma_{\tilde{\mathbf{g}}_k}^2+\sigma_k^2$, T_i and o_i are the weighting factor and ith zero, respectively, see [[1], (25.245)] for tabulated values of the Laguerre polynomials.

We can rewrite (7.3) to determine the ergodic rate for the confidential stream as

$$y_k^{\mathrm{p}} = \beta_{\mathrm{p}_k}\sqrt{P_{\mathrm{p}}}x_k+\sum_{i=1}^{K}\sqrt{P_{\mathrm{p}}}\tilde{\mathbf{g}}_k\mathbf{w}_i^{\mathrm{p}}x_i+n_k. \tag{7.12}$$

Thus, the SINR is [11, 13]

$$\gamma_k^{\mathrm{p}} = \frac{\frac{P_{\mathrm{p}}}{\left[\left(\hat{\mathbf{G}}\hat{\mathbf{G}}^{\mathrm{H}}\right)^{-1}\right]_{k,k}}+P_{\mathrm{p}}\sigma_{\tilde{\mathbf{g}}_k}^2}{(K-1)P_{\mathrm{p}}\sigma_{\tilde{\mathbf{g}}_k}^2+\sigma_k^2}. \tag{7.13}$$

For the confidential part, the ergodic rate can be obtained by perform the following expectation,

$$\mathbb{E}\left[r_k^{\mathrm{p}}\right] = \mathbb{E}\left[\log_2\left(1+\frac{\frac{P_{\mathrm{p}}}{\left[\left(\hat{\mathbf{G}}\hat{\mathbf{G}}^{\mathrm{H}}\right)^{-1}\right]_{k,k}}+P_{\mathrm{p}}\sigma_{\tilde{\mathbf{g}}_k}^2}{(K-1)P_{\mathrm{p}}\sigma_{\tilde{\mathbf{g}}_k}^2+\sigma_k^2}\right)\right], \tag{7.14}$$

From which using [12], we can obtain

$$\mathbb{E}\left[r_k^{\mathrm{p}}\right] = \sum_{i=1}^{n}\mathrm{T}_i\frac{1}{o_i}\log_2\left(1+\frac{P_{\mathrm{p}}o_i+P_{\mathrm{p}}\sigma_{\tilde{\mathbf{g}}_k}^2}{(K-1)P_{\mathrm{p}}\sigma_{\tilde{\mathbf{g}}_1}^2+\sigma_k^2}\right)\frac{o_i^{(N-K)}\left(\Psi_k\right)^{N-K+1}e^{-\Psi_k o_i}}{\Gamma(N-K+1)}, \tag{7.15}$$

If user j wishes to eavesdrop on a message intended for user k, it would be possible to do this by, first, detecting the shared element and its confidential part, then detecting the kth message using SIC. At user j, the received signal to get the confidential message of user k is

$$y^{\mathrm{p}}_{j\to k} = \sqrt{P_{\mathrm{p}}}\tilde{\mathbf{g}}_j\mathbf{w}^{\mathrm{p}}_k x_k + \sum_{i=1,i\neq j,k}^{K} \sqrt{P_{\mathrm{p}}}\tilde{\mathbf{g}}_j\mathbf{w}^{\mathrm{p}}_i x_i + n_j. \tag{7.16}$$

It is then possible to write the SINR as

$$\gamma^{\mathrm{p}}_{j\to k} = \frac{P_{\mathrm{p}}\sigma^2_{\tilde{\mathbf{g}}_j}}{(K-2)P_{\mathrm{p}}\sigma^2_{\tilde{\mathbf{g}}_j} + \sigma^2_j}. \tag{7.17}$$

We can now calculate the ergodic rate using

$$\mathbb{E}\left[\left\{r^{\mathrm{p}}_{j\to k}\right\}\right] = \log_2\left(1 + \frac{P_{\mathrm{p}}\sigma^2_{\tilde{\mathbf{g}}_j}}{(K-2)P_{\mathrm{p}}\sigma^2_{\tilde{\mathbf{g}}_j} + \sigma^2_j}\right). \tag{7.18}$$

Finally, the lower bound of the ergodic secrecy rate may be determined using [12]

$$\begin{aligned}\mathbb{E}\left[r_{\mathrm{s}}\right] = \frac{1}{\ln(2)}\sum_{i=1}^{n}\frac{\mathrm{T}_i}{o_i}\left(1 - \left(e^{-o_i P\sigma^2_{\tilde{\mathbf{g}}_k}}\left(1 + P_{\mathrm{p}}\Psi_k\, o_i\right)^{K-N_t-1}\right)\right) \\ \times e^{o_i\left(1-\left((K-2)P_{\mathrm{p}}\sigma^2_{\tilde{\mathbf{g}}_k}+\sigma^2_k\right)\right)} - \max_j \log_2\left(1 + \frac{P_{\mathrm{p}}\sigma^2_{\tilde{\mathbf{g}}_j}}{(K-2)P_{\mathrm{p}}\sigma^2_{\tilde{\mathbf{g}}_j} + \sigma^2_j}\right).\end{aligned} \tag{7.19}$$

7.3.2 Power Allocation Strategy for Secure RSMA Transmission

As explained above, we may increase the sum rate if the power allocated to the shared portion of the message is increased, but revealing part of the signals can result in a degraded secrecy rate. How the power is shared between the two streams, confidential and shared, should be assessed very carefully. An efficient power allocation scheme to attain the required secrecy rate while maximizing the ergodic sum rate can be obtained by considering the following optimization problem:

$$\begin{aligned}&\max_{0<t\leq 1} \mathbb{E}\left\{r_{\mathrm{c}}\right\} + \sum_{k=1}^{K}\left(\mathbb{E}\left\{r^{\mathrm{p}}_k\right\}\right), \\ &\text{s.t.}\ \mathbb{E}\left[r_{\mathrm{s}}\right] > r_{\mathrm{th}}, \forall_{j,k}, \\ &KP_{\mathrm{p}} + P_{\mathrm{c}} \leq P,\end{aligned} \tag{7.20}$$

where r_{th} is the required secrecy rate. We solve this problem using a low-complexity heuristic approach. First, we find a value for t, (t_{s}), that gives the required secrecy rate, r_{th}. Then, we know that any value of t greater than t_{s} will meet the required secrecy condition. Thus, the optimum t will have a value in the range $\left[t_{\text{s}}, 1\right]$, see Figure 7.3.

It is possible to obtain the value of t_{s} numerically by incrementally varying t across the range from zero to one. In order to gain deep insights into the process, we derive here an approximate value of t_{s}. For the secrecy constraint in (7.20) to hold, we have

$$r_{\text{th}} = \Xi \log_2\left(1 + \frac{tPo_1 + tP\sigma^2_{\tilde{\mathbf{g}}_k}}{(K-1)tP\sigma^2_{\tilde{\mathbf{g}}_k} + \sigma^2_k}\right) - \max_j \log_2\left(1 + \frac{tP\sigma^2_{\tilde{\mathbf{g}}_j}}{(K-2)tP\sigma^2_{\tilde{\mathbf{g}}_j} + \sigma^2_j}\right), \tag{7.21}$$

where $\Xi = \mathrm{T}_1 \frac{1}{o_1} \frac{o_1^{(N_t-K)} (\Psi_k)^{N_t-K+1} e^{-\Psi_k o_1}}{\Gamma(N_t-K+1)}$. Then, the value of t_{s} can be obtained as

$$t_{\text{s}} = \frac{(2^{\varrho}-1)\left(\sigma^2_k\right)}{Po_1 + P\sigma^2_{\tilde{\mathbf{g}}_k} - (2^{\varrho}-1)(K-1)P\sigma^2_{\tilde{\mathbf{g}}_k}}. \tag{7.22}$$

We can now reformulate the optimization problem as,

$$\max_{t_s < t \leq 1} \mathbb{E}\left\{r_{\text{c}}\right\} + \sum_{k=1}^{K} \left(\mathbb{E}\left\{r^{\text{p}}_k\right\}\right),$$
$$\text{s.t. } KP_{\text{p}} + P_{\text{c}} \leq P. \tag{7.23}$$

We can now determine the optimum value of t via a suitable linear search technique over the range $ts \leq t \leq 1$. Simple single-dimensional techniques such as the golden section can be used.

To illustrate the proposed scheme, Figure 7.4 shows how the ergodic secrecy rate and ergodic sum rate vary with the power fraction t, with SNR = 5 dB, $N_t = 5$, $K = 5$, and $\sigma^2_{\tilde{\mathbf{g}}_k} = 0.7$. We assume that the required secrecy rate is $r_{\text{th}} = 0.43$ (bits/s/Hz) as in Figure 7.4a. Interestingly, it is noted from the latter that the optimum value for the sum rate occurs when $t = 0.6$, while the maximum value of secrecy occurs for a value of $t = 1$. This observation is a clear demonstration of the trade-off between secrecy and sum rates. Accordingly, the best overall value of t is within the range $t^* \in [0.9, 1]$, see Figure 7.4b, which shows that, $t = 0.9$, is the optimum value. Thus, to attain the desired secrecy rate while optimizing the sum rate, a greater proportion of the power should be allocated to the confidential messages. Without any constraint on secrecy, the optimum t value is seen to be about 0.6. Thus, more power should be allocated to the shared part of the message to optimize the sum rate for RSMA transmissions with no

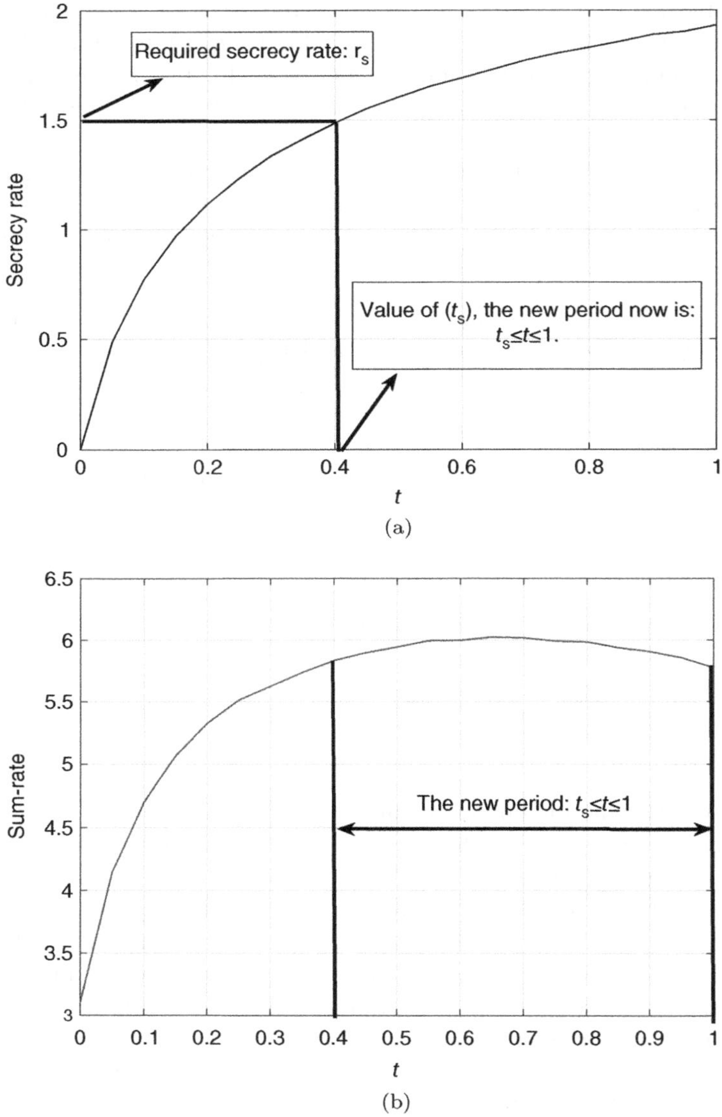

Figure 7.3 The proposed power-allocation scheme. (a) Secrecy rate versus t and (b) achievable sum rate versus t.

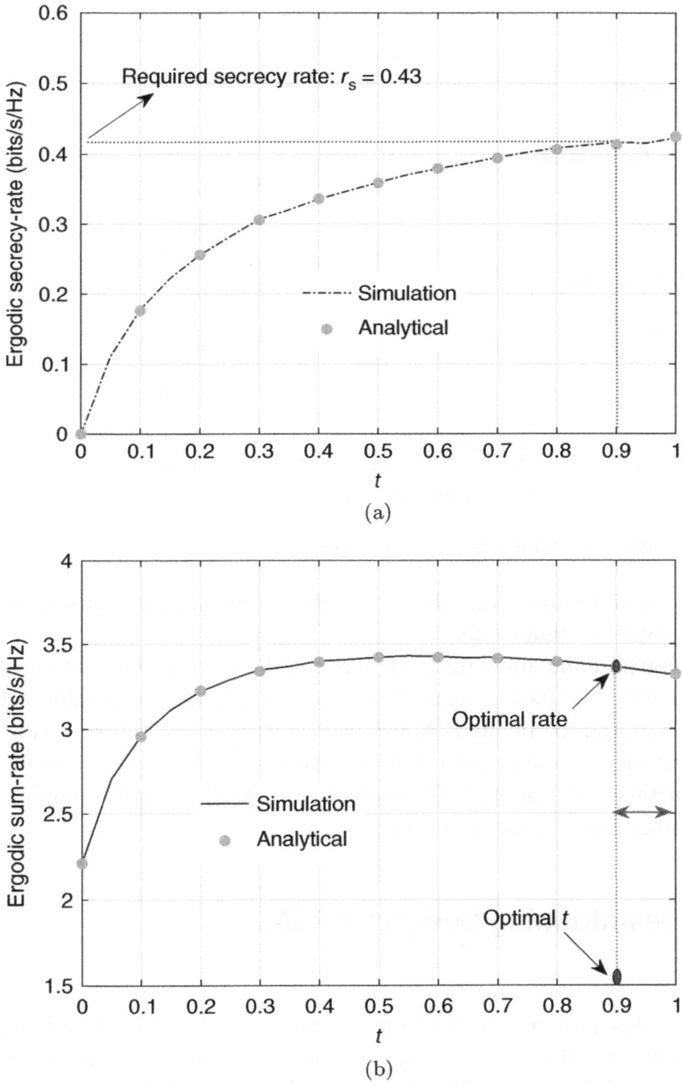

Figure 7.4 Ergodic secrecy rate and sum rate versus t. (a) Ergodic secrecy rate as a function of t and (b) ergodic sum rate as a function of t.

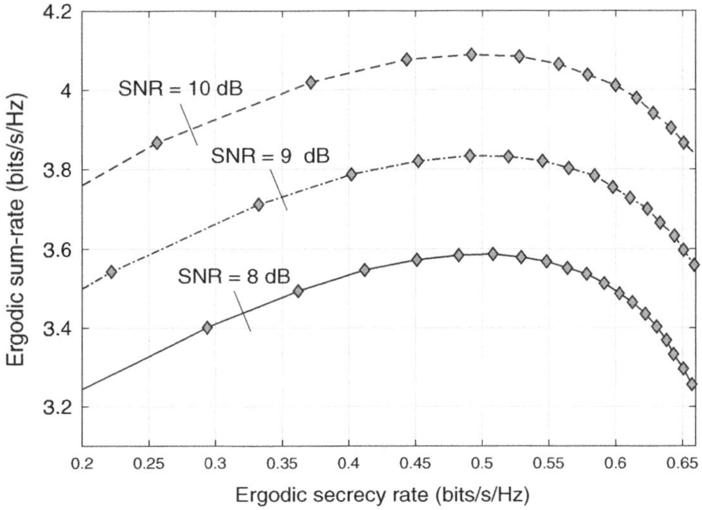

Figure 7.5 Ergodic sum rate versus ergodic secrecy rate.

secrecy constraint, which shows the substantial effect of the security constraints on the optimum power fraction t value.

To further demonstrate the direct trade-off between the secrecy and sum rates, Figure 7.5 shows sum rate plotted against the secrecy rate for three values of the SNR, SNR $= 8, 9, \text{and } 10$ dB, with $N = K = 3$, and $\sigma^2_{\tilde{\mathbf{g}}_k} = 0.7$. These results demonstrate the existence of an optimal secrecy rate value that maximizes the sum rate. This further confirms that the secrecy rate threshold value requires careful selection to ensure a higher sum-rate.

7.4 Secure Beamforming Design for RSMA Transmission

This section aims to design a precoder that maximizes the weighted sum rate while simultaneously achieving the required target for the user's secrecy rate. Let us study a typical downlink MU-MISO RSMA system, where each of the K users has one antenna, and all are served concurrently by a BS equipped with N_t antennas. The BS sends a confidential message to each of the k users, such that the message destined for the kth user should be kept secret from every other users. Following the practices of RSMA, the BS-transmitted signal can be presented as,

$$\mathbf{s} = \mathbf{W}\mathbf{x} = \mathbf{w}_c x_c + \sum_{i=1}^{K} \mathbf{w}_i^{\mathrm{p}} x_i. \tag{7.24}$$

The constraint on the transmit power may be defined as $\mathrm{tr}(\mathbf{W}\mathbf{W}^{\mathrm{H}}) \leq P$. For user k, the received message is

$$y_k = \mathbf{g}_k^{\mathrm{H}}\mathbf{s} + n_k. \tag{7.25}$$

The users now detect and remove the shared message using SIC scheme, then each user can detect its confidential message. Hence, the corresponding SINRs of the shared and confidential streams are given, respectively, by

$$\gamma_k^{\mathrm{c}} = \frac{\left|\mathbf{g}_k^{\mathrm{H}}\mathbf{w}_{\mathrm{c}}\right|^2}{\sum_{j\in K}\left|\mathbf{g}_k^{\mathrm{H}}\mathbf{w}_j^{\mathrm{p}}\right|^2 + \sigma_k^2}. \tag{7.26}$$

and

$$\gamma_k^{\mathrm{p}} = \frac{\left|\mathbf{g}_k^{\mathrm{H}}\mathbf{w}_k^{\mathrm{p}}\right|^2}{\sum_{j\neq k}\left|\mathbf{g}_k^{\mathrm{H}}\mathbf{w}_j^{\mathrm{p}}\right|^2 + \sigma_k^2}. \tag{7.27}$$

After decoding the confidential message, an additional SIC layer could be used by an eavesdropper to cancel the private stream in the received signal, and detect the confidential messages for any of the other $K-1$ users. Thus, the SINR at the jth user attempting to detect the kth private stream is

$$\gamma_{j\to k} = \frac{\left|\mathbf{g}_j^{\mathrm{H}}\mathbf{w}_k^{\mathrm{p}}\right|^2}{\sum_{k'\neq j,k}\left|\mathbf{g}_j^{\mathrm{H}}\mathbf{w}_{k'}^{\mathrm{p}}\right|^2 + \sigma_k^2}. \tag{7.28}$$

Now, we can write the secrecy rate as

$$r_{\mathrm{s},k} = \left[r_k^{\mathrm{p}} - \max\left\{r_{j\to k}, j\in K, j\neq k\right\}\right]^+. \tag{7.29}$$

To guarantee that the shared part is successfully detected and canceled by the users, the transmission rate for the shared message, r_{c}, should be the lowest rate of the shared parts for all users.

7.4.1 Optimization Framework

The focus here is on designing a secure precoder that maximizes the users' weighted sum rate under secrecy rate constraints. An optimal precoder is designed for two cases according to whether the CSI is perfectly or imperfectly estimated at the BS.

7.4.1.1 Perfect CSIT

The first case is when the channel state information at the transmitter (CSIT) is fully known. Depending on the secrecy constraints on each user, the instantaneous weighted sum rate (WSR) is to be maximized. The weight vector is $\mathbf{u} = [u_1, \dots, u_K]$, thus, secure RSMA precoder can be designed by considering the following optimization problem:

$$\max_{\mathbf{c},\mathbf{W}} \sum_{k\in K} u_k \left(C_k + r_k^{\mathrm{p}}\right),$$
$$\text{s.t. } r_{\mathrm{s},k} \geq r_{\mathrm{th},k} \quad \forall k,$$
$$\sum_{j\in K} C_j \leq r_k^{\mathrm{c}}, \mathbf{c} \geq 0, \quad \forall k,$$
$$\mathrm{Tr}\left(\mathbf{W}\mathbf{W}^{\mathrm{H}}\right) \leq P, \tag{7.30}$$

where $r_{\mathrm{th},k}$ is the target secrecy rate for user k, and $\mathbf{c}$ is the allocation vector of the shared parts, $\mathbf{c} = [C_1, \dots, C_K]$.

The condition on the security performance of each user is presented in the first constraint in (7.30), while the second constraint is to ensure the successful detection of the shared message at the users, and the last constraint is for the total transmit power.

It is seen that under these constraints and the non-convexity of the objective function, (7.30) is intractable and hard to solve. However, this problem is solvable using a successive convex approximation (SCA)-based algorithm as explained in [14].

7.4.1.2 Imperfect CSIT

A more realistic scenario is when an imperfect CSIT is available at the BS. The channel may be estimated as $\hat{\mathbf{G}} = [\hat{\mathbf{g}}_1, \dots, \hat{\mathbf{g}}_K]$, and channel error as $\tilde{\mathbf{G}} = [\tilde{\mathbf{g}}_1, \dots, \tilde{\mathbf{g}}_K]$. Every element of $\tilde{\mathbf{g}}_k$ independently follows an identical complex Gaussian distribution, characterized by zero-mean and $\mathbb{E}\left\{\tilde{\mathbf{g}}_k\tilde{\mathbf{g}}_k^{\mathrm{H}}\right\} = \sigma_{\tilde{\mathbf{g}}_k}^2 \mathbf{I}$, $\sigma_{\tilde{\mathbf{g}}_k}^2 \backsim O\left(P^{-\delta}\right)$ with scaling factor $\delta \in [0, \infty)$ describing the variance of the error and the CSIT quality for high SINR regimes. If the instantaneous WSR is maximized using the estimated channel states, the resulting transmission rates may be undecodable, since such a process neglects the errors due to multi-user interference. To describe the long-term performance when the CSIT is imperfectly known, the aim would be optimizing the weighted ergodic sum rate (WESR) for all users which is defined as

$$\mathrm{WESR}\left(\mathbf{G}, \hat{\mathbf{G}}\right) = \sum_{k\in K} u_k \left\{\mathbb{E}_{\{\mathbf{G},\hat{\mathbf{G}}\}}\left\{C_k + r_k^{\mathrm{p}}\right\}\right\}.$$

Accordingly, the WESR optimization problem may be formulated as

$$\max_{\mathbf{c},\mathbf{W}} \sum_{k\in K} u_k \left\{\mathbb{E}_{\{\mathbf{G},\hat{\mathbf{G}}\}}\left\{C_k + r_k^{\mathrm{p}}\right\}\right\}$$
$$\text{s.t. } \mathbb{E}_{\{\mathbf{G},\hat{\mathbf{G}}\}}\left\{r_{\mathrm{s},k}\right\} \geq r_{\mathrm{th},k} \quad \forall k \in K$$

$$\sum_{j \in K} \mathbb{E}_{\{\mathbf{G},\hat{\mathbf{G}}\}} \{C_j\} \leq \mathbb{E}_{\{\mathbf{G},\hat{\mathbf{G}}\}} \{r_c\}$$
$$\mathbb{E}_{\{\mathbf{G},\hat{\mathbf{G}}\}} \{C_j\} \geq 0, \forall j \in K$$
$$\mathbb{E}_{\{\mathbf{G},\hat{\mathbf{G}}\}} \{\mathrm{Tr}(\mathbf{W}\mathbf{W}^{\mathrm{H}})\} \leq P. \tag{7.31}$$

The problem expressed in (7.31) is intractable since only the conditional density $f_{\mathbf{G}|\hat{\mathbf{G}}}\left(\mathbf{G}\,\middle|\,\hat{\mathbf{G}}\right)$ can be known, and the joint probability density distribution $f_{\mathbf{G}|\hat{\mathbf{G}}}\left(\mathbf{G},\hat{\mathbf{G}}\right)$ cannot be known. Thus, when CSIT is imperfectly known, the problem expressed in (7.31) may be reduced to a two-step optimization problem. First, the WASR for a given channel estimate, $\hat{\mathbf{G}}$, is maximized, then the WSR is averaged over all possible channel estimates [14].

Figure 7.6 presents the WSR obtained using the RSMA scheme versus secrecy rate thresholds, r_{th}. We can see that, for both imperfect and perfect CSIT, the designed precoders of RSMA are able to provide secure transmissions by optimizing the private streams. With an increase in the required secrecy rate, r_{th}, additional power ought to be provided to the confidential streams to increase the security, though this will cause a loss in WSR. This phenomenon is more obvious with an imperfect CSIT case. Generally, the RSMA scheme is more sensitive to the changes in the required secrecy rate, r_{th}, because the power provided to the shared streams should be updated based on the value of the required secrecy rate.

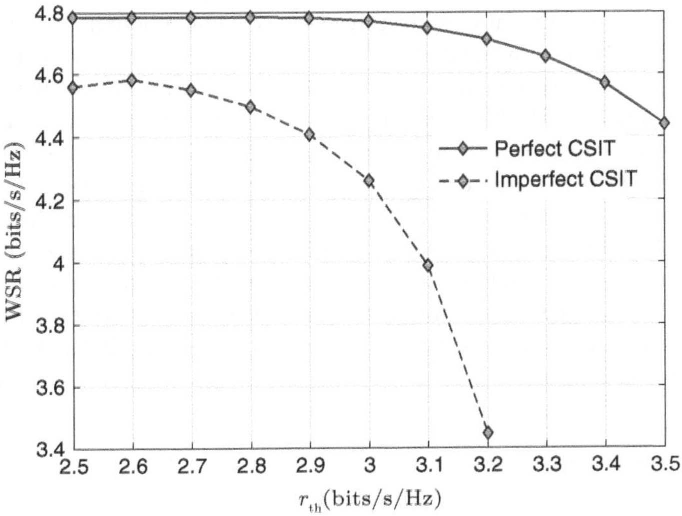

Figure 7.6 WSR versus desired secrecy rate, r_{th}, with $N = 2$, $K = 2$, and SNR = 20 dB.

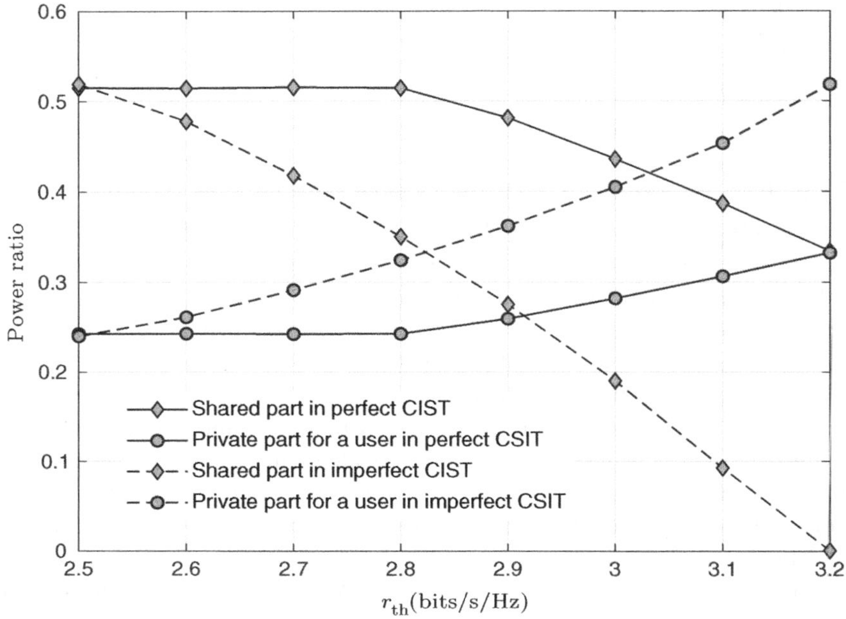

Figure 7.7 Power allocation versus desired secrecy rate, r_{th}, with $N = 2$, $K = 2$, and SNR = 20 dB.

Figure 7.7 illustrates how the ratio of the power provided to both confidential and shared streams of two users varies with secrecy rate, r_{th}, under both imperfect and perfect CSIT. Clearly, the power allocated to the confidential streams of both users overlaps with an increase in r_{th} value, because both users have a similar channel strength. With perfect CSIT, the power allocated to the shared element of the signal initially remains constant with an increase in the desired secrecy rate and then decreases with a further increase in r_{th}. With imperfect CSIT, the power allocated to the shared element of the signal decreases continuously with an increase in r_{th} and approaches zero when r_{th} becomes sufficiently large because most of the power is directed to the confidential stream. Interestingly enough, in imperfect CSIT, the power allocated to the confidential streams is greater than that for perfect CSIT, because under poor channel conditions, a higher power is required to achieve the same security level.

7.5 Conclusion

This chapter concerned security challenges and solutions for RSMA. We first introduced the concept of RSMA, and the benefits of RSMA over conventional communication systems have been explained. We then described why the security issue in RSMA technique is unique and challenging. We also showed that the typical physical-layer security techniques can be implemented to face any passive external eavesdropping attack, and the real threat occurring when a user within the system acts as an eavesdropper. To tackle this security issue, we discussed two techniques. In the first, an efficient power allocation scheme has been described in detail to achieve a given secrecy rate. In the second one, a secure RSMA precoder design has been presented and discussed.

References

1 M. Abramowitz and I. A. Stegun. *Handbook of Mathematical Functions With Formulas, Graphs, and Mathematical Tables*. Washington, DC: U.S. Dept. Commerce, 1972.

2 H. Bastami, M. Letafati, M. Moradikia, A. Abdelhadi, H. Behroozi, and L. Hanzo. On the physical layer security of the cooperative rate-splitting-aided downlink in UAV networks. *IEEE Transactions on Information Forensics and Security*, 16:5018–5033, 2021. doi: 10.1109/TIFS.2021.3122989.

3 B. Clerckx, H. Joudeh, C. Hao, M. Dai, and B. Rassouli. Rate splitting for MIMO wireless networks: A promising PHY-layer strategy for LTE evolution. *IEEE Communications Magazine*, 54(5):98–105, May 2016. ISSN 0163-6804. doi: 10.1109/MCOM.2016.7470942.

4 O. Dizdar and B. Clerckx. Rate-splitting multiple access for communications and jamming in multi-antenna multi-carrier cognitive radio systems. *IEEE Transactions on Information Forensics and Security*, 17:628–643, 2022. doi: 10.1109/TIFS.2022.3149145.

5 C. Hao, Y. Wu, and B. Clerckx. Rate analysis of two-receiver MISO broadcast channel with finite rate feedback: A rate-splitting approach. *IEEE Transactions on Communications*, 63(9):3232–3246, Sept 2015. ISSN 0090-6778. doi: 10.1109/TCOMM.2015.2453270.

6 H. Joudeh and B. Clerckx. Sum-rate maximization for linearly precoded downlink multiuser MISO systems with partial CSIT: A rate-splitting approach. *IEEE Transactions on Communications*, 64(11):4847–4861, Nov 2016. ISSN 0090-6778. doi: 10.1109/TCOMM.2016.2603991.

7 H. Joudeh and B. Clerckx. Robust transmission in downlink multiuser MISO systems: A rate-splitting approach. *IEEE Transactions on Signal Processing*, 64(23):6227–6242, Dec 2016. ISSN 1053-587X. doi: 10.1109/TSP.2016.2591501.

8 P. Li, M. Chen, Y. Mao, Z. Yang, B. Clerckx, and M. Shikh-Bahaei. Cooperative rate-splitting for secrecy sum-rate enhancement in multi-antenna broadcast channels. In *2020 IEEE 31st Annual International Symposium* *on Personal, Indoor and Mobile Radio Communications*, pages 1–6, 2020. doi: 10.1109/PIMRC48278.2020.9217123.

9 Y. Mao, O. Dizdar, B. Clerckx, R. Schober, P. Popovski, and H. V. Poor. Rate-splitting multiple access: Fundamentals, survey, and future research trends. *IEEE Communications Surveys & Tutorials*, 24(4):2073–2126, 2022. doi: 10.1109/COMST.2022.3191937.

10 A. Papazafeiropoulos and T. Ratnarajah. Rate-splitting robustness in multi-pair massive MIMO relay systems. *IEEE Transactions on Wireless Communications*, 17(8):5623–5636, Aug 2018. ISSN 1536-1276. doi: 10.1109/TWC.2018.2847668.

11 M. Sadek, A. Tarighat, and A. H. Sayed. A leakage-based precoding scheme for downlink multi-user MIMO channels. *IEEE Transactions on Wireless Communications*, 6(5):1711–1721, May 2007. ISSN 1558-2248. doi: 10.1109/TWC.2007.360373.

12 A. Salem, C. Masouros, and B. Clerckx. Secure rate splitting multiple access: How much of the split signal to reveal? *IEEE Transactions on Wireless Communications*, 22(6):4173–4187, 2022. doi: 10.1109/TWC.2022.3223961.

13 T. X. Tran and K. C. Teh. Spectral and energy efficiency analysis for SLNR precoding in massive MIMO systems with imperfect CSI. *IEEE Transactions on Wireless Communications*, 17(6):4017–4027, June 2018. ISSN 1558-2248. doi: 10.1109/TWC.2018.2819184.

14 H. Xia, Y. Mao, B. Clerckx, X. Zhou, S. Han, and C. Li. Weighted sum-rate maximization for rate-splitting multiple access based secure communication. In *2022 IEEE Wireless Communications and Networking Conference (WCNC)*, pages 19–24, 2022. doi: 10.1109/WCNC51071.2022.9771854.

8

End-to-End Autoencoder Communications with Optimized Interference Suppression*

Kemal Davaslioglu, Tugba Erpek, and Yalin Sagduyu

Nexcepta, Gaithersburg, MD, USA

8.1 Introduction

In conventional communication systems, the design of transmitters and receivers is usually divided into separate communication blocks, which are either based on analytical models or derived from expert knowledge. Sophisticated physical-layer (PHY) techniques are required for 6G to meet the stringent demands in terms of data rate, throughput, and interoperability [10, 26]. With the recent advancements in algorithmic and computational abilities, deep learning learns from and adapts to complex representations of spectrum data and offers a new approach to wireless communications, including physical-layer waveform design [14, 56] for 6G communications.

Autoencoder (AE) comprises two deep neural networks (DNNs): one acts as the encoder, tasked with learning a compressed representation of a dataset, primarily for dimension reduction purposes, while the other functions as the decoder, aiming to recover the original data samples. An AE communications (AECs) system can be built with an encoder hosted at the transmitter and a decoder hosted at the receiver. This encoder–decoder pair is jointly trained, while adapting to the channel conditions, to minimize the reconstruction loss, which directly impacts the bit error rate (BER). The BER performance of AEC was shown in [23] to either match or surpass that of conventional communication systems. The feasibility of AEC implementation was demonstrated with software-defined radios communicating

* Distribution Statement A (Approved for Public Release, Distribution Unlimited). This research was supported under funding from Defense Advanced Research Projects Agency (DARPA). The views, opinions, and/or findings expressed are those of the authors and should not be interpreted as representing the official views or policies of the Department of Defense or the US Government.

Physical-Layer Security for 6G, First Edition. Edited by Parthajit Mohapatra, Nikolaos Pappas, Arsenia Chorti, and Stefano Tomasin.

over wireless channels [11]. Beyond using a single carrier, AEC was expanded in [2, 16] to incorporate orthogonal frequency division multiplexing (OFDM).

In this chapter, we study OFDM-based AEC, where the encoder performs the coding and modulation functions of the transmitter and the decoder performs the demodulation and decoding functions of the receiver, as shown in Figure 8.1. Unlike previous works that have primarily focused on adapting AEC to channel effects, we formulate AEC to suit a more practical scenario, wherein we account for the mitigation of external interference like jamming effects, alongside channel effects. Our design also accounts for the limitations of the training data and the impact of embedded hardware implementation. The system performance is assessed using the BER and error vector magnitude (EVM) metrics.

We demonstrate that the performance of AEC heavily hinges upon the availability of training samples. When there is a shortage of training data, either because of a restricted sensing duration or a reduced sampling rate, we explore the possibility of utilizing a generative adversarial network (GAN) to produce synthetic data samples, thereby augmenting the training dataset for AEC. As the duration of data collection increases, there is a corresponding decrease in time available for data communications. We explore methods for identifying how many synthetic data samples to use optimally in order to improve the BER performance. Only a few synthetic samples do not effectively enhance the BER performance with the GAN assistance. On the other hand, too many synthetic data samples can lead to confusion during training, as synthetic artifacts introduced during the data generation process may interfere.

Subsequently, we take into account the *limitations imposed by hardware* on AEC. The deployment of these systems on edge devices, such as radio transceivers, would necessitate an *embedded implementation* on platforms such as

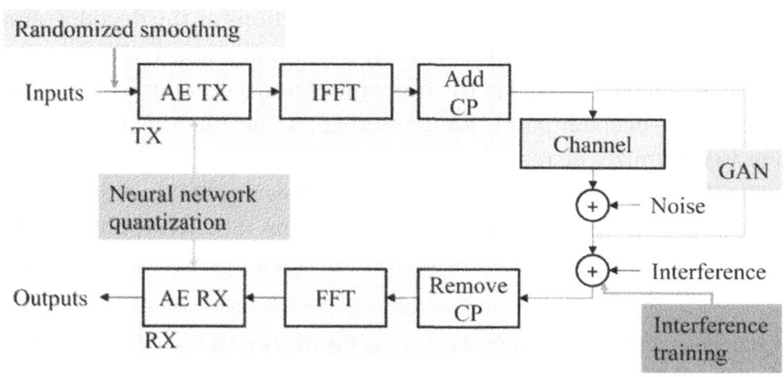

Figure 8.1 Interfaces of different system model components for the proposed AEC approach.

field-programmable gate arrays (FPGAs) or embedded graphics processing units (GPUs), where computations are performed using limited precision. To fulfill this requirement, we perform quantization on DNN models and illustrate that 8-bit quantized model (8QM) has minimal impact on BER performance compared to a floating-point model (FPM) implementation. On the other hand, quantization results in a considerable reduction in the memory demand for the model's size.

Interference management is a critical capability needed in wireless communication systems to sustain performance in case of in-network and out-of-network interference as well as jamming effects. 6G envisions the support of massive number of devices connected in heterogeneous network settings with dynamic spectrum-sharing capabilities, where interference to 6G needs to be carefully mitigated [3, 22, 28, 55]. In case of jamming attacks, novel techniques are needed to protect 6G communications by suppressing interference due to jamming. We demonstrate how potential performance degradation due to *unforeseen and dynamically shifting interference (jamming) effects*, alongside channel effects, can be effectively mitigated by a novel training approach referred to as *interference training* (IT) that helps mitigate the impact of interfering signals. Additionally, we utilize the *randomized smoothing* (RS) technique [7] to convert a classifier that functions effectively in the presence of Gaussian noise into a new classifier to make it certifiably robust with respect to adversarial disturbances. We initially showcase that AEC outperforms conventional communications when we consider only noise without external interference. Subsequently, we demonstrate that IT along with RS can improve the BER and EVM performance. We consider scenarios in which redundancy is introduced into transmitted signal by employing the channel multiple times for transmission. Incorporating external interference in the form of a jamming signal, we assess the maximum jammer-to-signal ratio (JSR) that can be accommodated when maintaining a target BER at a specified signal-to-noise ratio (SNR). The proposed approach *suppresses the interference up to* 36 *dB*.

Additionally, we expand the scope of AEC from the single input single output (SISO) to the multiple-input multiple-output (MIMO) configuration. By taking channel effects into account across both the transmit and receive antennas, we demonstrate that MIMO communications with AE notably enhances the BER relative to conventional communications for MIMO with spatial multiplexing. With the use of IT and RS, we also show that AEC can effectively suppress the interference and reduce the BER for the MIMO case.

The structure of the remaining section is outlined in the following. Section 8.2 overviews the prior work. Section 8.3 describes the system model for AEC. Section 8.4 assesses the performance under different effects of channel, quantization, and implementation on embedded platforms. Section 8.5 discusses the application of a GAN for augmenting the AE training data. Section 8.6 studies interference suppression. Section 8.7 extends the capabilities to encompass the MIMO scenario. Finally, Section 8.8 concludes this chapter.

8.2 Related Work

AEC was first explored in [23] and it was demonstrated that learning the communication models for both the transmitter and receiver is feasible using the provided channel model. This concept was subsequently extended to encompass MIMO and massive MIMO capabilities in [20, 24, 49]. In [54], a conditional GAN was used to learn channel distributions for AEC. Our study distinguishes itself from prior research by considering interference suppression along with channel adaptation, quantization, and embedded implementation aspects.

The use of DNN-based AEs was considered in [16] for OFDM systems. The effectiveness of AEC was assessed in comparison to an OFDM system in [2] by taking into account multiple channel uses for transmitting a single message as a form of redundancy similar to channel coding. A DNN was applied in [17] for high-dimensional modulation to mitigate channel effects rather than provide interference resistance. Additionally, Erpek et al. [15] considered how to design AEC with beamforming and reconfigurable intelligent surface (RIS), taking into account channel effects only. AEC was extended to task-oriented communications [35–38] by optimizing the communication pipeline with respect to the task to be completed at the receiver. In addition to reconstruction of messages, the AE can be also optimized to sustain the meaning in communications such as studied in semantic communications [32–34, 39].

In [12, 51], interference channel was studied for single carrier communications. The two pairs of transmitters and receivers were optimized jointly to reduce the symbol error rate (SER) subject to the self-generated interference. The receiver-side interference management was also addressed in [50, 52] through the training of DNNs. This chapter aims to tackle the problem of suppressing unknown and dynamic external interference such as jamming for AEC.

Due to the shared medium, wireless communications is susceptible to jamming attacks. Hence, it is crucial to investigate the viability of jamming attacks and design corresponding countermeasures [30, 53]. To that end, deep learning has been effectively used to launch jamming attacks [13, 43] as well as to detect and mitigate jamming effects [1, 9]. On the other hand, adversaries can also attempt to jam the deep learning engine as studied for AEC [29], for reinforcement learning [45, 47] and for federated learning [41, 46] in wireless systems.

Prior work on AEC has not taken into account hardware limitations, such as computing limitations at edge devices, during model training. Specifically, prior work has not addressed the ramifications of limited training data and quantization effects for embedded implementation (e.g., on FPGA and embedded GPU). Supervised machine-learning techniques require a sufficient amount of training data, but it is hard to collect the required training since sensing time and sampling rate can be low in practical cases. To overcome this challenge, a GAN can produce synthetic data samples and supplement the training process, as studied for spectrum sensing [8, 42], jamming [13, 44], modulation classification [27], infiltration of signal authentication [31], signal camouflage [19], and wireless anomaly detection [4]. The GAN was also utilized in channel modeling and estimation [5, 6, 25, 54]. In this chapter, we strengthen the training process of AEC with the GAN,

8.3 System Model

Using two different DNNs, the transmitter implements the encoding and the receiver decodes the data, together forming the AE system that is illustrated in Figure 8.2.

The transmitter encodes $n_{\text{FFT}} \times k$ bits (represented as $\mathbf{b}$) into $\mathbf{X}$-modulated symbols, where the number of subcarriers and bits per symbol are denoted by n_{FFT} and k, respectively, as illustrated in Figure 8.2. The concept of *number of channel uses* (as in [16]) is utilized by adding redundant symbols in the transmitter to increase the error correction performance through joint modulation and coding. This system is tailored for OFDM systems that divide the available frequency spectrum into multiple narrowband subcarriers that are orthogonal to each other.

For AEC, it is possible to transmit different symbols in successive channel utilizations. We capture this effect by introducing a parameter n_{ch} that signifies the frequency of symbol transmissions across the channel. To ensure orthogonality, we take the inverse fast Fourier transform (IFFT) of the encoded symbols that have the dimensions of $n_{\text{FFT}} \times k \times n_{\text{ch}}$. The IFFT outputs are represented as $\mathbf{X}_{\text{OFDM}}$.

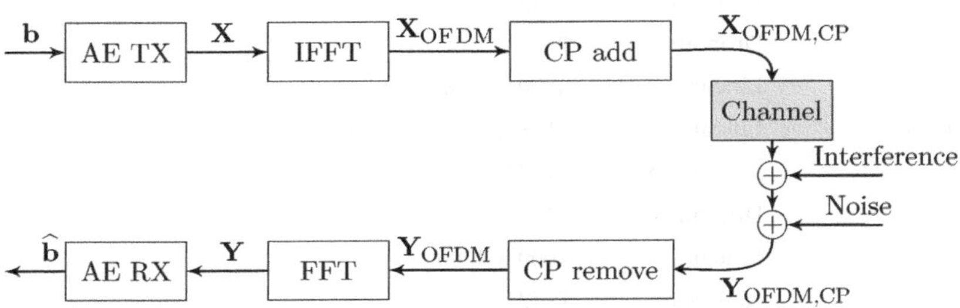

Figure 8.2 System model.

We add cyclic prefix (CP) for the reduction of inter-symbol interference (ISI) which is denoted by $\mathbf{X}_{\mathrm{OFDM,CP}}$.

Wireless channel varies the phase and frequency of the output symbols. Moreover, we account for the additive white Gaussian noise (AWGN), along with an interference signal at a specified JSR. We represent received symbols with $\mathbf{Y}_{\mathrm{OFDM,CP}}$. The CP is removed from the received symbols and fast Fourier transform (FFT) is taken to obtain $\mathbf{Y}$. This data stream is input to the AE's decoder which estimates $\hat{\mathbf{b}}$ bits. The encoder–decoder pair undergoes joint training to minimize the BER, taking channel and interference effects into account. We train the AE to minimize the dissimilarity between $\mathbf{b}$ and $\hat{\mathbf{b}}$. This can be formulated as a multi-label classification for which the loss function is

$$\begin{aligned}\ell(\hat{\mathbf{b}}, \mathbf{b}) &= \frac{1}{N}\sum_{n=0}^{N} \ell(\hat{b}_n, b_n) = \frac{1}{N}\sum_{n=0}^{N} \ell\left(h(b_n), b_n\right) \\ &= -\frac{1}{N}\sum_{n=0}^{N}\left(b_n \log(h(b_n)) + (1-b_n)\log(1-h(b_n))\right),\end{aligned} \tag{8.1}$$

where b_n stands for the nth input bit of the encoder and $\hat{b}_n = h(b_n)$ denotes the nth output bits of the decoder. The function $h(\mathbf{b})$ encapsulates the transmitter–receiver pair operations, encompassing the effects of the noise and channel. It follows that $h(\mathbf{b}) = \hat{\mathbf{b}}$. In this formulation, a total of $N = (n_{\mathrm{FFT}} + n_{\mathrm{cp}}) \times n_{\mathrm{ch}}$ symbols are transmitted in n_{ch} channel uses, where n_{cp} stands for the number of subcarriers allocated for the CP. The DNN architectures are shown in Table 8.1. For numerical results, we use 12 bits for FFT and 3 bits for CP. The number of channel uses is adjusted while maintaining 2 bits per symbol. The signal constellation does not

Table 8.1 The architecture of the AE.

Network name	Layer	Properties
Encoder	Linear	$(k \cdot n_{\mathrm{FFT}}) \times 4000$
	Activation	ReLU
	Dropout	25%
	Linear	$4000 \times 2 \cdot n_{\mathrm{FFT}} \cdot n_{\mathrm{ch}}$
Decoder	Linear	$(2 \cdot n_{\mathrm{FFT}} \cdot n_{\mathrm{ch}}) \times 4000$
	Activation	ReLU
	Dropout	25%
	Linear	4000×4000
	Activation	ReLU
	Dropout	25%
	Linear	$4000 \times (k \cdot n_{\mathrm{FFT}})$

have fixed points, unlike in the case of quadrature phase shift keying (QPSK), but they are instead determined by AE to consider the effects of channel and any possible interference.

8.4 Performance Evaluation of AEC Considering the Effects of Channel, Quantization, and Embedded Implementation

Figure 8.3 illustrates the BER performance under AWGN channel. "AE TX" from Figure 8.2 is substituted in conventional communications with traditional modulation which typically depends on n_b and n_{ch}. Similarly, instead of "AE RX," respective demodulation and decoding operations would be used. At a 10^{-3} BER, AEC demonstrates an improvement of up to 3 dB with increasing n_{ch}.

Figure 8.4 illustrates the BER versus SNR in the presence of channel impairments, such as 10° phase offset and 30 Hz frequency offset. The advantage of AEC becomes more pronounced when these impairments are present. This is because while conventional communications suffer from degradation due to channel impairments distorting the waveform, AEC is trained to recognize and counteract these effects. Furthermore, the error improvement that can be achieved by AEC becomes more significant when the channel use is increased

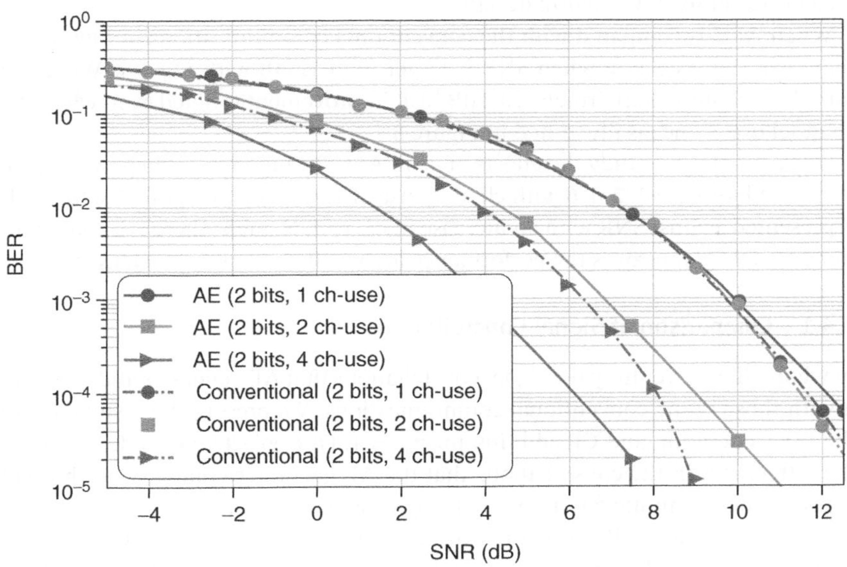

Figure 8.3 BER in an AWGN channel.

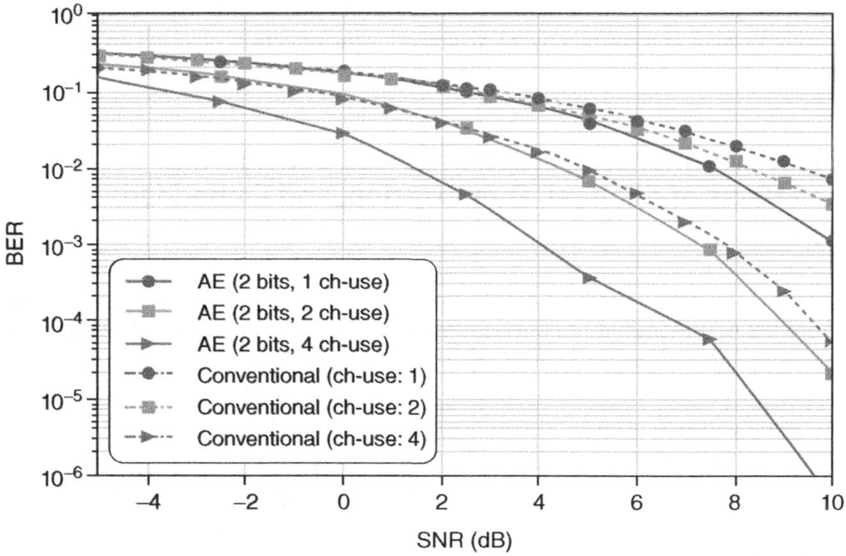

Figure 8.4 BER in an AWGN channel when considering channel impairments where the phase and frequency offsets are taken as 10° and 30 Hz, respectively.

as AE simultaneously optimizes channel utilization and modulation, providing greater flexibility in waveform design.

When AEC is compared with the conventional communications in Figure 8.3, there is a slight decline when SNR is larger than 12 dB with $n_{\text{ch}} = 1$. When we introduce channel impairments, a 2 dB BER improvement is noted even when the channel use is one, as illustrated in Figure 8.4.

Subsequently, we study the effects of n_{b} in AEC. In Figure 8.5, the performance is assessed for $n_{\text{b}} = 1, 2, 4, 6$ with channel use of 1. Moving to Figure 8.6, the BER is presented against SNR with an increase in n_{ch} to 4. Notably, there is a consistent trend that AEC improves the performance across all cases as n_{ch} increases.

8.4.1 Comparison of Signal Constellations

In what follows, we study the signal constellations that AE generates and compare it with those of the conventional communications. We present the results before the IFFT operation and CP addition for $n_{\text{ch}} = 1$. Our objective is to demonstrate the differences in the constellations that the AE generates compared to those of conventional communications, specifically when QPSK is used.

Figures 8.7 and 8.8 illustrate the constellations when $SNR = -5$ dB and $SNR = 30$ dB, respectively. When we compare the two constellations, we observe that the constellation generated by AE is designed to minimize the BER with taking the channel effects into account and it is different compared to the fixed signal constellation points that the conventional communication system uses.

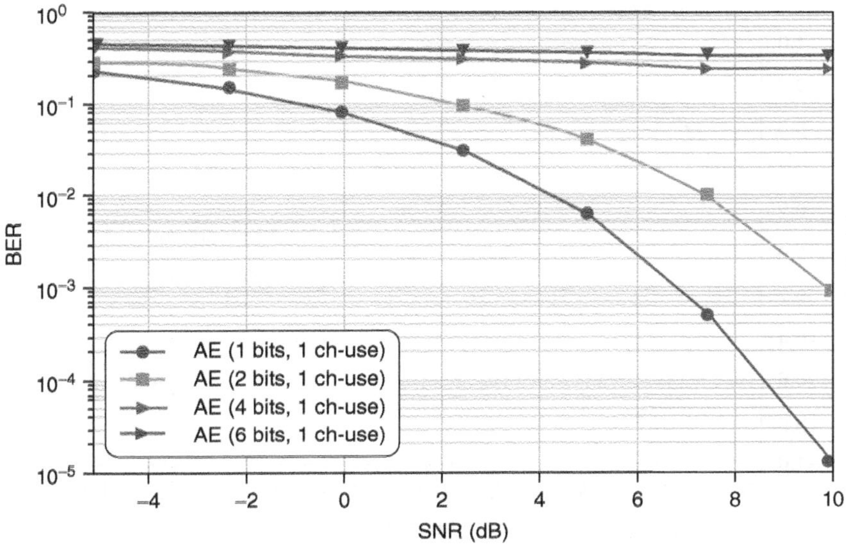

Figure 8.5 BER of different number of bits for channel use of 1.

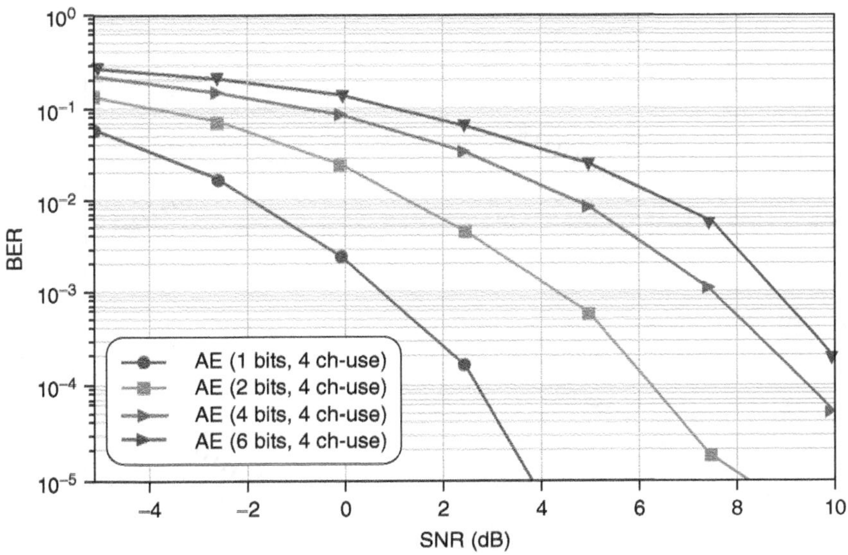

Figure 8.6 BER of different number of bits for channel use of 4.

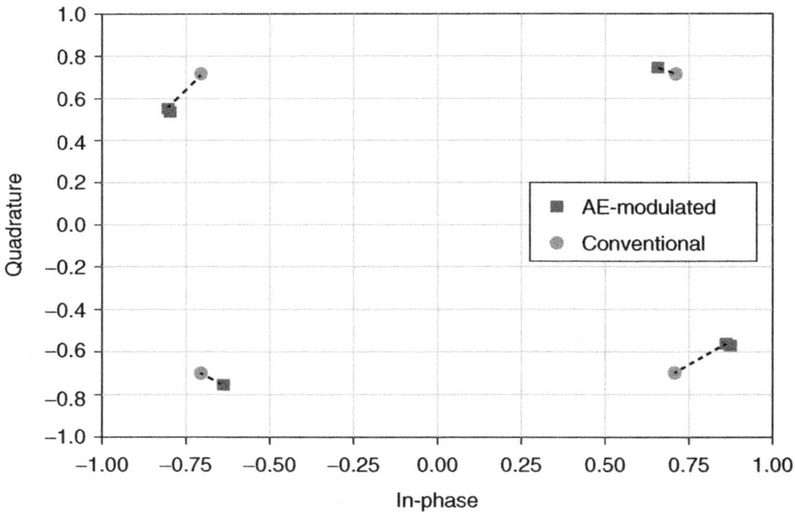

Figure 8.7 Signal constellation comparison at −5 dB SNR.

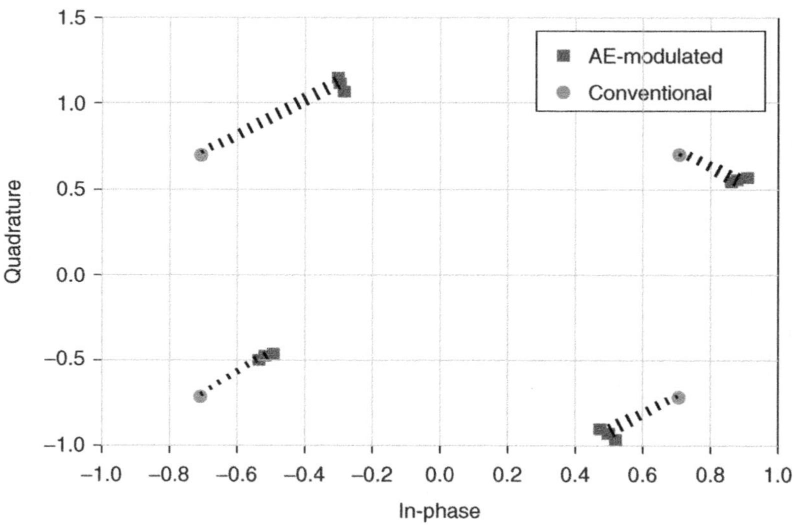

Figure 8.8 Signal constellation comparison at 30 dB SNR.

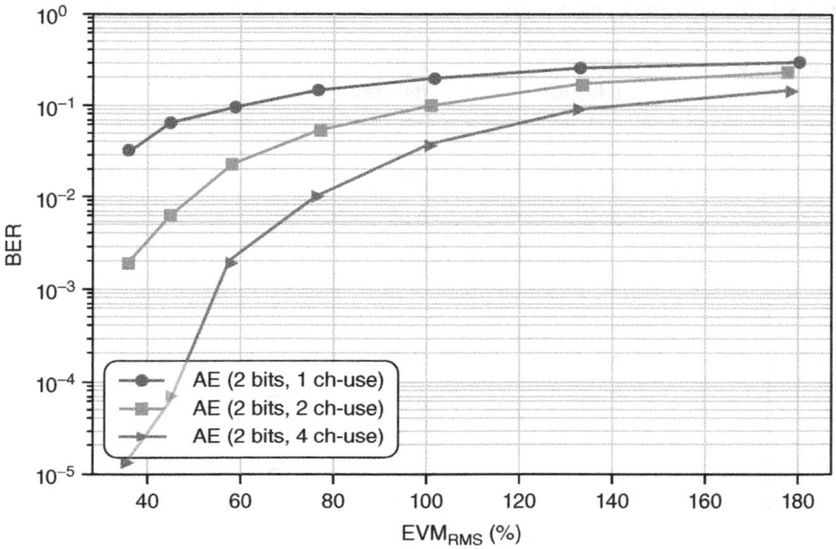

Figure 8.9 EVM results for different channel uses keeping the bits per symbol the same at $n_b = 2$.

8.4.2 Effects of EVM

EVM is typically used to measure the modulation quality in wireless channels [40]. We will use the root mean square (RMS) of the error vector normalized with respect to the ideal signal. We can extend the error definition to AEC as the error vector between the output of the AE encoder and the input of AE decoder. Figure 8.9 presents the BER versus EMV for different channel uses keeping the bits per symbol as $n_b = 2$. Note that a large EVM value shows a larger difference between the ideal transmitter signal and received signal and often yields to an increase in BER which we also observe in Figure 8.9. However, when the channel use is increased, the coding gains overtake the effects of EVM and the BER decreases.

8.4.3 Effects of Quantization

Quantization is used to reduce the bit resolution for embedded devices. However, this operation is prone to rounding errors which yield to error propagation. To study quantization effects on the BER, we apply dynamic quantization which uses different bit resolution in each layer. We have conducted our experiments on an NVIDIA GPU (GeForce RTX 2080 Ti).

To study the effects of quantization, we measured the error performance, model size, and inference time as key metrics. The DNN model sizes for different channel uses are shown in Table 8.2, which shows the size of the DNN model is reduced by a factor of 4 with quantization. The inference time of models for different channel

Table 8.2 Comparison of model size.

Channel use n_{ch}	**FPM** (Mbytes)	**8QM** (Mbytes)
1	2.88	0.73
2	3.03	0.77
4	3.34	0.85

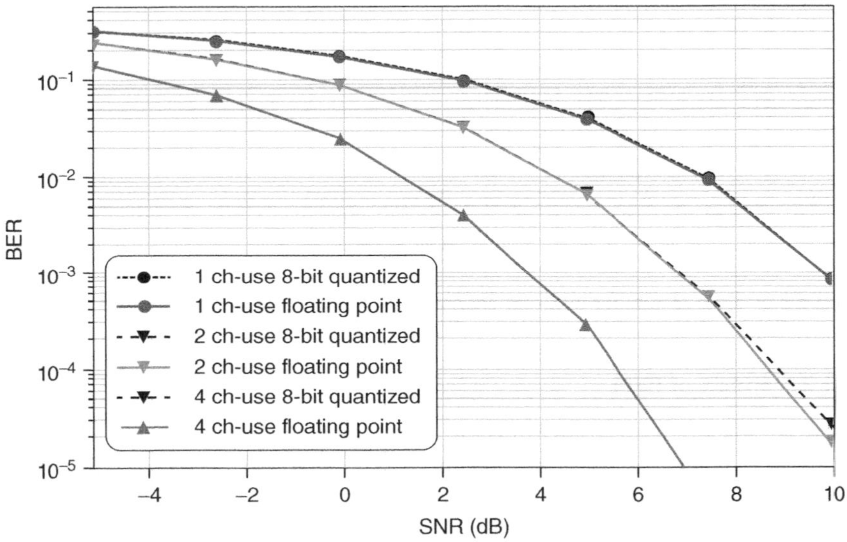

Figure 8.10 Comparison of the BER performance for $n_{ch} = 1, 2, 4$ channels, between FPM and 8QM.

uses with and without quantization are compared in Table 8.3 where we normalize the results for each sample. In our evaluations, we repeat the experiments 10,000 times with a batch size of 50. Our results show that the models with and without quantization have very similar inference time. Figure 8.10 presents the BER of the models with and without quantization for different channel uses and the error performance at different SNR levels is presented in Tables 8.4–8.6. These results show that the effects of quantization error gradually diminish as the SNR increases.

8.4.4 Practical Considerations for Embedded Devices

When we consider practical consideration for embedded devices, the size of the model and its latency on the device need to be studied. For this purpose, we

Table 8.3 Comparison of inference time.

Channel use n_{ch}	FPM (s)	8QM (s)
1	0.000109	0.000114
2	0.000333	0.000335
4	0.000354	0.000358

Table 8.4 Comparison of BER for $n_{ch} = 1$.

SNR (dB)	FPM	8QM
−5.0	0.3033	0.3037
−2.5	0.2393	0.2394
0.0	0.1664	0.1665
2.5	0.0951	0.0952
5.0	0.0390	0.0392
7.5	0.0091	0.0093
10.0	0.0009	0.0009

Table 8.5 Comparison of BER for $n_{ch} = 2$.

SNR (dB)	FPM	8QPM
−5.0	0.2238	0.2242
−2.5	0.1527	0.1529
0.0	0.0831	0.0833
2.5	0.0308	0.0309
5.0	0.0064	0.0065
7.5	0.0005	0.0005
10.0	0.0000	0.0000

Table 8.6 Comparison of BER for $n_{ch} = 4$.

SNR (dB)	FPM model	8QM model
−5.0	0.1376	0.1377
−2.5	0.0696	0.0697
0.0	0.0239	0.0240
2.5	0.0040	0.0041
5.0	0.0003	0.0003
7.5	0.0000	0.0000
10.0	0.0000	0.0000

Table 8.7 The network sizes and the total number of parameters for the AEC's FPGA implementation.

Model	Size of the model	Total number of parameters
AEC transmitter	0.16 M	39.2K
AEC receiver	2.72 M	680.0K

measure the latency of the DNN model on an FPGA embedded implementation where we use Xilinx Zynq Ultrascale MPSoC FPGA (XCZU7EV) model. We have converted the encoder network and the decoder network to FPGA binaries and ported these files to the FPGA. We have leveraged our prior work in [48] in this step. We present the size of the model for the AE-TX and AE-RX in Table 8.7 along with the number of parameters in each model. We observe that it takes 8.69 μs for the transmitter to encode the bits to symbols and 16.74 μs for the decoder to decode the symbols back to bits, which demonstrate the low-latency properties of the proposed method.

8.5 Data Augmentation to Train the AE Model Using GANs

Deep-learning models typically suffer when there is limited training data. The same can be expected for the AE training. In such cases, the training data size can be increased by training a conditional GAN that learns the data distribution

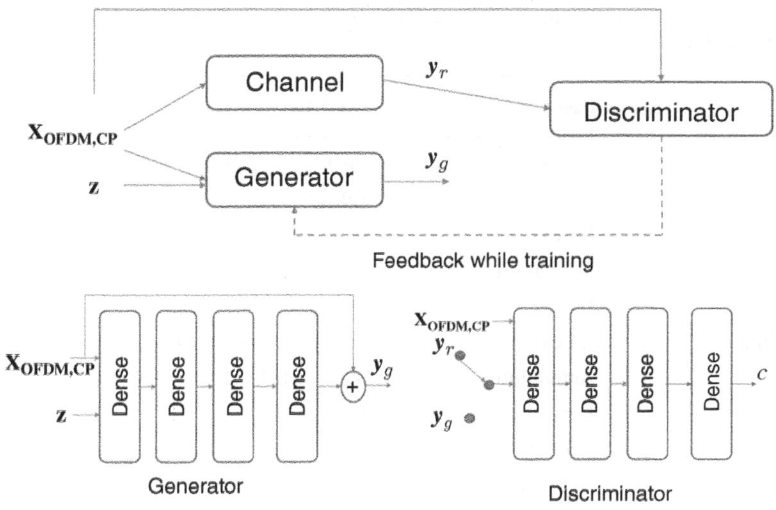

Figure 8.11 System model of the conditional WGAN-GP.

of the data, namely, received signals in our case, and effectively generates training data samples that incorporate channel and other impairments for AEC.

The system model to learn the received signal distributions with the GAN is shown in Figure 8.11. In this model, the generator network and discriminator network are trained jointly. We use $\mathbf{y}_r$ to represent the real data which takes into account the impairments due to channel such as phase and frequency shifts. Similarly, $\mathbf{y}_g$ represents the data synthesized by the generator model. In general, GAN models are trained with the minimax objective which is given by

$$\min_{G}\max_{D} \quad \mathbb{E}_{\mathbf{y}_r\sim\mathbb{P}_{\mathbf{y}_r}}[\log(D(\mathbf{y}_r))] + \mathbb{E}_{\mathbf{y}_g\sim\mathbb{P}_{\mathbf{y}_g}}[\log(1 - D(\mathbf{y}_g))], \tag{8.2}$$

where the distributions are denoted by $\mathbb{P}_{\mathbf{y}_r}$ and $\mathbb{P}_{\mathbf{y}_r}$, respectively, for the real data and synthetic data. The generator uses the noise vector $\mathbf{z}$ which follows $\mathbf{z} \sim p(\mathbf{z})$ distribution. The discriminator output is shown by $D(\cdot)$ and the expectation function over $\mathbf{y}$ is represented by $\mathbb{E}_{\mathbf{y}}[\cdot]$. Equation (8.2) attains the objective to minimize the Jensen–Shannon divergence between real and synthetic data distributions if the discriminator reaches optimality before generator updates. It is shown that the objective in (8.2) can lead to the vanishing gradient problem if the discriminator saturates [18] and the conditional Wasserstein GAN with gradient penalty (WGAN-GP) loss can be used to mitigate this problem. We can formulate the loss of the conditional WGAN-GP as

$$\begin{aligned} L &= \mathbb{E}_{\mathbf{y}_g\sim\mathbb{P}_{\mathbf{y}_g}}[D(\mathbf{y}_g|\mathbf{x})] - \mathbb{E}_{\mathbf{y}_r\sim\mathbb{P}_{\mathbf{y}_r}}[D(\mathbf{y}_r|\mathbf{x})] \\ &\quad + \lambda \cdot \mathbb{E}_{\hat{\mathbf{y}}\sim\mathbb{P}_{\hat{\mathbf{y}}}}\left[\left(||\nabla_{\hat{\mathbf{y}}}D(\hat{\mathbf{y}}|\mathbf{x})||_2 - 1\right)^2\right], \end{aligned} \tag{8.3}$$

Table 8.8 Architecture of the generator and discriminator networks of the conditional WGAN-GP.

	Layer	Properties
Generator	Linear	$(20 + k \cdot n_{\text{FFT}}) \times 800$
	Activation	ReLU
	Linear	800×800
	Activation	ReLU
	Linear	$800 \times 2(n_{\text{FFT}} + n_{\text{cp}})n_{\text{ch}}$
Discriminator	Linear	$4(n_{\text{FFT}} + n_{\text{cp}})n_{\text{ch}} \times 64$
	Activation	LeakyReLU, 0.2
	Linear	64×64
	Dropout	40%
	Activation	LeakyReLU, 0.2
	Linear	64×64
	Dropout	40%
	Activation	LeakyReLU, 0.2
	Linear	64×1

where $\hat{\mathbf{y}}$ is drawn from $\mathbb{P}_{\hat{\mathbf{y}}}$, indicating evenly spaced points along the line connecting pairs from the generator and data distributions, and λ stands for gradient penalty coefficient. Table 8.8 presents the generator and discriminator network architectures.

8.5.1 BER Performance with GAN-Based Data Augmentation

To study the effects of data augmentation on the BER, we consider two cases:

1. *AEC in few-shot settings*: This case considers limited training data condition. In our analysis, we use 100 real samples.
2. *AEC in many-shot setting*: In this case, we assume that there are enough data samples that can be used to train the GAN. In our analysis, we use 1000 real samples.

In our approach, we first train the GAN to learn the data distribution, followed by the AE. During the AE training, we use the real received signals as well as the samples synthesized by the generator. During this process, determining the appropriate number of synthetic samples is very crucial; an excess number of samples can lead to overfitting, while few samples may not enhance performance. Therefore, we study the error performance with varying numbers of synthetic samples. A noise vector $\mathbf{z}$ serves as input to the GAN generator during training,

Table 8.9 Evaluation of BER as a function of $N_{synthetic}$.

$N_{synthetic}$	BER
No augmentation	0.025
2×	0.015
4×	0.006
6×	0.008
8×	0.012

with a dimension of 50, which follows our prior work in [8] for spectrum sensing augmentation. Table 8.9 depicts the BER for $n_{ch} = 1$ at 10 dB SNR, employing 100 real samples (N_{real}) and adjusting the number of synthetic samples ($N_{synthetic}$) in increments of N_{real}. Increasing synthetic samples generally decreases the BER. For example, the BER drops from 0.025 to 0.006 when 100 real and 400 synthetic samples are used for training. Excessive use of synthetic samples can cause the BER to rise, as there is the potential overfit to GAN artifacts, diverging from the true distribution.

To establish a systematic approach for assessing how many synthetic training samples are optimally used, the AE undergoes training with increments of N_{real} until $10 \times N_{real}$. We select the configuration that demonstrates the best performance. The BER in the low-shot learning scenario is illustrated in Figure 8.12 for different n_{ch}. Notably, when we use $n_{ch} = 4$, data augmentation increases the BER performance by 4 dB at 10^{-4}. Additionally, Figure 8.13 presents the effects of data augmentation in the many-shot learning scenario for various n_{ch} values. In this case, the error performance improvements are more moderate as the training sample size increases. Specifically, we observe marginal improvements at low SNRs, while slightly greater gains are observed at high SNRs, where the BER improves by 2 dB when we consider a channel use of 4.

8.6 Methods to Suppress the Effects of Interference

The subsequent phase of our procedure involves delineating the training protocol for interference suppression. The IT strategy is devised to accommodate interference to a certain extent during the training phase. This is achieved by training the AE using samples that incorporate both instances with and without interference, which aims to minimize the dissimilarity between the distributions of training and testing data in the presence of interference. We note that during the testing phase, the AE lacks prior knowledge regarding the level or timing

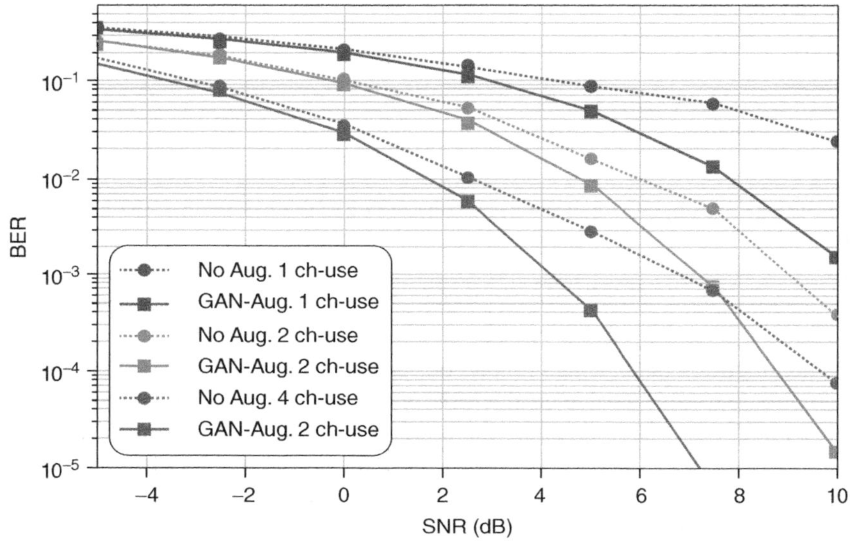

Figure 8.12 BER with 100 real samples.

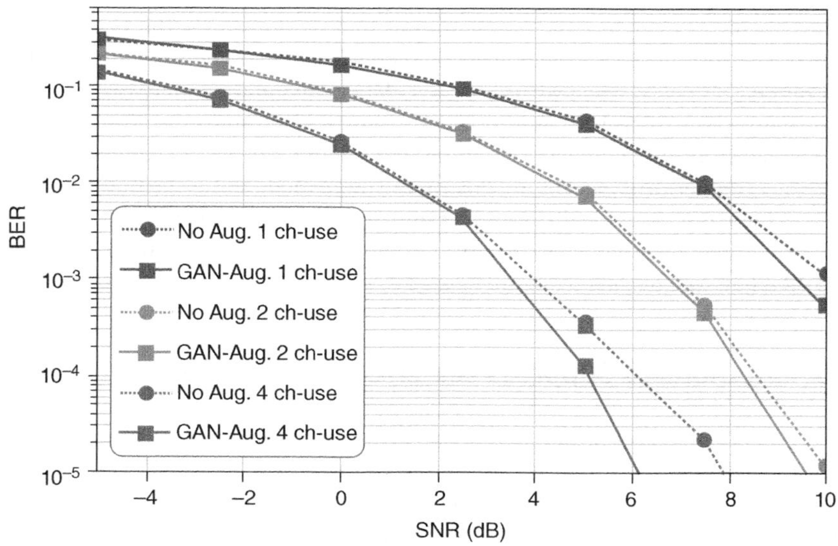

Figure 8.13 BER with 1000 real samples.

of interference. Initially, we conduct a conventional training step, involving a forward pass and backpropagation, using OFDM symbols and not considering the effects of interference. We can represent this loss function as

$$\ell_{\text{AE}} = \frac{1}{N}\sum_{n=0}^{N}\ell\left(h(b_n), b_n\right). \tag{8.4}$$

We train AEC taking the channel and noise effects into account in this step, but the effects of interference are not included.

In the next step, we take the effects of interference into account and conduct multiple iterations of forward pass and backpropagation. To express the contributions of IT in the loss function, we derive the following loss expression:

$$\ell_{\text{IT}} = \frac{1}{N_{\text{it}}N}\sum_{k=0}^{N_{\text{it}}}\sum_{n=0}^{N}\ell\left(g(b_n), b_n\right), \tag{8.5}$$

where the function $g(\mathbf{b}) = [g(b_1), \ldots, g(b_N)]$ includes channel impairments. Since we do not have any prior knowledge on the degree and type of interfering waveform, we consider that the OFDM symbols may encounter random interference at any point during the reception. To account for such effects, we examine various scenarios where the AE system is trained considering an interference waveform that effects $N_{\text{tr-jam}}$ symbols but we consider $N_{\text{te-jam}}$ symbol interference during testing, where we make no assumptions about $N_{\text{tr-jam}}$ and $N_{\text{te-jam}}$.

In the third step, we apply the RS method (or Gaussian smoothing) which was first introduced as a countermeasure for adversarial machine learning attacks in computer vision. This technique enhances the training dataset by introducing Gaussian or Laplacian noise, thereby increasing the model resilience against ℓ_2 or ℓ_1 attacks [7]. In our analysis, the noise has Gaussian distribution, that is, $\mathbf{n} \sim \mathcal{N}(\mathbf{0}, \sigma^2\mathbf{I})$, where σ^2 denotes the noise variance. We use RS to increase the classifier's robustness to suppress interference that can be transmitted from different emitters. We formulate the loss function that takes the effects of RS into account as

$$\ell_{\text{RS}} = \frac{1}{N_{\text{rs}}N}\sum_{k=0}^{N_{\text{rs}}}\sum_{n=0}^{N}\ell\left(h(b_n + n_{k,n}), b_n\right), \tag{8.6}$$

where we use $n_{k,n}$ to represent the noise that is introduced to the nth bit for kth realization. The noise vector $\mathbf{n}_k = [n_{k,1}, \ldots, n_{k,N}]$ represents AWGN noise with zero mean and variance σ^2. In this setup, it needs to be noted that we need to adjust N_{rs} and σ^2 to find a suitable trade-off between accuracy and robustness. Combining expressions in (8.4)–(8.6), we can express the final loss function as

$$\ell_{\text{Total}} = \ell_{\text{AE}} + \ell_{\text{IT}} + \ell_{\text{RS}}. \tag{8.7}$$

Algorithm 8.1 presents the overall process to train the AEC system to suppress interference. We use N_{ep} training cycles, each comprising N_{batches} batch iterations. N random bits are input to the encoder neural network where they are encoded to $\mathbf{b}$ for each batch. We incorporate the channel effects, noise, and hardware impairments, before being decoded by the decoder network, which is represented by $h(\mathbf{b})$. During the RS step, we add AWGN noise $\mathbf{n}$ to $\mathbf{b}$, and compute the loss in (8.6). Subsequently, we compute the function $g(\mathbf{b})$ that considers the introduced interference effects for IT. Finally, we compute the loss function in (8.5). The model is updated with Adam optimizer [21], where α is the learning rate and β_1 and β_2 are the hyperparameters governing the exponential decay of the first and second moments.

Algorithm 8.1 Algorithm to train the AE system to suppress interference

---

Input: Parameters for training without interference: N and $h(b_n)$; RS: N_{it} and $g(b_n)$; AE optimization: N_{rs} and σ^2; AE model parameters: $\boldsymbol{\theta}_0$, and Hyperparameters of Adam optimizer: $(\alpha, \beta_1, \beta_2)$.
Initialize: First and second moment vectors $\mathbf{m}_0$ and $\mathbf{v}_0$ as $\mathbf{0}$, respectively.
for epoch $t = 1, 2, \dots, N_{\text{ep}}$ **do**
  for $j = 1, 2, \dots, N_{\text{batches}}$ **do**
    Generate input bits $\mathbf{b}$ and calculate $h(\mathbf{b})$.
    Use (8.4) to compute ℓ_{AE}.
    for $k = 1, \dots, N_{\text{it}}$ **do**
      Generate interference.
      Compute $g(\mathbf{b})$.
      Use (8.5) to compute ℓ_{IT}.
    end for
    for $k = 1, \dots, N_{\text{rs}}$ **do**
      Generate AWGN noise following $\mathbf{n}_k \sim \mathcal{N}(0, \sigma^2\mathbf{I})$ distribution.
      Compute $h(\mathbf{b} + \mathbf{n}_k)$.
      Use (8.6) to compute ℓ_{RS}.
    end for
  end for
  Calculate the loss function gradient $\nabla\ell_{\text{Total}}(\boldsymbol{\theta}_{t-1})$.
  Calculate the first and second moment estimates to calculate the updates for the Adam optimizer, $\widehat{\mathbf{m}}_t, \widehat{\mathbf{v}}_t$
  Update the neural network parameters, $\boldsymbol{\theta}_t$.
end for
Output: Trained AE network parameters $\boldsymbol{\theta}_t$.

We make the assumption that in each data sample $N_{\text{te-jam}}$ randomly selected symbols are affected by interference. A random phase shift is added to $N_{\text{te-jam}}$ symbols. In this analysis, we sample the phase shift uniformly between $[0, \pi]$ in $\pi/50$ increments. To calculate the interfering signal strength, we use the JSR term.

To assess the AEC system's ability to handle interference, different jammer types can be considered. In this study, we consider a tone jammer interfering with the transmitted signal at a JSR of 15 dB. We incorporate channel and receiver noise and consider the following cases:

1. *Baseline model*: This model represents the regular AE training without taking interference effects into account.
2. *Improved model*: This model applies RS during training process.
3. *Best model*: This model enhances the improved model by incorporating IT during the training stage.

Cases 1 and 2 demonstrate that RS effectively suppresses interference while preserving the BER performance. The BER versus SNR performance is illustrated in Figures 8.14–8.16 for different channel use cases of $n_{\text{ch}} = 1, 2, 4$, respectively. Despite a large interference signal with a JSR of 15 dB, we observe that the interference is effectively attenuated, with the BER performance matching or even surpassing that of the conventional scheme even when there is no interference.

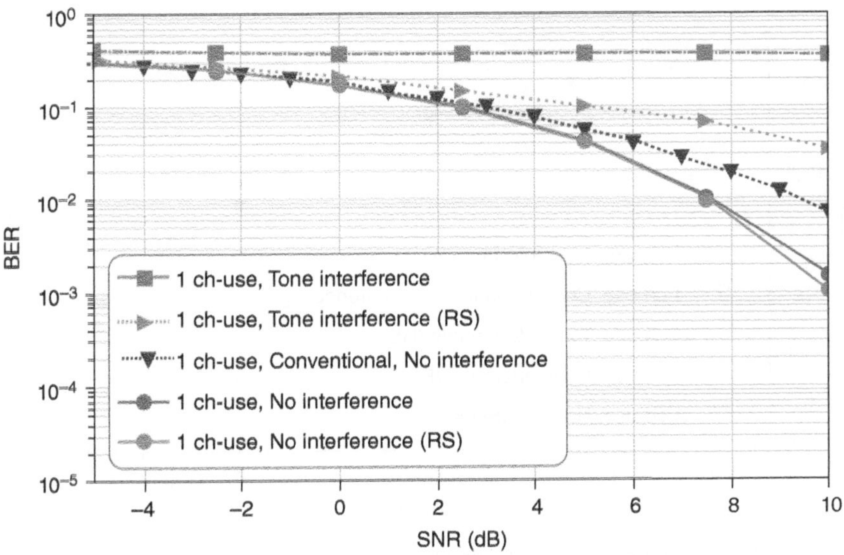

Figure 8.14 Error performance evaluation for channel use of 1.

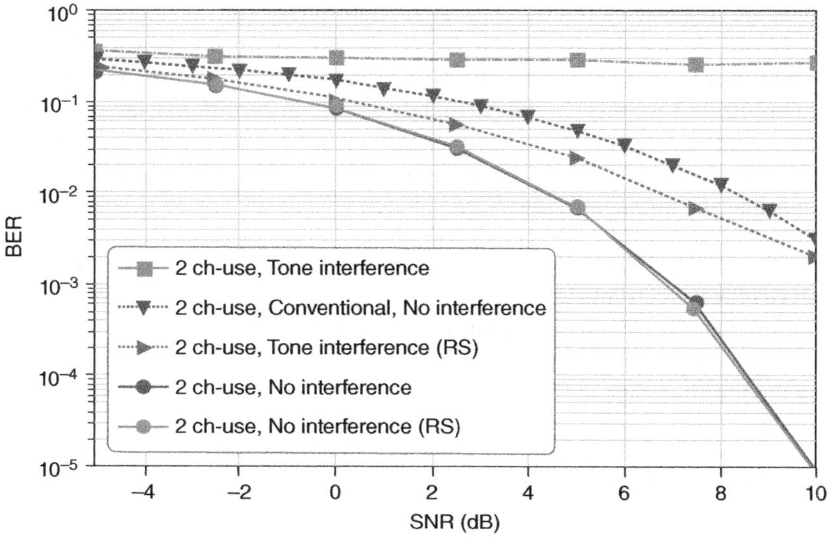

Figure 8.15 Error performance evaluation for channel use of 2.

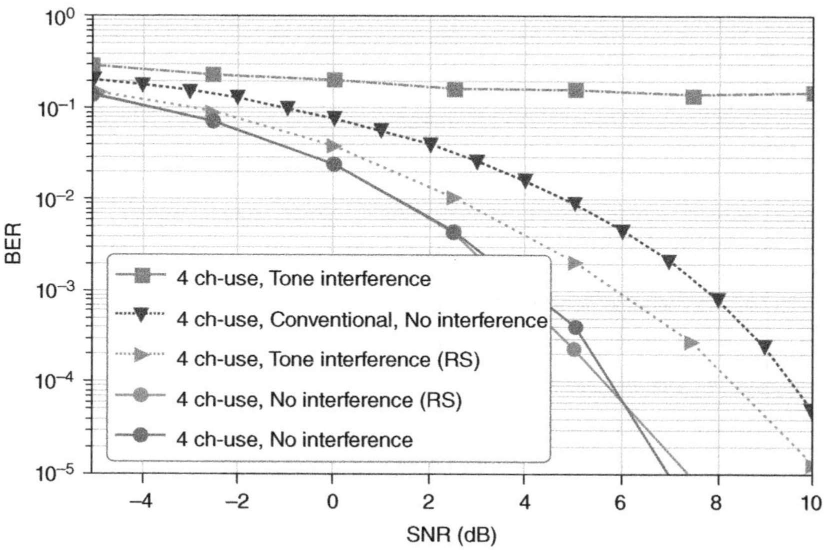

Figure 8.16 Error performance evaluation for channel use of 4.

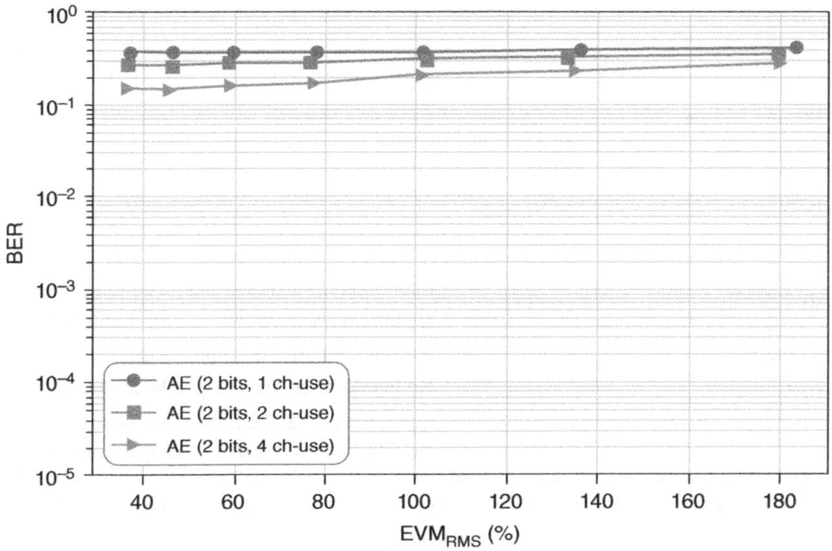

Figure 8.17 Assessing the EVM versus BER for AEC without interference suppression in the presence of jamming.

Figure 8.17 illustrates the BER versus error vector magnitude (EVM) for AEC without interference suppression. Noticeably, even at low EVM values, we observe a poor error performance. In contrast, Figure 8.18 depicts the BER versus EVM for AEC when RS is used for training. Notably, for $n_{\text{ch}} = 2$ and 4, the BER remains at 10^{-2} at EVM levels of 50% and 78%, respectively, despite the presence of 15 dB single-symbol interference. These findings underscore the effectiveness of training the AEC system using the proposed RS approach to suppress interference.

For the best model, the AE is trained under specific interference scenarios, such as training a model to mitigate single-symbol interference. Upon evaluation, we gauge its ability to suppress interference across a spectrum of conditions ranging from 1-symbol to 12-symbol interference. To maintain fairness, we ensure that the total interference power remains constant for each symbol count. For instance, when we consider the 1-symbol interference case, only one symbol is subject to interference per channel use, whereas the interference affects all symbols for the 12-symbol interference case. Figure 8.19 shows how these two interference scenarios differ, with power allocation being uniform across symbols for simplicity.

To further study the effects of suppressing the interference, an achievable SNR–JSR plot is constructed by determining the SNR at which a BER of 10^{-2} or lower is attained for each JSR value. We also evaluate the capability of the conventional system to suppress interference and compare it with the proposed

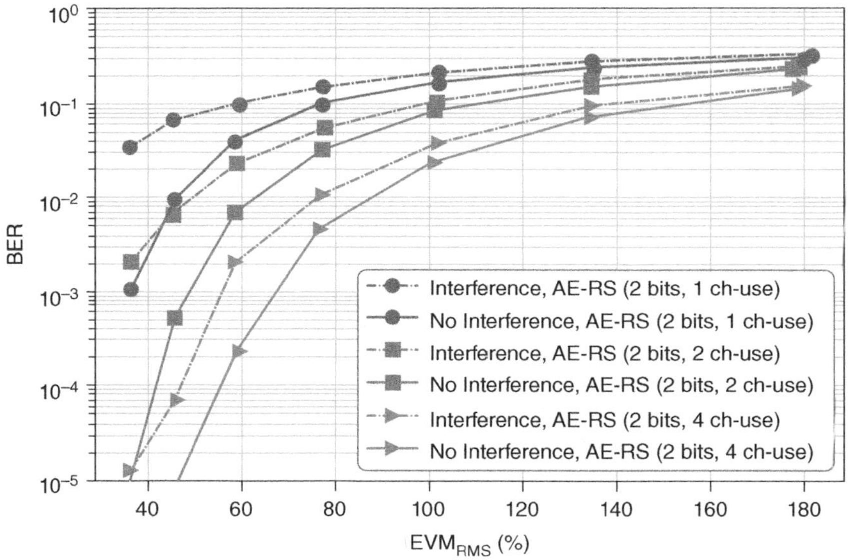

Figure 8.18 Assessing the EVM versus BER for AEC that is trained with RS to suppress interference.

Figure 8.19 Visualization of jamming signal interfering with only one symbol (a) versus all symbols (b).

approach. Results depicted in Figure 8.20 demonstrates the SNR versus JSR relationship for the AE models that are trained with 1 symbol ($N_{\text{tr-jam}} = 1$) and tested under varying interference conditions where we adjust $N_{\text{te-jam}}$. It is observed that the interference up to 24–26 dB can be mitigated. With an increase in the number of symbols exposed to interference energy, the proposed approach demonstrates enhanced performance against higher jamming power. Notably, AE models trained with both IT and RS suppress interference over 36 dB, outperforming conventional methods.

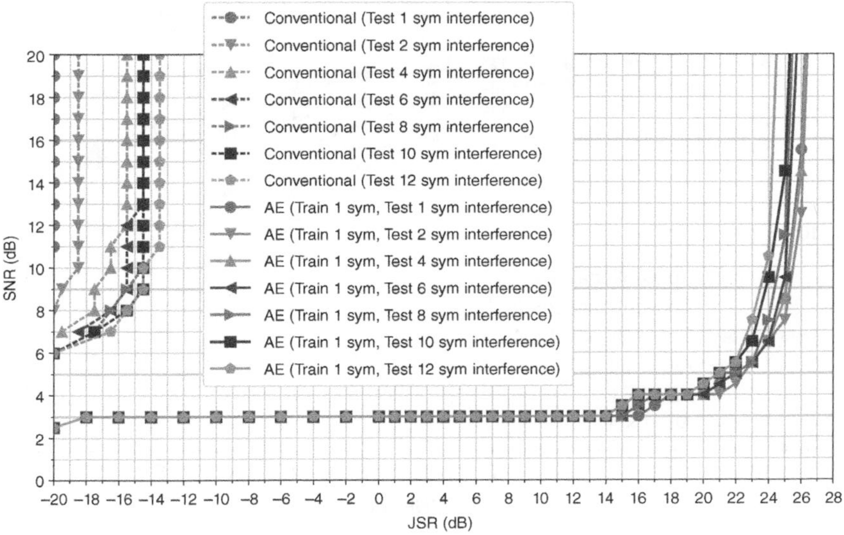

Figure 8.20 Assessing the AE models trained on single-symbol interference for their performance in handling multi-symbol interference.

8.7 AE Communications with Interference Suppression for MIMO Systems

Next, AEC that is developed for SISO systems is extended to MIMO communications systems. We assume spatial multiplexing where each antenna transmits different information such that n_T and n_R represent how many antennas there are for the transmitter and the receiver, respectively. The input and output dimensions for the encoder and decoder are shown in Table 8.10.

The encoder feedforward neural network (FNN) architecture includes input, 2 dense layers with 4000 neurons each, and output layers. The decoder FNN architectures includes input, 3 dense layers with 4000 neurons each, and

Table 8.10 The size of the inputs and outputs for the encoder and decoder in the AEC for MIMO.

	Encoder input	**Encoder output**	**Decoder input**	**Decoder output**
Dimensions	$n_b \times n_T$	$2n_T$	$2n_R$	$n_b \times n_R$

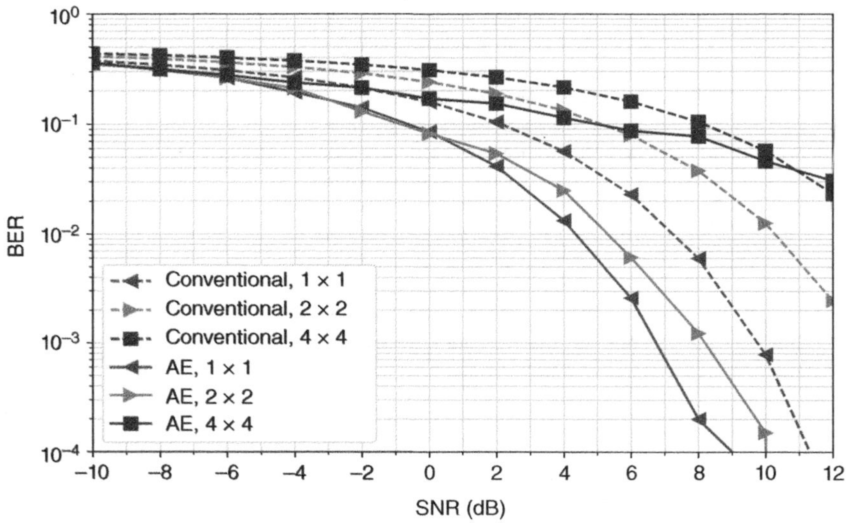

Figure 8.21 The BER performance comparison between the AEC and conventional MIMO communications for different antenna configurations (1 × 1, 2 × 2, and 4 × 4).

output layer. A ReLU activation layer followed by a dropout layer (with dropout probability of 25%) are used between each fully connected layer.

The BER performance of AEC for MIMO is compared with that of conventional MIMO communications in the absence of any interference in Figure 8.21. The MIMO rate for spatial multiplexing is increased by adding more antennas (1 × 1, 2 × 2, and 4 × 4). We apply RS to enhance the performance when there are channel impairments including 10° phase offset and 30 Hz frequency offset.

Table 8.11 BER of a MIMO system taking interference into account.

MIMO Scheme	BER for 0 dB SNR	BER for 5 dB SNR	BER for 10 dB SNR
Conventional 2 × 2 (No interference)	0.2395	0.1048	0.0125
AE 2 × 2 (interference)	0.1327	0.0489	0.0028
Conventional 4 × 4 (No interference)	0.3084	0.1867	0.0569
AE 4 × 4 (Interference)	0.1929	0.1244	0.0405

Subsequently, we examine how interference affects the BER performance of AEC for MIMO in contrast to conventional systems. AEC undergoes training using RS and IT, and it is subsequently tested under interference involving 1-symbol at 0 dB JSR, with a rate of 2 bits per symbol and utilizing 1 channel use.

Table 8.11 demonstrates how AEC improves the BER beyond conventional communications for the MIMO case subject to interference effects. Specifically, there is a reduction in the BER by an average of 59% for 2×2 MIMO and 33% for 4×4 MIMO.

8.8 Conclusion

In this chapter, a novel communication approach was presented for 6G by utilizing an AE model, wherein the transmitter and receiver were modeled by the DNN architectures serving as the encoder–decoder pair. AEC was shown to outperform conventional communications across various practical scenarios including limited training data, embedded implementation limitations, and the presence of both channel and interference effects such as in the case of jamming attacks. A GAN was designed to enhance the AE training, resulting in an improvement in the BER performance. Additionally, the impact of quantization on the DNN model was shown to be minimal on the BER performance, while reducing memory requirements for embedded implementation. IT and RS were introduced for AEC and they are shown to be highly effective in handling unknown and dynamic jamming conditions, achieving up to 36 dB of interference suppression compared to conventional schemes. Finally, the MIMO setting with spatial multiplexing was studied and the application of the AE was shown to improve the BER across different antenna configurations. These results highlight how effective the proposed approach is in suppressing interference in SISO and MIMO communications and protecting 6G communications against interference effects.

References

1 N. Abuzainab et al. QoS and jamming-aware wireless networking using deep reinforcement learning. In *IEEE Military Communications Conference (MILCOM)*, 2019.

2 F. Ait Aoudia and J. Hoydis. Trimming the fat from OFDM: Pilot-and CP-less communication with end-to-end learning. In *IEEE International Conference on Communications Workshops (ICC Workshops)*, pages 1–6, 2021.

3 R. Askar, J. Chung, Z. Guo, H. Ko, W. Keusgen, and T. Haustein. Interference handling challenges toward full duplex evolution in 5G and beyond cellular networks. *IEEE Wireless Communications*, 28(1):51–59, 2021.

4 E. Ayanoglu, K. Davaslioglu, and Y. E. Sagduyu. Machine learning in NextG networks via generative adversarial networks. *IEEE Transactions on Cognitive Communications and Networking*, 8(2):480–501, 2022.

5 E. Balevi and J. G. Andrews. Wideband channel estimation with a generative adversarial network. *IEEE Transactions on Wireless Communications*, 20(5):3049–3060, 2021.

6 E. Balevi, A. Doshi, A. Jalal, A. Dimakis, and J. G. Andrews. High dimensional channel estimation using deep generative networks. *IEEE Journal on Selected Areas in Communications*, 39(1):18–30, 2021.

7 J. M. Cohen, E. Rosenfeld, and Z. Kolter. Certified adversarial robustness via randomized smoothing. In *International Conference on Machine Learning*, pages 1310–1320, 2019.

8 K. Davaslioglu and Y. E. Sagduyu. Generative adversarial learning for spectrum sensing. In *IEEE International Conference on Communications (ICC)*, 2018.

9 K. Davaslioglu, S. Soltani, T. Erpek, and Y. E. Sagduyu. DeepWiFi: Cognitive WiFi with deep learning. *IEEE Transactions on Mobile Computing*, 20(2):429–444, 2021.

10 A. F. Demir, M. Elkourdi, M. Ibrahim, and H. Arslan. Waveform design for 5G and beyond. *5G Networks: Fundamental Requirements, Enabling Technologies, and Operations Management*, pages 51–76, 2018.

11 S. Dörner, S. Cammerer, J. Hoydis, and S. T. Brink. Deep learning based communication over the air. *IEEE Journal of Selected Topics in Signal Processing*, 12(1):132–143, 2018.

12 T. Erpek, T. J. O'Shea, and T. C. Clancy. Learning a physical layer scheme for the MIMO interference channel. In *IEEE International Conference on Communications (ICC)*, 2018.

13 T. Erpek, Y. E. Sagduyu, and Y. Shi. Deep learning for launching and mitigating wireless jamming attacks. *IEEE Transactions on Cognitive Communications and Networking*, 5(1):2–14, 2018.

14 T. Erpek, T. J. O'Shea, Y. E. Sagduyu, Y. Shi, and T. C. Clancy. Deep learning for wireless communications. In *Development and Analysis of Deep Learning Architectures*, pages 223–266. Springer, 2020.

15 T. Erpek, Y. E. Sagduyu, A. Alkhateeb, and A. Yener. Autoencoder-based communications with reconfigurable intelligent surfaces. In *IEEE International Symposium on Dynamic Spectrum Access Networks (DySPAN)*, 2021.

16 A. Felix, S. Cammerer, S. Dörner, J. Hoydis, and S. T. Brink. OFDM-autoencoder for end-to-end learning of communications systems. In *IEEE International Workshop on Signal Processing Advances in Wireless Communications (SPAWC)*, 2018.

17 M. Goutay, F. A. Aoudia, J. Hoydis, and J.-M. Gorce. End-to-end learning of OFDM waveforms with PAPR and ACLR constraints. In *IEEE Globecom Workshops (GC Wkshps)*, 2021.

18 I. Gulrajani, F. Ahmed, M. Arjovsky, V. Dumoulin, and A. Courville. Improved training of Wasserstein GANs. In *NeurIPS*, 2017.

19 T. Hou, T. Wang, Z. Lu, Y. Liu, and Y. Sagduyu. IoTGAN: GAN powered camouflage against machine learning based IoT device identification. In *2021 IEEE International Symposium on Dynamic Spectrum Access Networks (DySPAN)*, 2021.

20 Q. Hu, Y. Cai, K. Kang, G. Yu, J. Hoydis, and Y. C. Eldar. Two-timescale end-to-end learning for channel acquisition and hybrid precoding. *IEEE Journal on Selected Areas in Communications*, 40(1):163–181, 2021.

21 D. P. Kingma and J. Ba. Adam: A method for stochastic optimization. In *International Conference on Learning Representations (ICLR)*, 2015.

22 N. H. Mahmood, O. A. Lopez, H. Alves, and M. Latva-Aho. A predictive interference management algorithm for URLLC in beyond 5G networks. *IEEE Communications Letters*, 25(3):995–999, 2020.

23 T. J. O'Shea and J. Hoydis. An introduction to deep learning for the physical layer. *IEEE Transactions on Cognitive Communications and Networking*, 3(4):563–575, 2017.

24 T. J. O'Shea, T. Erpek, and T. C. Clancy. Physical layer deep learning of encodings for the MIMO fading channel. In *Allerton Conference on Communication, Control, and Computing (Allerton)*, 2017.

25 T. J. O'Shea, T. Roy, and N. West. Approximating the void: Learning stochastic channel models from observation with variational generative adversarial networks. In *2019 International Conference on Computing, Networking and Communications (ICNC)*, 2019.

26 B. Ozpoyraz, A. T. Dogukan, Y. Gevez, U. Altun, and E. Basar. Deep learning-aided 6G wireless networks: A comprehensive survey of revolutionary PHY architectures. *IEEE Open Journal of the Communications Society*, 3:1749–1809, 2022.

27 M. Patel, X. Wang, and S. Mao. Data augmentation with conditional GAN for automatic modulation classification. In *ACM Workshop on Wireless Security and Machine Learning (WiseML)*, 2020.

28 D. Peng, D. He, Y. Li, and Z. Wang. Integrating terrestrial and satellite multibeam systems toward 6G: Techniques and challenges for interference mitigation. *IEEE Wireless Communications*, 29(1):24–31, 2022.

29 M. Sadeghi and E. G. Larsson. Physical adversarial attacks against end-to-end autoencoder communication systems. *IEEE Communications Letters*, 23(5):847–850, 2019.

30 Y. E. Sagduyu, R. A. Berry, and A. Ephremides. Jamming games in wireless networks with incomplete information. *IEEE Communications Magazine*, 49(8):112–118, 2011.

31 Y. E. Sagduyu, T. Erpek, and Y. Shi. Adversarial machine learning for 5G communications security. In C. A. Kamhoua, C. D. Kiekintveld, F. Fang, and Q. Zhu, editors, *Game Theory and Machine Learning for Cyber Security*, pages 270–288. Wiley, 2021.

32 Y. E. Sagduyu, T. Erpek, S. Ulukus, and A. Yener. Is semantic communications secure? A tale of multi-domain adversarial attacks. *IEEE Communications Magazine*, 61(11):50–55, 2023.

33 Y. E. Sagduyu, T. Erpek, S. Ulukus, and A. Yener. Vulnerabilities of deep learning-driven semantic communications to backdoor (Trojan) attacks. In *Conference on Information Sciences and Systems (CISS)*, 2023.

34 Y. E. Sagduyu, T. Erpek, A. Yener, and S. Ulukus. Joint sensing and semantic communications with multi-task deep learning. *arXiv preprint arXiv:2311.05017*, 2023.

35 Y. E. Sagduyu, T. Erpek, A. Yener, and S. Ulukus. Joint sensing and task-oriented communications with image and wireless data modalities for dynamic spectrum access. *arXiv preprint arXiv:2312.13931*, 2023.

36 Y. E. Sagduyu, T. Erpek, A. Yener, and S. Ulukus. Multi-receiver task-oriented communications via multi-task deep learning. *arXiv preprint arXiv:2308.06884*, 2023.

37 Y. E. Sagduyu, S. Ulukus, and A. Yener. Task-oriented communications for NextG: End-to-end deep learning and AI security aspects. *IEEE Wireless Communications*, 30(3):52–60, 2023.

38 Y. E. Sagduyu, S. Ulukus, and A. Yener. Age of information in deep learning-driven task-oriented communications. In *IEEE INFOCOM Age of Information Workshop*, 2023.

39 Y. E. Sagduyu, T. Erpek, A. Yener, and S. Ulukus. Will 6G be semantic communications? Opportunities and challenges from task oriented and secure communications to integrated sensing. *arXiv preprint arXiv:2401.01531*, 2024.

40 R. A. Shafik, M. S. Rahman, A. R. Islam, and N. S. Ashraf. On the error vector magnitude as a performance metric and comparative analysis. In *International Conference on Emerging Technologies*, pages 27–31, 2006.

41 Y. Shi and Y. E. Sagduyu. How to launch jamming attacks on federated learning in NextG wireless networks. In *IEEE Globecom Workshops (GC Wkshps)*, 2022.

42 Y. Shi and Y. E. Sagduyu. Sensing-throughput tradeoffs with generative adversarial networks for NextG spectrum sharing. In *IEEE Military Communications Conference (MILCOM)*, 2022.

43 Y. Shi, Y. E. Sagduyu, T. Erpek, K. Davaslioglu, Z. Lu, and J. H. Li. Adversarial deep learning for cognitive radio security: Jamming attack and defense strategies. In *IEEE International Conference on Communications Workshops (ICC Workshops)*, 2018.

44 Y. Shi, K. Davaslioglu, and Y. E. Sagduyu. Generative adversarial network for wireless signal spoofing. In *ACM Workshop on Wireless Security and Machine Learning (WiseML)*, 2019.

45 Y. Shi, Y. E. Sagduyu, T. Erpek, and M. C. Gursoy. Jamming attacks on NextG radio access network slicing with reinforcement learning. In *IEEE Future Networks World Forum (FNWF)*, 2022.

46 Y. Shi, Y. E. Sagduyu, and T. Erpek. Jamming attacks on decentralized federated learning in general multi-hop wireless networks. *arXiv preprint arXiv:2301.05250*, 2023.

47 Y. Shi, Y. E. Sagduyu, T. Erpek, and M. C. Gursoy. How to attack and defend NextG radio access network slicing with reinforcement learning. *IEEE Open Journal of Vehicular Technology*, 4:181–192, 2023.

48 S. Soltani, Y. E. Sagduyu, R. Hasan, K. Davaslioglu, H. Deng, and T. Erpek. Real-time and embedded deep learning on FPGA for RF signal classification. In *IEEE Military Communications Conference (MILCOM)*, 2019.

49 J. Song, C. Häger, J. Schröder, T. O'Shea, and H. Wymeersch. Benchmarking end-to-end learning of MIMO physical-layer communication. In *IEEE Global Communications Conference (GLOBECOM)*, 2020.

50 H. Sun, X. Chen, Q. Shi, M. Hong, X. Fu, and N. D. Sidiropoulos. Learning to optimize: Training deep neural networks for interference management. *IEEE Transactions on Signal Processing*, 66(20):5438–5453, 2018.

51 D. Wu, M. Nekovee, and Y. Wang. Deep learning-based autoencoder for m-user wireless interference channel physical layer design. *IEEE Access*, 8:174679–174691, 2020.

52 J. Xia, K. He, W. Xu, S. Zhang, L. Fan, and G. K. Karagiannidis. A MIMO detector with deep learning in the presence of correlated interference. *IEEE Transactions on Vehicular Technology*, 69(4):4492–4497, 2020.

53 W. Xu, W. Trappe, Y. Zhang, and T. Wood. The feasibility of launching and detecting jamming attacks in wireless networks. In *ACM International Symposium on Mobile Ad Hoc Networking and Computing*, 2005.

54 H. Ye, L. Liang, G. Y. Li, and B.-H. Juang. Deep learning-based end-to-end wireless communication systems with conditional GANs as unknown channels. *IEEE Transactions on Wireless Communications*, 19(5):3133–3143, 2020.

55 Q. Zhang, W. Ma, Z. Feng, and Z. Han. Backhaul-capacity-aware interference mitigation framework in 6G cellular internet of things. *IEEE Internet of Things Journal*, 8(12):10071–10084, 2021.

56 S. Zhang, J. Liu, Ti. K. Rodrigues, and N. Kato. Deep learning techniques for advancing 6G communications in the physical layer. *IEEE Wireless Communications*, 28(5):141–147, 2021.

9

AI/ML-Aided Processing for Physical-Layer Security

Muralikrishnan Srinivasan[1], *Sotiris Skaperas*[2], *Mahdi Shakiba Herfeh*[2], *and Arsenia Chorti*[2,3]

[1] *Department of Electronics Engineering, Indian Institute of Technology (BHU), Varanasi, India*
[2] *ETIS UMR8051, CY University, ENSEA, CNRS, Cergy, France*
[3] *Wireless Connectivity Group, Barkhausen Institut, Dresden, Germany*

9.1 Introduction

The field of reliable communications focuses on creating efficient and effective communication systems. Researchers take a bottom-up approach, starting with idealized models of sources and channels. They use this information to develop powerful theorems that describe the limits of rates that can be achieved. These theorems then guide the design of real-world wireless systems, which must accommodate the fact that there will always be some connection outages and errors.

In contrast, information security requires much tighter guarantees. Cryptographic proofs are used to ensure that the communication systems are secure and that sensitive information is protected from unauthorized access. In these proofs, "adversarial advantage" is used to quantify the potential for unauthorized access. The advantage gained by an adversary should be limited within a specific, computationally feasible timeframe, reflecting the constraints posed by available computational resources [1].

Mathematical constructions such as pseudorandom number generators (PRGs), pseudorandom functions (PRFs), permutations (PRPs), and random oracles (RO) are utilized in cryptographic proofs. These abstractions serve as the foundation for the fundamental cryptographic primitives [1]. To ensure that practical cryptosystems reflect these idealized mathematical structures as closely as possible, they are designed to resemble them. To put it simply, when creating secure stream ciphers, it is beneficial if they share characteristics with PRGs and, for block ciphers, they should resemble PRFs or PRPs. This alignment with

Physical-Layer Security for 6G, First Edition. Edited by Parthajit Mohapatra,
Nikolaos Pappas, Arsenia Chorti, and Stefano Tomasin.

idealized mathematical structures is a common practice in designing practical cryptographic systems, and it helps achieve semantic security in real-world scenarios.

In simpler terms, in the realm of physical-layer security (PLS), proofs for perfect secrecy or creating secure keys without leaks are built using theoretical channel models that represent idealized conditions. It has been demonstrated that semantic security for wiretap channels can be proven [26]. Additionally, recent advances have been made in the area of finite blocklength for analysis and code design [19, 21, 25, 30]. However, despite these achievements, there is still a challenge in translating the theory into practice. The reason for this is that real communication channels often do not match the perfect models assumed in the proofs.

To deal with inconsistencies in 6G systems, a crucial first step is to learn and understand the specific characteristics of the communication channel at each site [29]. This involves using real-time data to create secrecy maps [25]. However, there are challenges in directly assessing secrecy from field data due to limited precision in estimating mutual information from measured channel information, mainly because of dataset size [13]. To overcome this, it is essential to adjust the transmission based on the learned channel model. The goal is to make the end-to-end transmission resemble an idealized channel with explicit security guarantees. One approach is to use machine learning (ML) for dimensionality reduction, transforming the actual channel into a simpler form with less correlation across dimensions. Additionally, any preprocessing methods proposed should be carefully evaluated for their potential impact on achievable secrecy or secret key rates.

9.1.1 Facilitating the Incorporation of PLS in 6G

Integrating PLS into 6G networks involves three important steps:

1. Trusted entities like base stations should actively learn about the communication channel during the network's actual operation. This learning needs to be ongoing, adapting to the specific conditions of each site [3].
2. Ensure that end-to-end security in communication links is well engineered. ML techniques, like dimensionality reduction, can be helpful tools in achieving this goal.
3. PLS can play a role in creating and sharing authentication or symmetric keys in hybrid PLS-crypto systems. The desired key rates depend on the cryptographic suite or network properties.

Incorporating PLS into 6G security protocols can introduce new security measures at all layers for the first time [7]. ML-based preprocessing algorithms can help to bridge the gap in designing practical PLS systems. These algorithms can process observed channel information to create or share secret keys in hybrid security systems. They can also be employed for tasks like determining location or verifying identity through radio frequency (RF) fingerprints [10, 14, 16]. The design criteria for these algorithms include separability of signatures (fingerprints) and unpredictability of shared randomness in secret key generation (SKG).

The wireless channel can be seen as a combination of deterministic and stochastic components [6]. The suggested preprocessing idea is that the large-scale fading, influenced by location, can act as a source for authentication, while the small-scale fading can act as a source of randomness for SKG [15, 22, 23]. This chapter introduces a power-domain decomposition-based preprocessing of the channel state information (CSI) into two components, which are then further analyzed through unsupervised learning techniques such as principal component analysis (PCA) and autoencoders (AEs). The decomposition should meet the criteria of maximum separability of RF fingerprints and minimum dependence between the sources of shared randomness at different nodes. The suggested approach is put to the test using synthetic datasets and the Quadriga models [8].

Section 1.2 proposes metrics for RF fingerprinting and SKG. The reciprocity and mismatch probability (MP) are discussed as trade-offs between reducing dependencies and preserving the reciprocity of SKG seeds. Section 1.3 proposes two approaches, PCA and AE, to separate predictable and unpredictable CSI components.

9.2 Proposed Metrics for RF Fingerprinting and SKG

In this section, the metrics that can validate the security notions after the preprocessing stage is defined.

9.2.1 Total Variation Distance for Radio Frequency Fingerprinting

In PLS, establishing authenticity necessitates a verifiable source of distinctiveness, such as node positioning and fingerprinting. The fingerprints utilized for authentication should exhibit statistical distinctiveness at each location and showcase gradual variations, enhancing the reliability of the authentication process [20]. Ideally, these fingerprints should also manifest statistical separability at each location, though complete lack of correlation is not a strict requirement.

To measure the separability of fingerprints, e.g., between nodes at neighboring locations, one can use the Total Variation Distance (TVD), which measures the L1 distance between the empirical measures.

Definition 9.1 The TVD between two probability distributions μ and ν on a countable configuration space $\mathcal{X}^N$ is defined by [11]

$$\text{TVD} = \|\mu - \nu\|_{\text{TV}} = \frac{1}{2} \sum_{x \in \mathcal{X}^N} |\mu(x) - \nu(x)| .$$

As the TVD increases, the distinguishability of the measures increases and vice versa.

9.2.2 Cross Correlation for SKG

The SKG from wireless fading coefficients relies on three crucial factors: (i) the reciprocal nature of the channel between nodes Alice and Bob within the channel coherence time, (ii) spatial independence, often assessed through decorrelation, theoretically existing at distances comparable to the wavelength, and (iii) temporal variations, such as those resulting from node mobility [32].

According to Jakes' model, a channel is uncorrelated when a third party is positioned at a distance equal to half the wavelength [6]. However, real-world experiments have demonstrated that this distance is applicable mainly in environments with abundant scattering, and attacks using ray tracing have proven successful in other scenarios [4, 5, 9, 27, 31].

In the current literature, it is common to conduct SKG without eliminating predictable elements from wireless channel coefficients and without considering them in the privacy amplification step. This oversight can lead to reduced energy efficiency, as larger quantizers and heavier hashing are needed [12, 17, 28].

It is vital to pay attention to spatial correlations near potential attackers (referred to as Eve) who are in close proximity to legitimate users. Without preprocessing the observed CSI, an attacker can distill highly correlated sequences, significantly reducing the effective key space. Although these correlations can be removed through privacy amplification, this approach requires larger quantizers and heavier hashing.

Despite the introduction of various polynomial-time statistical tests like the National Institute of Standards and Technology (NIST) suite, these tests do not consider dependencies and cross correlations between legitimate and adversarial nodes, which may exist in SKG scenarios. To tackle this, it is essential to use a metric that gauges the predictability or unpredictability of observed CSI vectors by assessing spatial dependencies between them. This approach can be expanded to include time, frequency, and antenna domains.

9.2.3 Statistical Independence Metric

In the PLS literature, the cross-correlation coefficient (CC) has been universally used as an indirect measure of unpredictability. Nevertheless, this method has constraints because relying solely on decorrelation may not be sufficient to confirm unpredictability. Nonlinear dependencies and deviations from Gaussian distributions could still be present. Introducing the concept of spatial independence of SKG seeds is crucial to ensuring that the design of the SKG process aligns with the definition of PRG. This concept serves as a criterion for achieving semantically secure key distillation from wireless fading coefficients and provides a more robust measure of unpredictability. Ensuring independence between observations at legitimate and adversarial entities is enough to conclude unpredictability, making it an essential condition for SKG design.

The d-parameter Hilbert–Schmidt independence criterion (d-HSIC), introduced in recent research [18], offers a method for assessing dependencies among observed CSI vectors. This hypothesis test employs a positive-definite kernel on N-dimensional random variables (RVs) and transform their distribution into the reproducing kernel Hilbert space, determining whether the N dimensions are independent. More specifically, consider $\tilde{\boldsymbol{H}} = \left(\tilde{\boldsymbol{h}}^1, \dots, \tilde{\boldsymbol{h}}^N\right)$, an $M \times N$ matrix comprising observation vectors $\tilde{\boldsymbol{h}}^i = [\tilde{h}_1^i, \dots, \tilde{h}_M^i]^T$ for $i \in 1, \dots, N$. The null hypothesis H_0 assumes independence among the vectors $\tilde{\boldsymbol{h}}^i$, where i ranges from 1 to N. Accordingly, the joint probability density function of these vectors is represented as the product of their individual marginals: $H_0 : F_{\tilde{\boldsymbol{h}}^1,\dots,\tilde{\boldsymbol{h}}^N} = F_{\tilde{\boldsymbol{h}}^1} \dots F_{\tilde{\boldsymbol{h}}^N}$. Conversely, the alternative hypothesis H_A indicates that $\tilde{\boldsymbol{H}}$ includes at least two dependent vectors. An estimator (test statistic) dHSIC of the statistical functional was defined as follows [[18], Def 2.6]:

$$\begin{aligned} d\text{HSIC}(\tilde{\boldsymbol{H}}) = {} & \frac{1}{M^2}\sum_{i,j=1}^{M}\prod_{l=1}^{N}\left(\mathbf{1}_{M\times M} \circ K_{ij}^l\right) + \frac{1}{M^{2N}}\prod_{l=1}^{N}\sum_{i,j=1}^{M}K_{ij}^l \\ & - \frac{2}{M^{N+1}}\sum_{i,j=1}^{M}\prod_{l=1}^{N}\left(\mathbf{1}_{M\times 1} \circ K_{ij}^l\right), \end{aligned} \tag{9.1}$$

where the operator $\circ$ denotes the Hadamard product and $\mathbf{1}_{M\times M}$ is an $M \times M$ matrix of ones. Also, $\mathbf{K}^l = \left(\mathbf{K}_{ij}^l\right) = \left(k^l(x_i, x_j)\right) \in \mathbb{R}^{M\times M}$ is the Gram matrix of the positive semi-definite Gaussian kernel k^l, defined $\forall x_i, x_j \in \mathbb{R}$ by, $k^l = \exp\left(-\frac{\|x_i - x_j\|^2}{\sigma^2}\right)$, with bandwidth $\sigma = \sqrt{\frac{\text{med}\left(\|x_i - x_j\|^2\right)}{2}}$ and $\text{med}(\cdot)$ is the median heuristic.

In accordance with the hypothesis test outlined in [[18], Theorem 3.1], the critical value for a specific significance level α can be determined as follows:

$$\mathrm{CV}_{\alpha} = \left[\mathbf{D}^{d\mathrm{HSIC}'}\right]_{\lceil (B+1)(1-\alpha)\rceil + \sum_{i=1}^{B} \mathbb{1}_{\{d\mathrm{HSIC}'(\tilde{\boldsymbol{H}})=d\mathrm{HSIC}'(\tilde{\boldsymbol{H}}_i)\}}}.$$

Here, $\mathbf{D}^{d\mathrm{HSIC}'}$ comprises the B Monte-Carlo realizations of $d\mathrm{HSIC}'(\tilde{\boldsymbol{H}})$ in ascending order, and $d\mathrm{HSIC}'\left(\tilde{\boldsymbol{H}}\right)$, where $\tilde{\boldsymbol{H}}$ is constructed as $\tilde{\boldsymbol{H}} = \left(r_1(\tilde{\boldsymbol{h}}_1), \dots, r_N(\tilde{\boldsymbol{h}}_M)\right)$, involves $r_1, \dots, r_M$ random resamplings without replacement. The operators $\lceil . \rceil$ and $[.]_j$ denote the ceiling function and the jth element of a vector, respectively, and $\mathbb{1}_{\{\cdot\}}$ is the indicator function.

Utilizing the dHSIC test statistic alongside its corresponding critical value, we can formulate a normalized metric to quantify the degree of dependence as follows:

$$\overline{\Delta} = \frac{d\mathrm{HSIC}(\tilde{\boldsymbol{H}})}{\mathrm{CV}_{\alpha}} \mathbb{1}_{\{d\mathrm{HSIC}(\tilde{\boldsymbol{H}})>\mathrm{CV}_{\alpha}\}}. \tag{9.2}$$

$\overline{\Delta}$ is close to unity in case of low dependence between the variables and grows without bound in case of high dependence.

In Section 9.2.4, we will delve into the crucial balance required in managing the reduction of dependencies among observations from different spatial locations. This balance is essential to preserve, as much as possible, the reciprocity between observations made by Alice and Bob.

9.2.4 Reciprocity and Mismatch Probability

The preprocessing step has a trade-off effect on the security of SKG. On the one hand, it reduces the dependencies between the observations made by legitimate and adversarial entities, increasing unpredictability. On the other hand, it may compromise the reciprocity of the SKG seeds. To gauge the reciprocity of the SKG seeds in both the uplink (from Alice to Bob) and the downlink (from Bob to Alice), a one-bit quantizer centered around the median point in the time dimension is utilized. The MP, calculated as the ratio of the number of mismatched bits to the total number of bits, represents the degree of disagreement between Alice and Bob. Although information reconciliation is intended to handle discrepancies, a lower reconciliation decoder rate may be noticeable when there is a high MP, making the reconciliation challenging beyond a certain threshold, particularly for short blocklengths.

9.3 Power Domain Preprocessing

This section outlines the suggested methods to differentiate between predictable components of CSI, mainly attributed to large-scale fading, and unpredictable components, primarily influenced by small-scale fading. The separation is accomplished through the application of PCA and two different AEs. The importance of using TVD as a metric for measuring the statistical separability of RF fingerprints is discussed. The introduced metric $\overline{\Delta}$ is used to evaluate the spatial dependencies of SKG seeds [18]. The validation of the proposed methodology is carried out on a synthetic dataset, and is described in the current section.

Consider a scenario with single-antenna legitimate nodes, referred to as Alices, and a base station, referred to as Bob. The spatial locations of Alices are represented by $\mathbf{x}_{nn=1}^{N}$, where $n = 1, \ldots, N$ and $\mathbf{x}_{nn=1}^{N} \in \mathbb{R}^L$, where L is the number of spatial dimensions considered (usually $L = 2$). The channel responses are measured between Bob, located at $(x, y, z) = (0, 0, 10)$, and the Alices at $N = 400$ equidistant (1 m) spatial points within a square area on the ground, spanning from $x = 100$ to $x = 290$ and $y = -100$ to $y = 90$. These responses are obtained using the "Berlin-UMa-NLOS" configuration in the Quadriga channel model [8], as illustrated in Figure 9.1. In this scenario, it is assumed that any specific Alice can potentially act as an attacker (Eve) to any other Alice.

The function that maps spatial locations to the $M \times 1$ CSI vectors, denoted as $\mathbf{h}_{nn=1}^{N}$, is represented by $\mathcal{H} : \mathbb{R}^L \rightarrow \mathbb{R}^M$, where M indicates the number of time-domain snapshots (obtained by combining real and imaginary components into a single-column vector). The CSI observations at both Alice and Bob, acquired

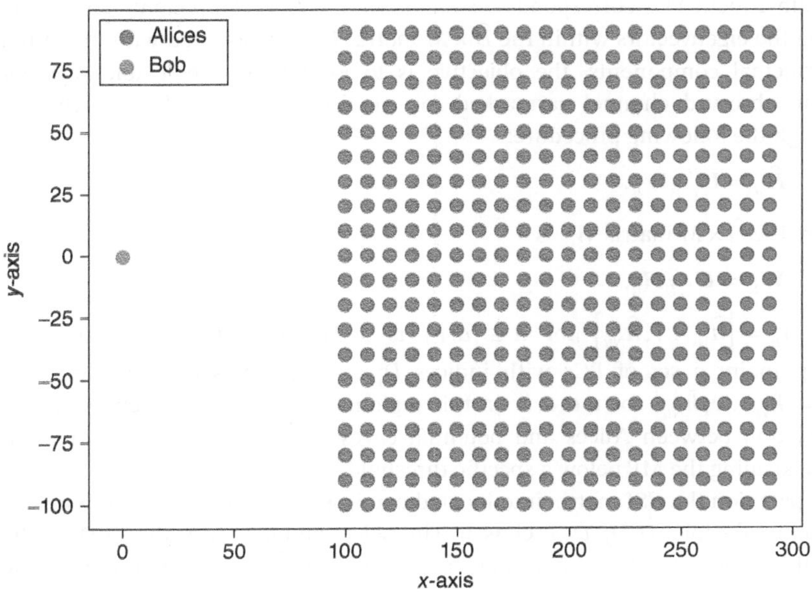

Figure 9.1 Positions of nodes in the synthetic Quadriga dataset.

through pilot signal exchanges, can be expressed as:

$$\mathbf{y}_{nu} = \mathbf{h}_n s + \mathbf{n}_{nu}, : n = 1, \dots, N, \; u \in a, b, \tag{9.3}$$

where the index a represents Alice and b represents Bob. The variables $\mathbf{n}_{na}$ and $\mathbf{n}_{nb}$ are circularly symmetric Gaussian noise, and the pilot symbols s are selected from the binary phase-shift keying (BPSK) constellation [2]. The CSI estimates at Alice and Bob are represented as $\mathbf{h}_{na} = \mathbf{y}_{na}$ and $\mathbf{h}_{nb} = \mathbf{y}_{nb}$, respectively, for $n = 1, \dots, N$.

In Sections 9.3.1 and 9.3.2, how the observed CSI matrices at Bob can be used to learn the functional mapping that separates the predictable and potentially spatially correlated components from the unpredictable and spatially decorrelated components by applying PCA and AEs will be demonstrated. While PCA is a linear approach, AEs have the ability to capture nonlinear dependencies.

9.3.1 Preprocessing Using PCA

Consider $\boldsymbol{H}_u = [\boldsymbol{h}_{1u}, \dots, \boldsymbol{h}_{Nu}]$ as the overall observed CSI matrix at Bob, consolidating CSI vectors from all Alices. Let $\boldsymbol{U}$ be an $M \times M$ matrix, with its rows containing eigenvectors obtained from the covariance matrix of $\mathbf{H}_u$, arranged in descending order of magnitude. In various scenarios, like those with Rician or generally line-of-sight setups, it's reasonable to assume that the initial principal components (PCs) align with significant large-scale fading elements, while the subsequent PCs correspond to smaller-scale fading components and noise. Using the eigenvectors within the $\widehat{D} \times M$ matrix $\boldsymbol{U}_{1:\widehat{D}}$, associated with the initial $\widehat{D}$ principal components, the objective is to isolate the predictable segment of the observed channel earmarked for RF fingerprinting. This is achieved through the following procedure:

$$\widehat{\boldsymbol{H}}_u = \boldsymbol{U}^H_{1:\widehat{D}} \boldsymbol{W}_u, \tag{9.4}$$

where the $\widehat{D} \times M$ matrix $\boldsymbol{W}_u$ is

$$\boldsymbol{W}_u = \boldsymbol{U}_{1:\widehat{D}} \boldsymbol{H}_u, \tag{9.5}$$

and $\widehat{\boldsymbol{H}}_u = \left[\widehat{\boldsymbol{h}}_{1u}, \dots, \widehat{\boldsymbol{h}}_{Nu}\right]$ for $u \in \{a, b\}$ is a $M \times N$ matrix. Moreover, one can pinpoint a set of PCs with indices $\tilde{D}_1, \dots, \tilde{D}_2$, representing components $\tilde{\boldsymbol{H}}_u = [\tilde{\boldsymbol{h}}_{1u}, \dots, \tilde{\boldsymbol{h}}_{Nu}]$ for $u \in a, b$. In this region, achieving low dependence and correlation between Alices and potential eavesdroppers (Eves) is possible, all while keeping the MP below a specific threshold.

To partition the CSI matrix into its predictable and unpredictable components, the optimal triplet $\widehat{D}, \tilde{D}_1, \tilde{D}_2$ is chosen. This selection maximizes the TVD for the initial $\widehat{D}$ principal components while minimizing both $\overline{\Delta}$ and MP within the range $\tilde{D}_1, \dots, \tilde{D}_2$. It is important to note that the principal components beyond $\tilde{D}_2 + 1$ are predominantly noise and should be disregarded.

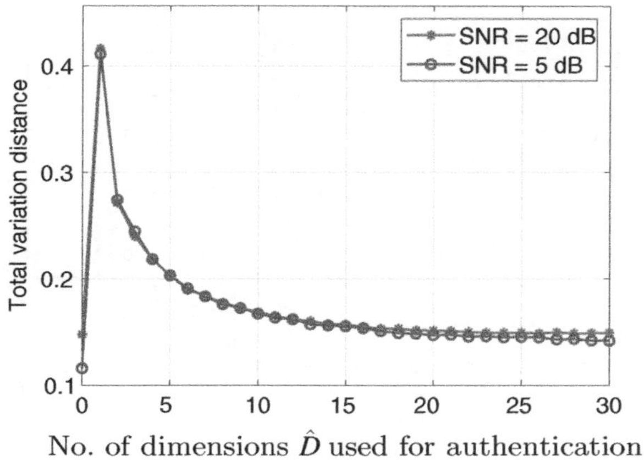

Figure 9.2 TVD versus $\hat{D}$.

In Figure 9.2, we present the average TVD between the initial $\hat{D}$ PCs at any Alice and her neighbors using the Quadriga dataset. The graph demonstrates that the maximum TVD is attained when $\hat{D} = 1$, with $\hat{D} = 0$ representing the original measurements. As the signal-to-noise ratio (SNR) increases, a slight uptick in TVD is observed due to the amplified variance of the first PCs amid reduced noise levels. This amplified TVD's impact is showcased in Figure 9.3, where we illustrate the fluctuation of the original CSI's amplitude alongside that of the first PC over time for four neighboring Alices. It is evident that the time series linked to the first PC in Figure 9.3b is more distinguishable compared to the original signal in Figure 9.3a.

In Figure 9.4a,b, we delve into the ramifications of maintaining various dimensions of the CSI matrix on two key metrics: the average CC among locations and

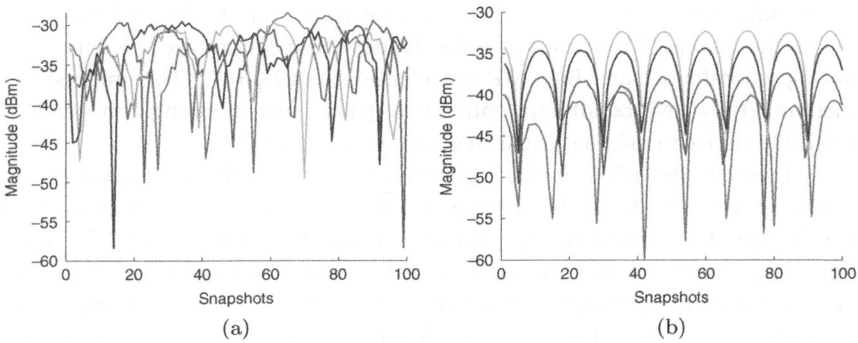

Figure 9.3 Separation of 6 neighbors for the original signal and the $\hat{D} = 1$ principal component for SNR of 20 dB. (a) Original signals and (b) predictable components.

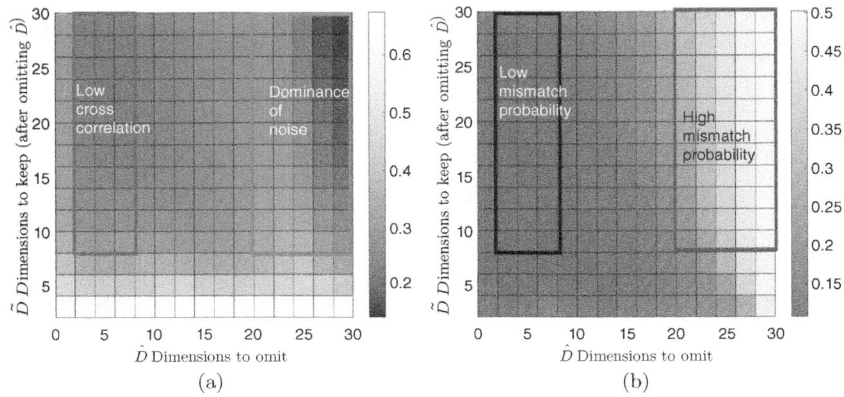

Figure 9.4 Trade-off between CC and MP for SNR = 20 dB for the Quadriga dataset. Darker colors indicate lower values. (a) Average CC (Original = 0.30) and (b) average MP.

their nearest neighbors, and the average MP between Alice and Bob, focusing on an SNR of 20 dB. Initially, the average CC stands at approximately 0.30 without any preprocessing. However, as we retain more dimensions, the CC steadily declines. Specifically, when $\tilde{D}_1 = 2$ and $\tilde{D}_2 = 20$, the CC drops to 0.18, accompanied by a negligible uptick in MP.

The prevalence of uncorrelated components, evident in the decline of CC, is most notable when a sufficient number of dimensions is retained. The corresponding region with low MP is labeled as the "Low Mismatch Probability" phase in Figure 9.4b. Conversely, the reduction in CC is less pronounced beyond $\tilde{D}_1 = 14$, where noise becomes predominant, leading to an uptick in MP. This phase is denoted as the "Dominance of Noise" in Figure 9.4a and "High Mismatch Probability" in Figure 9.4b.

An analogous examination is depicted in Figure 9.5 for an SNR of 5 dB, showcasing the increased influence of noise. Under these conditions, the domain of noise predominance and high MP is evident even at $\tilde{D}_1 = 10$. This underscores that, under low SNR conditions, opting to forgo any preprocessing steps remains a feasible choice to safeguard against compromising the MP.

In Figure 9.6, the behavior of $\overline{\Delta}$ is depicted in relation to the variation of $\tilde{D}_1, \tilde{D}_2$. A similar trend to CC can be observed for $\overline{\Delta}$ especially for higher values of $\tilde{D}_1$, which suggests independence between the retained PCs. However, for $\tilde{D}_1 < 8$, the trend in $\overline{\Delta}$ does not follow the decrease in CC, which highlights the limitations of CC in capturing dependence. This implies that simply retaining a low number of PCs, such as the first 6, may result in a low CC, but not necessarily a significant decrease in $\overline{\Delta}$ and independence between the retained PCs.

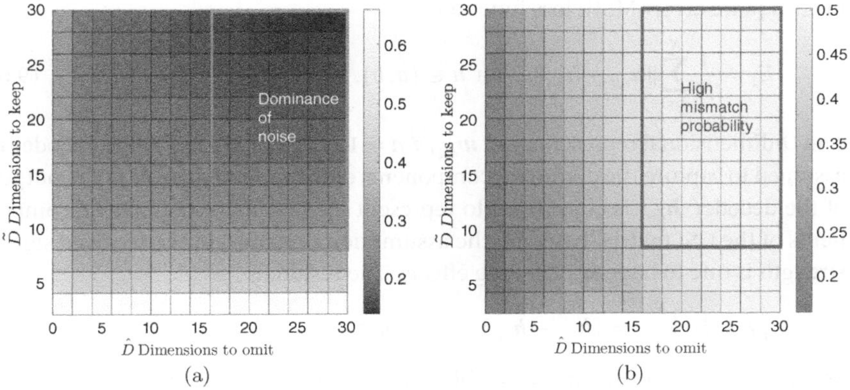

Figure 9.5 Trade-off between CC and MP for SNR = 5 dB for the Quadriga dataset. (a) Average CC (Original = 0.21) and (b) average MP.

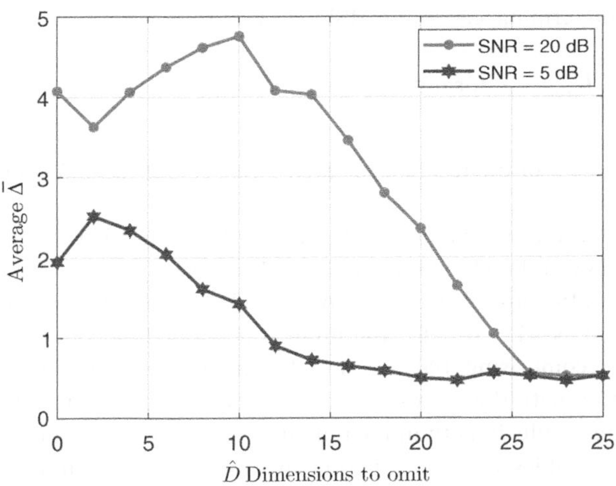

Figure 9.6 Evolution of $\overline{\Delta}$ with $\tilde{D}_1$ for $\tilde{D}_2 = 30$ for the Quadriga dataset.

9.3.2 Preprocessing Using AEs

AEs are unsupervised learning models that utilize two functions to learn from the data: an encoder, which maps the M-dimensional input matrix $\boldsymbol{h}_{nu}$ to a $\widehat{D}$-dimensional representation $\boldsymbol{w}_{nu}$ for all $n = 1, \dots, N$ and $u \in a, b$, and a decoder, which maps the encoded values back to an M-dimensional output $\widehat{\boldsymbol{h}}_{nu}$ for all $n = 1, \dots, N$ and $u \in a, b$. The primary objective of the AE is to minimize the mean square error (MSE) loss function:

$$E_1 = \frac{1}{N}\sum_{n=1}^{N} \|\boldsymbol{h}_{nu} - \widehat{\boldsymbol{h}}_{nu}\|_2^2, \text{ for } u \in \{a, b\}. \tag{9.6}$$

The $\widehat{D}$-dimensional representation $\boldsymbol{w}_{nu}, \forall\, n = 1, \ldots, N$ produced by the encoder is assumed to capture the dominant components of the input matrix $\boldsymbol{h}_{nu}$. The output of the decoder, $\widehat{\boldsymbol{h}}_{nu}$, is considered to represent the dominant predictable components of the CSI matrix, based on the assumption that most of the received signal strength is due to large scale fading effects. The residuals

$$\left\{\tilde{\boldsymbol{h}}_{nu}(\widehat{D})\right\}_{n=1}^{N} = \{\boldsymbol{h}_{nu} - \widehat{\boldsymbol{h}}_{nu}\}_{n=1}^{N}, \text{ for } u \in \{a, b\} \tag{9.7}$$

are considered to be the unpredictable components of the CSI matrix.

The hyperparameter $\widehat{D}$ can be optimized to emphasize SKG. This optimization involves creating an alternative loss function that balances the spatial correlation and reciprocity of the residuals in the uplink and downlink. The loss function may also include a correlation term instead of the MSE. For example,

$$E_2 = \frac{1}{N}\sum_{\substack{n_1=1 \\ n_2 \in \mathcal{U}(n_1)}}^{N} \tilde{\boldsymbol{h}}_{n_1 u}^{H} \tilde{\boldsymbol{h}}_{n_2 u}, \qquad \text{for } u \in \{a, b\}, \tag{9.8}$$

represents the inner product of the residuals at each location and those from its nearest neighbors $\mathcal{U}(n_1)$.

The architecture and activation functions of the AE are delineated in Table 9.1, following the methodology outlined in Studer et al. [24]. For clarity, we designate the AE employing the MSE loss function as AE1, and the AE utilizing the dot-product loss function as AE2. In AE2, the input is generated by combining the 200×1 CSI vector (comprising 100 real and 100 imaginary components) from each spatial location with the corresponding 200×1 CSI vectors from the nearest 8 neighbors. This results in both input and output dimensions of 400×1. This data arrangement enables the loss function to simultaneously minimize the correlation between users and the reconstruction error between the input and output. Two training methodologies are feasible: localized training, where Bob and the set of Alices each train separate AEs using their local datasets, or centralized training, where Bob trains a global AE, and its parameters are disseminated to the Alices.

Table 9.2 provides a comparison between the outcomes obtained using PCA and AE techniques. Similar to PCA, the CC diminishes and the MP increases as the SNR declines. Moreover, as the encoding dimensions $\widehat{D}$ escalate, the AE gains more adaptability to capture the predictable components, thereby causing a decline in CC. For the Quadriga dataset, AE2 employing a dot-product loss

Table 9.1 The layer configuration and activation function for AE1.

Layer	Dimensions	Activation
Input	200	Linear
1	100	tanh
2	50	Softplus
3	20	tanh
Intermediate	D	Linear
4	20	Relu
5	50	Softplus
6	100	tanh
Output	200	Linear

For AE2, the only alteration is that the dimensions of the input and output layers are 400.

Table 9.2 AE: key results for Quadriga.

$\hat{D}$	1				8			
SNR (dB)	5		20		5		20	
AE type	**AE1**	**AE2**	**AE1**	**AE2**	**AE1**	**AE2**	**AE1**	**AE2**
Original-CC	0.21	0.21	0.30	0.30	0.21	0.21	0.30	0.30
Residual-CC	0.20	0.19	0.27	0.24	0.18	0.15	0.25	0.21
Original-$\overline{\Delta}$	2.18	2.18	4.5	4.5	2.18	2.18	4.5	4.5
Residual-$\overline{\Delta}$	1.57	1.11	3.38	2.8	0.98	0.65	2.65	2.6
MP for localized	0.37	0.39	0.19	0.27	0.39	0.45	0.17	0.36
MP for centralized	0.35	0.36	0.14	0.15	0.34	0.40	0.15	0.15

function achieved a CC of 0.22 for $\hat{D} = 8$ and SNR= 20 dB for the residual components with centralized training, demonstrating comparable results to PCA with $\tilde{D}_1 = 2$ and $\tilde{D}_2 = 20$.

The average deviation of the residuals, denoted as $\overline{\Delta}$, exhibited a notable reduction from the original components, particularly noticeable for SNR = 20 dB. In comparison to AE1 utilizing an MSE loss function, AE2 displayed a lower CC for the residuals but a heightened MP, particularly evident in the case of local training. Centralized training yielded a lower MP than local training.

9.4 Conclusions

In conclusion, this chapter presents innovative methodologies for preprocessing CSI data in the context of RF fingerprinting and SKG for wireless communication systems. The study proposes the use of PCA and AE to effectively separate the CSI matrix into predictable and unpredictable components. The validation of these approaches is carried out on a synthetic dataset generated using the Quadriga channel model. The results demonstrate that both PCA and AE methods show promising results in achieving better separability and reducing correlation among the CSI components.

With PCA, the dominant components of the CSI matrix are extracted, and it is observed that the first PCA components are more distinguishable and exhibit higher TVD compared to the original measurements. However, as more PCA components are retained, the CC decreases, and the MP increases, indicating the increasing influence of noise. AEs are employed to separate the predictable and unpredictable components, with AE2 utilizing a dot-product loss function to balance spatial correlation and reciprocity. The results highlight that AE2, particularly with centralized training, achieves better performance in terms of CC and MP compared to AE1 with MSE loss function.

In summary, both PCA and AE methods hold great potential for enhancing the effectiveness of RF fingerprinting and SKG applications in wireless communication systems. The utilization of these preprocessing techniques can lead to improved separability of CSI components, thereby contributing to better security.

References

1 D. Boneh and V. Shoup. *A Graduate Course in Applied Cryptography*. Stanford, 2017.
2 A. Chorti. Optimal signalling strategies and power allocation for wireless secret key generation systems in the presence of a jammer. In *2017 IEEE International Conference on Communications (ICC)*, pages 1–6. IEEE, 2017.
3 A. Chorti, A. N. Barreto, S. Kopsel, M. Zoli, M. Chafii, P. Sehier, G. Fettweis, and H. V. Poor. Context-aware security for 6G wireless, the role of physical layer security. In *IEEE Communications Standards Magazine*, Special Issue on Emerging Security Technologies for 6G Security and Privacy.
4 R. Dautov and G. R. Tsouri. Effects of passive negative correlation attack on sensors utilizing physical key extraction in indoor wireless body area networks. *IEEE Sensors Letters*, 3(7):1–4, 2019.

5 M. Edman, A. Kiayias, Q. Tang, and B. Yener. On the security of key extraction from measuring physical quantities. *IEEE Transactions on Information Forensics and Security*, 11(8):1796–1806, 2016.

6 A. Goldsmith. *Wireless Communications*. Cambridge University Press, 2005.

7 IEEE INGR Security. Security and Privacy, International Network Generations Roadmap (INGR) - 2021 Edition, 2021.

8 S. Jaeckel, L. Raschkowski, K. Börner, and L. Thiele. QuaDRiGa: A 3-D multi-cell channel model with time evolution for enabling virtual field trials. *IEEE Transactions on Antennas and Propagation*, 62(6):3242–3256, 2014.

9 Z. Ji, Y. Zhang, Z. He, K. Lin, B. Li, P. L. Yeoh, and H. Yin. Vulnerabilities of physical layer secret key generation against environment reconstruction based attacks. *IEEE Wireless Communications Letters*, 9(5):693–697, 2020.

10 E. Jorswieck, S. Tomasin, and A. Sezgin. Broadcasting into the uncertainty: Authentication and confidentiality by physical-layer processing. *Proceedings of the IEEE*, 103(10):1702–1724, 2015. doi: 10.1109/JPROC.2015.2469602.

11 D. A. Levin, Y. Peres, and E. L. Wilmer. *Markov Chains and Mixing Times*. American Mathematical Society. ISBN 9780821886274. URL https://books.google.se/books?id=6Cg5Nq5sSv4C.

12 G. Li, A. Hu, J. Zhang, L. Peng, C. Sun, and D. Cao. High-agreement uncorrelated secret key generation based on principal component analysis preprocessing. *IEEE Transactions on Communications*, 66(7):3022–3034, 2018.

13 D. McAllester and K. Stratos. Formal limitations on the measurement of mutual information. In S. Chiappa and R. Calandra, editors, *Proceedings of the Twenty Third International Conference on Artificial Intelligence and Statistics*, volume 108 of *Proceedings of Machine Learning Research*, pages 875–884. PMLR, 26–28 Aug 2020.

14 M. Mitev, A. Chorti, M. J. Reed, and L. Musavian. Authenticated secret key generation in delay-constrained wireless systems. *EURASIP Journal on Wireless Communications and Networking*, 2020:122, 2020.

15 M. Mitev, M. Shekiba-Herfeh, A. Chorti, and M. Reed. Multi-factor physical layer security authentication in short blocklength communication. *arXiv preprint arXiv:2010.14457*, 2020.

16 W. Njima, M. Chafii, A. Chorti, R. M. Shubair, and H. V. Poor. Indoor localization using data augmentation via selective generative adversarial networks. *IEEE Access*, 9:98337–98347, 2021. doi: 10.1109/ACCESS.2021.3095546.

17 Y. Peng, P. Wang, W. Xiang, and Y. Li. Secret key generation based on estimated channel state information for TDD-OFDM systems over fading channels. *IEEE Transactions on Wireless Communications*, 16(8):5176–5186, 2017.

18 N. Pfister, B. Buhlmann, and J. P. Scholkopf. Kernel-based tests for joint independence. *Journal of the Royal Statistical Society, Series B: Statistical Methodology*, 80(1):5–31, 2018.

19 M. Shakiba-Herfeh and A. Chorti. Comparison of short blocklength Slepian-Wolf coding for key reconciliation. In *2021 IEEE Statistical Signal Processing Workshop (SSP)*, pages 111–115, 2021.

20 M. Shakiba-Herfeh, A. Chorti, and H. V. Poor. *Physical Layer Security: Authentication, Integrity, and Confidentiality*, pages 129–150. Springer International Publishing, 2021.

21 M. Shakiba-Herfeh, L. Luzzi, and A. Chorti. Finite blocklength secrecy analysis of Polar and Reed-Muller codes in BEC semi-deterministic wiretap channels. In *2021 IEEE Information Theory Workshop (ITW)*, pages 1–6, 2021.

22 M. Srinivasan, S. Skaperas, and A. Chorti. On the use of CSI for the generation of RF fingerprints and secret keys. *To appear in 25th International ITG Workshop on Smart Antennas*, 2021.

23 M. Srinivasan, S. Skaperas, M. S. Herfeh, and A. Chorti. Joint localization-based node authentication and secret key generation. In *ICC 2022-IEEE International Conference on Communications*, pages 32–37. IEEE, 2022.

24 C. Studer, S. Medjkouh, E. Gonultas, T. Goldstein, and O. Tirkkonen. Channel charting: Locating users within the radio environment using channel state information. *IEEE Access*, 6:47682–47698, 2018.

25 Z. Utkovski, M. Frey, P. Agostini, I. Bjelakovic, and S. Stanczak. Semantic security based secrecy maps for vehicular communications. In *WSA 2021; 25th International ITG Workshop on Smart Antennas*, pages 1–5, 2021.

26 A. Vardy, M. Bellare, and S. Tessaro. Semantic security for the wiretap channel. In *CRYPTO*. Springer, 2012.

27 P. Walther and T. Strufe. Inference attacks on physical layer channel state information. In *2020 IEEE 19th International Conference on Trust, Security and Privacy in Computing and Communications (TrustCom)*, pages 935–942, 2020. doi: 10.1109/TrustCom50675.2020.00125.

28 W. Xi, C. Qian, J. Han, K. Zhao, S. Zhong, X.-Y. Li, and J. Zhao. Instant and robust authentication and key agreement among mobile devices. In *Proceedings of the 2016 ACM SIGSAC Conference on Computer and Communications Security*, pages 616–627, 2016.

29 W. Xia, S. Rangan, M. Mezzavilla, A. Lozano, G. Geraci, V. Semkin, and G. Loianno. Millimeter wave channel modeling via generative neural networks. In *2020 IEEE Globecom Workshops (GC Wkshps*, pages 1–6, 2020. doi: 10.1109/GCWkshps50303.2020.9367420.

30 W. Yang, R. F. Schaefer, and H. V. Poor. Wiretap channels: Nonasymptotic fundamental limits. *IEEE Transactions on Information Theory*, 65(7):4069–4093, 2019.

31 C. Zenger, H. Vogt, J. Zimmer, A. Sezgin, and C. Paar. The passive eavesdropper affects my channel: Secret-key rates under real-world conditions. In *2016 IEEE Globecom Workshops*, pages 1–6. IEEE, 2016.

32 J. Zhang, G. Li, A. Marshall, A. Hu, and L. Hanzo. A new frontier for IoT security emerging from three decades of key generation relying on wireless channels. *IEEE Access*, 8:138406–138446, 2020.

10

Joint Secure Communication and Sensing in 6G Networks

Miroslav Mitev[1], Amitha Mayya[1], and Arsenia Chorti[1,2]

[1] *Wireless Connectivity Group, Barkhausen Institut, Dresden, Germany*
[2] *ETIS UMR8051, CY University, ENSEA, CNRS, Cergy, France*

10.1 Introduction

Historically, the evolution of communication and radar technologies has followed separate paths, each with its own distinct resources and design parameters. Despite their independent development, both services share the fundamental principle of transmitting and receiving electromagnetic waves. Recognizing this commonality, there is a growing interest in a codesigned approach as this can offer numerous advantages and enable a range of 6G applications [30].

One way to combine both services on a single terminal is to utilize single frequency band but sending distinct waveforms. A significant challenge for this approach is to identify an interference cancelation scheme able to successfully adapt to various situations [32]. Another strategy involves employing a unified signal for both communication and sensing, albeit requiring a coordinated waveform design. In this scenario, the waveform design is crucial: it can be communication centric (serving applications that demand high throughput and minimal sensing precision) or radar centric (serving applications that require high precision and low throughput).

Typically, signals that are designed for communications are more complex and exhibit modulation in multiple domains, such as time, spatial, and frequency. This complexity contributes to higher costs, as it necessitates advanced transmitter/receiver modules [32]. Conversely, radar signals have a simpler structure, e.g., short or continuous pulses. These signals are tailored toward proving sensing precision, as opposed to reliable data exchange, hence, the processing chain can be simplified. Although sensing waveforms do not offer the same data rates, they can still provide enough throughput for specific radar-centric applications [4, 11].

Physical-Layer Security for 6G, First Edition. Edited by Parthajit Mohapatra, Nikolaos Pappas, Arsenia Chorti, and Stefano Tomasin.

While communication-centric systems have well-established security mechanisms, an inevitable question arises: *How can we ensure the security of exchanged information in radar-centric applications and keep the complexity to minimum?* In fact, existing security solutions cannot meet the expected performance. For instance, Teniou and Bensaber [29] highlighted that verifying a digital signature over a processor with a clock speed of 400 MHz demands nearly 20 ms which is well above the low-latency expectations. Additionally, standard public-key encryption protocols fall short in meeting low-energy demands which makes it difficult to scale for extensive deployment [31]. Moreover, the development of quantum computers poses a threat to current security solutions, unless key sizes are significantly increased [22]. Consequently, there is a need for new lightweight security alternatives.

Building on the aforementioned points, we conduct a performance assessment of the physical-layer security (PLS)-based secret key generation (SKG). In fact, SKG is a quantum-secure secret key exchange scheme that is currently under consideration for 6G [16]. The SKG protocol enables a pair of devices to derive a shared secret [3, 17]. The SKG steps rely on the reciprocal and random behavior of the wireless medium. Few of the advantages of PLS solutions are the independence from complex infrastructure (unlike public key encryption [PKE]), can be applied using arbitrary physical-layer (PHY) waveform, and matched receivers are not necessity [33, 34].

In this current work, we evaluate the performance of SKG using a signal that is widely employed in radar systems, i.e., chirp. Chirps typically occupy large bandwidth, which helps in extracting the randomness present in the fading coefficients. In this work, we obtain channel measurements, by employing a filterbank-based method. The method was initially proposed in [33] and subsequently analytically/numerically examined in [19, 20, 34]. Here, we take a step further and assess the approach's performance in a set of real-life setup experiments. The experiments are conducted indoor under multiple scenarios. To determine the final key rate we implement the complete SKG protocol. Advantage distillation is carried out using a filterbank-based approach. Subsequently, to perform information reconciliation, Polar coders are used. At last, the rate of key generation is evaluated based on conditional min-entropy estimation, where eavesdropper knowledge is taken into account.

This chapter is structured as follows: Section 10.2 provides the motivation for our research, Section 10.3 offers an outline of our system model, Section 10.4 delves into the steps within the SKG scheme, Section 10.5 introduces our experimental setup, Section 10.6 encapsulates key findings, and finally, provides concluding remarks.

10.2 Related Work and Motivation

Sensing technologies are widely used across various sectors such as smart factories, car manufacturing, and navigation systems [15]. It is evident that adopting a codesigned approach for communication and sensing can offer substantial benefits in these applications. Specifically, vehicles have the potential to continuously sense their surroundings, while simultaneously sharing this information with other vehicles on the road [8]. Likewise, autonomous machinery can engage in communication to enhance production processes while remaining aware of nearby individuals. It is important to note that in many instances where communication and sensing work jointly, the device will exchange sensitive information, such as location and identity. Hence, safeguarding the transmission against potential information leakage and ensuring confidentiality is of paramount importance.

Related to this topic, the 3GPP highlighted in [2] that security remains a weak point for communication systems, particularly for Internet of Things (IoT) networks. Traditional cryptographic solutions make use of computationally heavy operations (modulo arithmetics), rendering them impractical for devices with power constraints. This challenge is also a subject of research by the National Institute of Standards and Technology (NIST). A 10-year (2013–2023) investigation on the topic of cryptography for constrained environments has been recently finalized and a single winner was announced (the family Ascon ciphers) [23, 25]. While these ciphers can protect information exchange among IoT devices, it is still not clear how to obtain keys that can run these mechanisms. All approved key distribution mechanisms make use of longer keys [24], which introduces additional latency and complexity.

Consequently, the use of SKG emerges as a post quantum security mechanism, offering key distribution with minimal overhead. With its reciprocity and stochastic behavior, the wireless channel serves as a valid source of randomness [18, 21, 31]. The SKG allows nodes to harvest entropy directly from the wireless medium, eliminating the need for complex modulo arithmetic operations in large fields. However, it is crucial to acknowledge the fact that the rates at which keys are generated are determined by the channel behavior. Specifically, SKG rates are considerably lower in static channels compared to fast changing ones.

The exploration of security for joint communication and sensing applications has been scarce, with only a handful of studies addressing this aspect [8, 10, 28, 35]. In [10], an information-theoretic viewpoint is offered regarding information leakage. The work considers a transmitter that aims to conduct channel estimation and simultaneously perform reliable information exchange with another device. The study in [28] introduces a multi-carrier chirp approach to integrate communication and sensing. The objective is to optimize the secrecy rate considering channel with multiple eavesdroppers. The study provides numerical results and shows that eavesdropping nodes can be identified through sensing. Lastly, the works presented in [8] and [35] evaluate SKG using chirps through simulations. Specifically, Dwivedi et al. [8] focus on assessing of the bit mismatch while Zoli et al. [35] evaluate the similarity of the bits generated at eavesdropping locations.

The aforementioned papers present encouraging findings, affirming the potential benefits of employing PLS for future networks. Building upon this, we further investigate the SKG protocol considering a filterbank-based approach. This approach proves particularly relevant for scenarios as those outlined in the 3GPP report [1] (e.g., collective situational awareness). Given their wide deployment in sensing applications, wideband chirp signals emerge as a viable waveform candidate for scenarios as in the 3GPP report. Employing channel estimation using chirps the measurements, which are typically used mainly for equalization purposes, can be reused for SKG (i.e., without introducing additional overhead). The core concept of employing filterbank is to divide the received bandwidth into multiple parallel sub-bands (through filtering) and measure the power at the output of each filter. Using this method as opposed to SKG schemes that rely on a single power measurement per transmission (e.g., received signal strength indicator [RSSI]), the filterbank approach provides multiple measurements that can better capture channel entropy.

10.3 System Model

The system model in this study involves two authorized users, Alice and Bob, and an eavesdropper, Eve. Each transceiver has one antenna that is used to transmit and receive in a time division duplex manner. The eavesdropper is strategically located in close proximity to one of the legitimate users (Bob) and listens to the communication. Alice and Bob exchange series of chirp signals to obtain channel measurements which are then converted into secret key. Their goal is to leave Eve oblivious of the generated sequence.

The transmitted signal, C, has bandwidth B. Given that we represent the received signals at Alice, Bob, and Eve by:

$$Y_{\mathrm{A}} = CH + Z_{\mathrm{A}}, \tag{10.1}$$

$$Y_{\mathrm{B}} = CH + Z_{\mathrm{B}}, \tag{10.2}$$

$$Y_{\mathrm{E}} = CH_{\mathrm{E}} + Z_{\mathrm{E}}, \tag{10.3}$$

respectively. As mentioned above, Eve is located close to Bob; therefore, she is interested only at measurements from the channel Alice–Eve, given here as H_{E}.

The channel response between Alice and Bob is given by H. Both channels may have high or low correlation factor, which is determined by the location of the eavesdropping device. The noise variables are given by Z_A, Z_B, and Z_E, for at Alice, Bob, and Eve, respectively. The overall distribution of the channel and received measurements will differentiate depending on multiple factors, i.e., the number of resolvable paths, their delay, their power, and the presence of line-of-sight. This has been evaluated numerically in our previous work [20]. To extract the randomness present in the channel, all parties execute the SKG. The steps of the protocol are detailed in Section 10.4.

10.4 Secret Key Generation Protocol

The SKG protocol consists of three fundamental steps: advantage distillation, information reconciliation, and privacy amplification. Figure 10.1 depicts an outline of the protocol as employed in this work. Alice and Bob utilize a quantizer to convert their channel estimation into bit vectors $\mathbf{d}_A$ and $\mathbf{d}_B$, respectively. Discrepancies in bit vectors are corrected by one-way exchange of helper data on a public channel $\mathbf{g}_A$ (e.g., Alice to Bob). As a last step, both users employ privacy amplification on the reconciled information to generate secret keys with maximum entropy. A comprehensive explanation of the specific protocol implementation in this work is provided below.

10.4.1 Advantage Distillation

To derive randomness from (10.1) to (10.3), each participant processes its received signal using a filterbank comprising N filters. A series of raised cosine filters

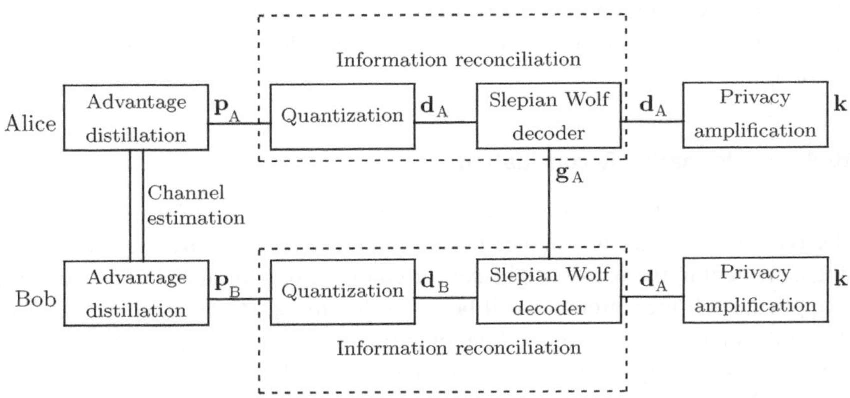

Figure 10.1 Physical-layer-based SKG between Alice and Bob.

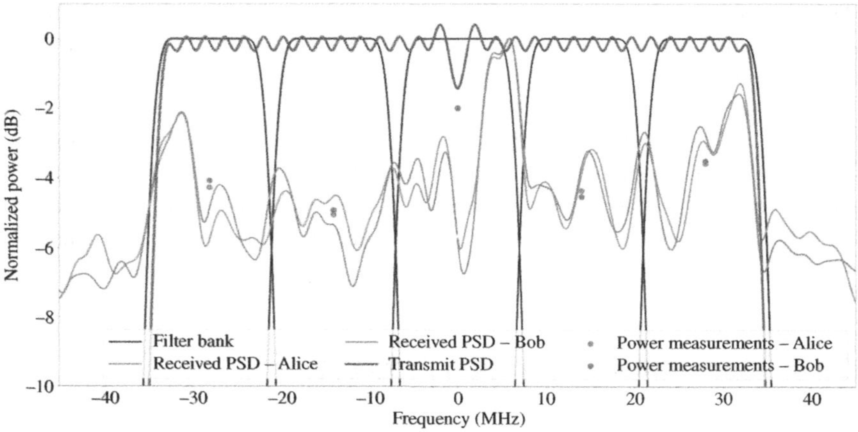

Figure 10.2 A filterbank comprising five filters. The transmit signal has a bandwidth of 70 MHz.

are employed in this work. The filters have identical bandwidth of B/N and are spread uniformly over the total bandwidth of B MHz. A general representation of the method is given in Figure 10.2. Each party measures the power at the band of each filter, which is in fact the source of randomness in the considered scheme. The example in the figure shows power spectrum densities (PSDs) of signals at transmitter and receiver with $B = 70$ MHz. Additionally, the figure shows the responses of the filters ($N = 5$) and the measured power at each frequency band. Alice's measurements are denoted as $\mathbf{p}_\mathrm{A} = \left[\hat{p}_{A,1} \cdots \hat{p}_{A,N}\right]$. Similar expressions can be used for the measurements at both Bob and Eve.

In our earlier work [20], the probability distribution of one measurement, under Rayleigh channel conditions, has been derived. It is a Gamma variable composed of M Gamma variables with M being the number of resolvable multipath components. Interested readers are referred to [20] for the derivation of the final power distribution.

10.4.2 Information Reconciliation

As noted earlier, the measurements are quantized and transformed into bits. The type of quantizer is a choice depending on the system architecture, i.e., a single element in the vector can be converted to one or several bits. If gray coding is applied, each measurement will be converted to $\log_2(Q)$ bits, where Q is the amount of quantization regions. After quantizing Alice, Bob, and Eve have binary sequences $\mathbf{d}_\mathrm{A}, \mathbf{d}_\mathrm{B}, \mathbf{d}_\mathrm{E} \in \{0, 1\}^{N \log_2(Q)}$. Depending on the channel conditions, the generated sequences at the legitimate users may not be identical, $\mathbf{d}_\mathrm{A} \neq \mathbf{d}_\mathrm{B}$.

To arrive at the same sequence (e.g., $\mathbf{d}_A$), Alice transmits helper data, $\mathbf{g}_A$. The helper data can be generated through different coding techniques, i.e., $\mathbf{g}_A$ is a decoding syndrome, which when used together with $\mathbf{d}_B$ can give $\mathbf{d}_A$. It is important to note that, $\mathbf{g}_A$ is transmitted over a public channel, hence, it is available at Eve, too. Therefore, to minimize the leakage the SKG protocol has one final step described below.

10.4.3 Privacy Amplification

Keys are fundamental part of cryptography and their quality is vital for security. In this regard, the shared key $\mathbf{k} \in \mathcal{K}$ at the output of the protocol must comply with the requirement of randomness, i.e., [5] $|\mathbf{k}| \leq H_\infty(\mathbf{d}_A)$, with

$$H_\infty(\mathbf{d}_A) = -\log_2 \max_{\mathbf{d}_A \in \mathcal{D}_A} p(\mathbf{d}_A), \tag{10.4}$$

being min-entropy [9]. We note that the inequality above does assume eavesdropper-free communication. In this sense, a stricter metric is needed for our work. To assess the generated keys in the presence of an eavesdropper, we use conditional min-entropy which is defined as [26]:

$$H_\infty(\mathbf{d}_A | \mathbf{d}_E, \mathbf{d}_A) = -\log_2 \max_{\mathbf{d}_A \in \mathcal{D}_A, \mathbf{d}_E \in \mathcal{D}_E, \mathbf{g}_A \in \mathcal{G}_A} p(\mathbf{d}_A | \mathbf{d}_E, \mathbf{g}_A). \tag{10.5}$$

In the above, $\mathbf{d}_A$ is conditioned on Eve's initial measurements of the leaked helper data. It is clear that the subtraction of (10.5) from (10.4) gives the leakage [27], i.e., $H_\infty(\mathbf{d}_A) - H_\infty(\mathbf{d}_A | \mathbf{d}_E, \mathbf{g}_A)$. Using conditional min-entropy, as opposed to min-entropy, the size of the key is now limited as $|\mathbf{k}| \leq H_\infty(\mathbf{d}_A | \mathbf{d}_E, \mathbf{g}_A)$.

Estimation of the quantities above is a challenging task, as it requires closed-form evaluation of the conditional distribution. Such derivations are beyond the scope of the current work. Here, we employ a numerical estimator, namely the fast black-box leakage estimator (FBLEAU) [6]. It uses the frequentist and nearest neighbor estimates to determine the conditional min-entropy [7]. Having this estimate Alice and Bob compress their sequence to the desired size using collision-resistant one-way compression (such as hashing).

10.5 Measurement Setup

Following the system model detailed in Section 10.3, the SKG protocol is performed using three USRPs (Alice, Bob, and Eve). This is illustrated in Figure 10.3. The USRPs used in this work are the 2974 units from National Instruments. Alice and Bob are at a distance of 4 m, with Alice positioned at a height of 2.3 m, and Bob and Eve situated are 0.85 m from the ground. During the campaign, Alice

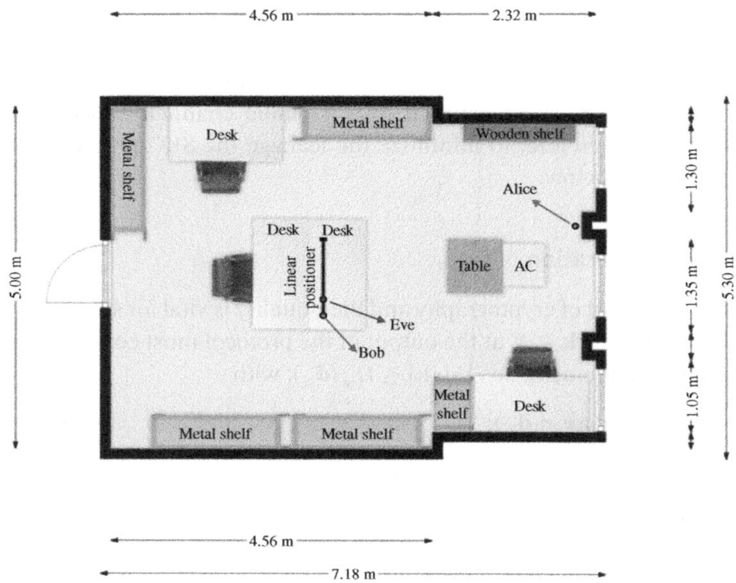

Figure 10.3 Diagram depicting the measurement setup (made using floorplaner.com). The separation between legitimate users is 4 m, with Alice positioned at a height of 2.3 m. Bob and Eve are situated at a height of 0.85 m. During the measurements, the eavesdropping node is placed on a linear positioner. The positioner incrementally adjusts its position relative to Bob's by one wavelength steps, ranging from 1λ to 10λ.

and Bob remain at static positions. On the other hand, 10 distinct positions are considered for Eve. This helps to better assess possible leakages. As it can be seen in the floorplan, Eve is placed on a positioner adjacent to Bob. The positioner moves Eve in one wavelength steps, 1λ to 10λ. All exchanged signals have a central frequency of 3.75 GHz, which results in $\lambda \approx 8$ cm.

During the exchange between the legitimate users Eve consistently operates in reception mode. For the setup considered here, each exchanged chirp signal has duration of 17.1875 ms, and 70 MHz bandwidth. Throughout the measurements, each of the legitimate users, Alice and Bob, transmits 10^5 chirp signals per single position of Eve. This procedure is reiterated across four distinct scenarios.

10.5.1 Scenarios

The following scenarios are considered for the experimental campaign: non-line-of-sight (NLoS) static, NLoS dynamic, LoS static, and LoS dynamic. Dynamic scenarios involve the movement of people and objects, whereas static measurements are conducted during the night with no movement occurring.

Figure 10.4 LoS measurement setup.

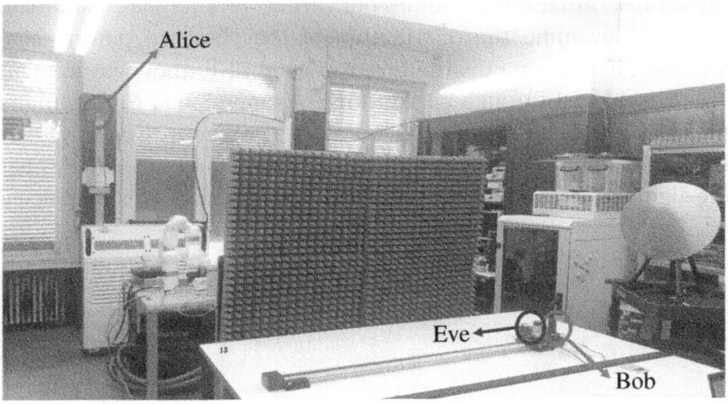

Figure 10.5 NLoS measurement setup.

The LoS scenario is depicted in Figure 10.4. The picture shows also a metal plate that has been moved in dynamic environments. To achieve NLoS, we have placed absorbers to obstruct the direct path between the transceivers (see Figure 10.5).

10.5.2 Implementation of the SKG Protocol

After a signal is received, each user performs SKG as detailed in Section 10.4. Initially, the received signals Y_A, Y_B, and Y_E undergo filtering. The protocol is evaluated considering varying numbers of filters, $N = 8, 16, 32,$ and 64. Power estimates after filtering are utilized for the SKG. The estimates are transformed into binary sequences, $\mathbf{d}_A$, $\mathbf{d}_B$, and $\mathbf{d}_E$ using a uniform quantization. Two different numbers for the quantization regions are considered, $Q = \{4, 16\}$. The binary vectors undergo reconciliation using Slepian–Wolf decoders. Here, the decoders are implemented with Polar codes (interested readers on the topic of Slepian–Wolf decoding with Polar codes are referred to [12]). Using her decoder, Alice creates helper data (e.g., a syndrome) $\mathbf{g}_A$ which is then sent to Bob. Due to the use of public channel Eve obtains the syndrome, too. Both utilize the syndrome along with their observations in an attempt to replicate Alice's bit sequence. Finally, leveraging conditional min-entropy $H\infty(\mathbf{d}_A|\mathbf{d}_E, \mathbf{g}_A)$, the legitimate users obtain a shared secret key $\mathbf{k}$. Section 10.6 provides a comprehensive evaluation of the SKG protocol, considering different N, Q, and coding rates.

10.6 Results and Discussion

Here, we provide an in-depth examination of the SKG performance under different system setups. Initially, the findings our earlier work [20] are examined. More specifically, we examine the distribution of the observed power measurements and check if a gamma distribution is a good fit to the measurement histograms. This is achieved through curve fitting. As mentioned earlier, filters are evenly spread across bandwidth of 70 MHz. In this sense, having higher amount of filters leads to narrower filters. As this alters the amount samples per subbands in the digital domain, the power distribution is affected. This is depicted in Figure 10.6.

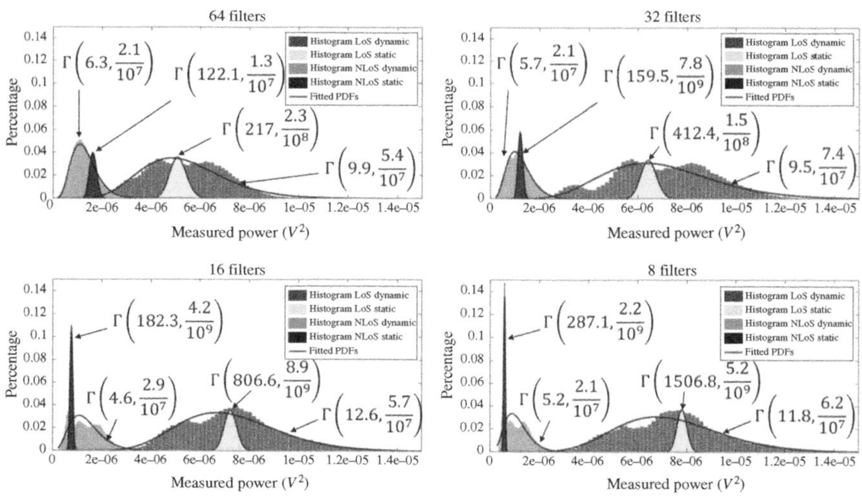

Figure 10.6 Histograms of power measurements captured across various environments each fitted to distinct Gamma distributions. The histograms compare different number of filters, $N = 8, 16, 32,$ and 64.

It can be seen that under static channel conditions, the histograms closely conform to our analytical findings, fitting well with the Gamma distribution. Another notable observation, also noted in [20], the lesser the filters, the smaller the variance of the observations. This is a consequence of the fact that, when less filters are used, they become wider which leads to higher number of samples in the subbands. This helps to better mitigate the impact of noise. This trend is evident in static scenarios.

Regarding gamma fitting, it can be seen that in dynamic scenarios, curves and histograms align only at the tails. We see that variations in dynamic scenarios are more pronounced and create several peaks. Differently from above, more variations are present for wider filters. This effect is a consequence of the fact that in dynamic conditions, the noise is only a small part of the variations; instead, multipath introduces additional degrees of freedom, leading to a wider spread of measurements with higher bandwidth. Overall, the figure shows that conventional statistical models are insufficient to fully characterize real-world scenarios, which necessitates further investigation.

After measurements are collected, the subsequent step is quantization. Depending on N and Q, the generated binary vectors can vary in length. Here, the average bit mismatch probability is assessed. This is shown in Figure 10.7. As anticipated, increasing the number of generated bits per channel leads to higher mismatch. This occurs because the extracted binary vectors become more susceptible to non-reciprocal noise variations. Furthermore, it is evident that in NLoS, more errors

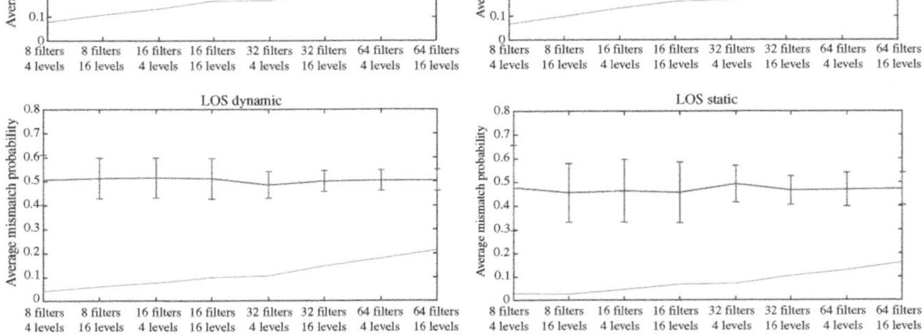

Figure 10.7 Bit mismatch probability at Bob and Eve after quantization. Different number of filters and quantization levels are compared.

are present. The difference, as compared to LoS scenarios, is due to the fact that the received power is typically less when LoS is absent.

An intriguing observation pertains to Eve's mismatch. Vertical lines denote her standard deviation for all possible positions on the linear positioner. We have seen in the experimental data that Eve's mismatch does not improve if she is closer to Bob; instead, she may have a more correlated sequence at positions further from Bob. This phenomenon could be attributed to the unique addition of multipath components (MPCs) at specific locations. This aligns with the theoretical finding in [13] where it is noted that decorrelation happens at distances approximately half a wavelength.

It is demonstrated that, across scenarios, the mean value at the eavesdropping node remains constant and is not affected by quantization or number of filters, hovering around 50%. However, it is apparent that the deviation decreases when less filters and levels are used, i.e., if fewer bits are generated per frame gives her higher probability to observe a similar to the legitimate bit sequence.

Next step is to perform information reconciliation. In the current work, this is achieved using Polar codes. Figures 10.8 and 10.9 show the frame error rate (FER), i.e., unsuccessful reconciliation, at Bob and Eve, respectively.

Figure 10.8 presents the FER observed at Alice and Bob considering different number of quantization levels and filters. Notably, changing Q has a more pronounced effect on the error rate if compared to changing N. Specifically, when Alice and Bob utilize 32 filters and 4 levels (64 bits), the FER is lower compared to the scenario with 16 filters and 16 levels (also 64 bits).

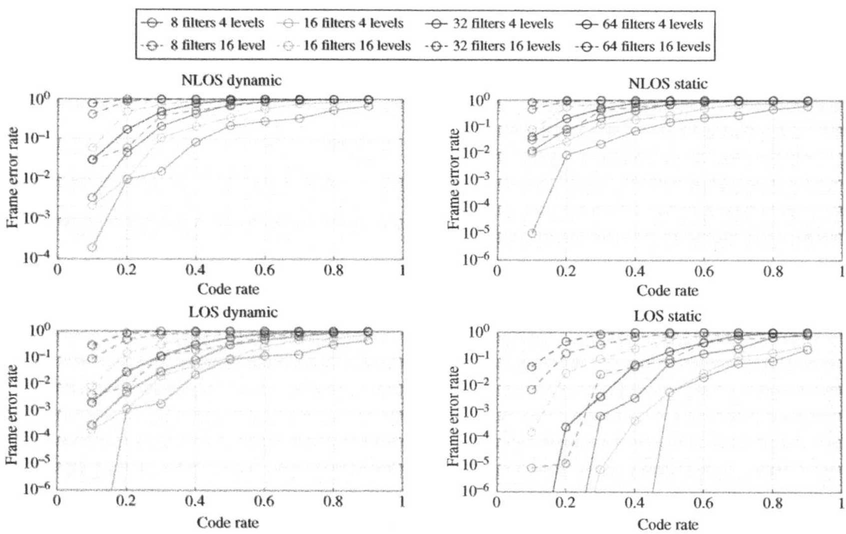

Figure 10.8 Rate of unsuccessful reconciliation at Alice and Bob.

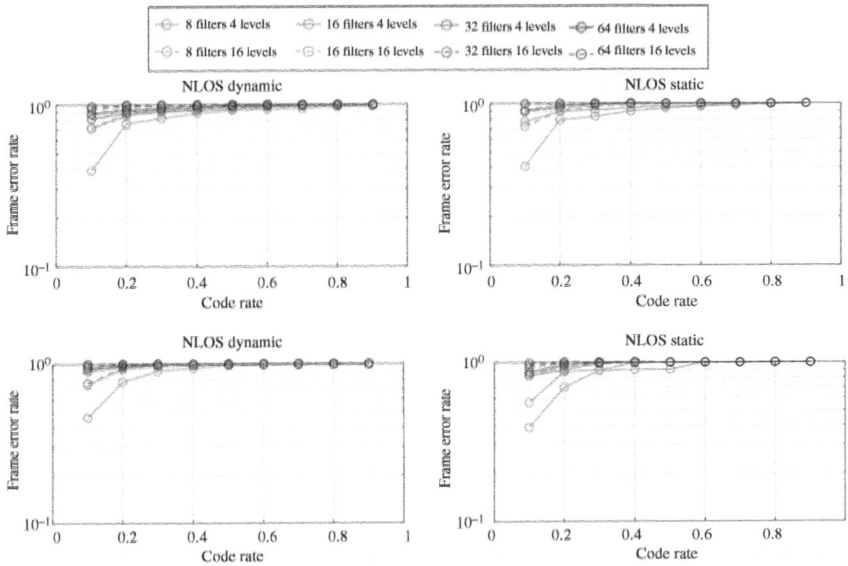

Figure 10.9 Rate of unsuccessful reconciliation at Eve.

Similar to observations in Figure 10.7, the performance of legitimate is worse in NLoS. This outcome is unsurprising, as the absence of LoS naturally reduces the signal-to-noise ratio (SNR) between communicating parties. All figures show that code rate can also strongly influence the success of reconciliation. Lower code rates result in negligible FER.

However, lowering the code rate requires larger syndrome sequence, $\mathbf{g}_A$, hence, more information is leaked. This can be seen in Figure 10.9 where the error rate at Eve is given considering she has access to the syndrome. It is clear that at low rates, her chance to arrive at the correct sequence after decoding increases. Similar to the legitimate users, Eve's performance improves with N and not Q. Hence, more filters can provide higher reconciliation rates for Alice and Bob but at the same time leak more information to Eve. Potential leakages must be accounted for in the last step of the protocol, i.e., privacy amplification.

We employ the FBLEAU numerical estimator [6] to calculate the conditional min-entropy $H_\infty(\mathbf{d}_A|\mathbf{d}_E, \mathbf{g}_A)$. The estimator takes the three binary sequences (Alice's sequence, Eve's sequence, and the syndrome) as inputs and yields an estimate of $H_\infty(\mathbf{d}_A|\mathbf{d}_E, \mathbf{g}_A)$. The results are given in Figure 10.10, where conditional min-entropy is estimated for different system parameters. Due to computational limitations, the results are restricted to a maximum of 32 filters and 4 quantization regions. The limitation is a consequence of the curse of dimensionality [14], (an ongoing area of research) meaning computational convergence time increases exponentially with additional dimensions.

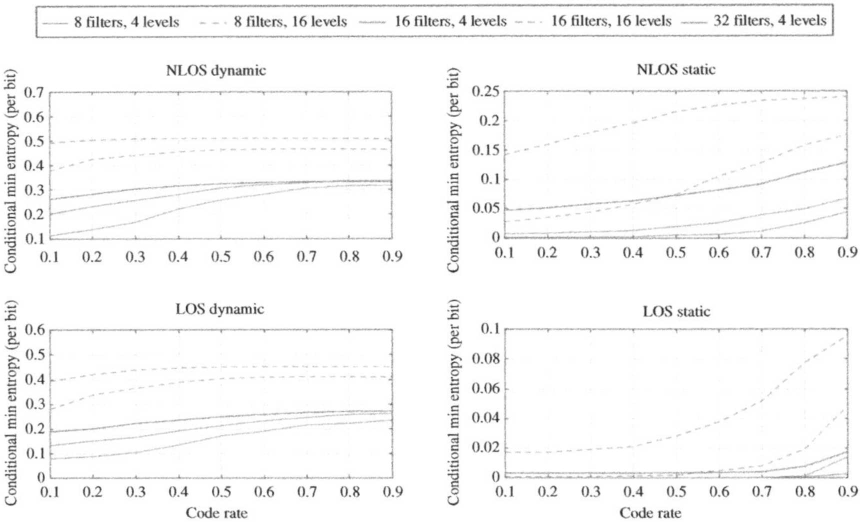

Figure 10.10 Evaluation of the conditional min-entropy for different numbers of quantization regions, filters, and different code rates.

The figure reveals that higher code rate gives rise in the conditional min-entropy. This is directly associated with the fact that higher rates require shorter syndromes which consequently reduces the information leakage to Eve. Moreover, since most of the randomness in the generated bits comes from channel variability, we note that the conditional min-entropy tends to be lower in static scenarios. This shows that movement can contribute to the randomness of the channel and to the quality of the generated keys. Interestingly, as opposed to what it was observed during reconciliation, increasing the number of quantization regions provides higher entropy as compared to increasing the number of filters. Nevertheless, for the successful execution of the complete SKG protocol, all steps highlighted in Section 10.4 should be taken into account.

Following from the above, the average SKG rate R_{SKG} (per channel use) can be defined using the parameters: (i) amount of bits generated per channel use L (it depends on the choice of filterbank and quantization); (ii) frame error rate after decoding at the reconciliation stage; (iii) conditional min-entropy $H_{\infty}(\mathbf{d}_{\text{A}}|\mathbf{d}_{\text{E}}, \mathbf{g}_{\text{A}})$. SKG rate can then be expressed as:

$$R_{\text{SKG}} = L \times (1 - \text{FER}) \times H_{\infty}(\mathbf{d}_{\text{A}}|\mathbf{d}_{\text{E}}, \mathbf{g}_{\text{A}}). \tag{10.6}$$

This quantity is evaluated and the results are shown in Figure 10.11.

The pattern in the figure is evident. To boost the SKG rate, more bits must be generated per channel use, preferably by having more quantization regions rather

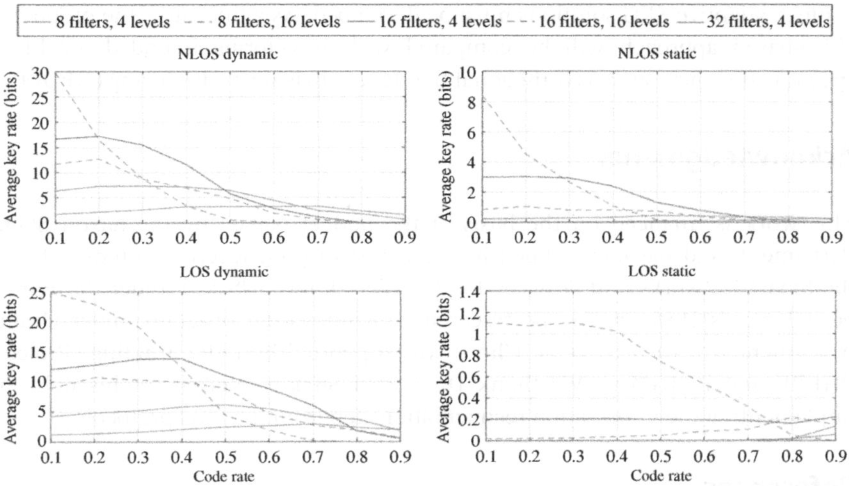

Figure 10.11 Average SKG rate for different numbers of quantization regions, filters, and different code rates.

than more filters. At the same time, lower code gives better performance. Given the system model employed here, we also calculate the rates in bits per second per Hertz (b/s/Hz). This is given in Table 10.1. Table 10.1 represents the maximum for each scenario from the curves in Figure 10.11. Note that a single exchange between the legitimate users takes around 35 ms. The values can also be converted to bit per second, e.g., in NLoS dynamic for the considered bandwidth, the key rate is 848, 220 bits per second.

In conclusion, this chapter has demonstrated that the amount of random bits that can be extracted for SKG is dependent not only on the channel behaviour but also on the parameters chosen at the system level. Further, it was shown that the chirp waveform allows to harvest randomness across wide bandwidth. With the advancements in wireless technology and integration of sensing at the edge of networks (which can offer environmental context), SKG becomes a practical technique for key distribution. In future research efforts, the authors intend to delve deeper into the findings presented in this chapter and explore multiple antenna scenarios. Furthermore, depending on the environmental

Table 10.1 SKG rates for different scenarios.

	NLoS dynamic	NLoS static	LoS dynamic	LoS static
Rate (b/s/Hz)	0.0121	0.0034	0.0102	0.0005

context, adaptive SKG solutions must be developed. As a final step, the proposed chirp-based approach will be compared with waveforms intended for high spectral efficiency, such as orthogonal frequency-division multiplexing (OFDM).

Acknowledgments

This work is financed on the basis of the budget passed by the Saxon State Parliament, and partially funded by Hexa-X-II project, receiving funding from the Smart Networks and Services Joint Undertaking (SNS JU) under the European Union's Horizon Europe research and innovation program under Grant Agreement No. 101095759. A. Chorti was supported by INEX funding (Project PHEBE and ENIGMA). We thank Dr. Mahdi Shakiba-Herfeh for his help in implementing the information reconciliation step from the SKG protocol.

References

1 3GPP TR 22.886 V2.0.0. Study on enhancement of 3GPP support for 5G V2X services (Release 16). 3rd Generation Partnership Project; Technical Specification Group Services and System Aspects, 2018.
2 3GPP TR 33.825 V2.0.0. Study on the Security for 5G URLLC (Release 16). 3rd Generation Partnership Project; Technical Specification Group Services and System Aspects, 2019.
3 R. Ahlswede and I. Csiszar. Common randomness in information theory and cryptography. I. Secret sharing. *IEEE Transactions on Information Theory*, 39(4):1121–1132, 1993.
4 A. N. Barreto, T. M. Pham, S. George, P. Sen, and G. Fettweis. Analysis of a chirp-based waveform for joint communications and radar sensing (JC&S) using non-linear components. In *15th European Conference on Antennas and Propagation (EuCAP)*, 2021.
5 D. Brown. Formally assessing cryptographic entropy. *IACR Cryptology ePrint Archive*, 2011.
6 G. Cherubin, K. Chatzikokolakis, and C. Palamidessi. F-BLEAU: Fast black-box leakage estimation. In *2019 IEEE Symposium on Security and Privacy (SP)*, 2019.
7 T. Chothia, Y. Kawamoto, and C. Novakovic. A tool for estimating information leakage. In *Computer Aided Verification*. Springer-Verlag, Berlin Heidelberg, 2013.
8 S. Dwivedi, M. Zoli, A. N. Barreto, P. Sen, and G. Fettweis. Secure joint communications and sensing using chirp modulation. In *6G Wireless Summit*, 2020.

9 B. Espinoza and G. Smith. Min-entropy as a resource. *Information and Computation*, 226:57–75, 2013. Special Issue: Information Security as a Resource.

10 O. Günlü, M. Bloch, R. F. Schaefer, and A. Yener. Secure joint communication and sensing. In *IEEE International Symposium on Information Theory (ISIT)*, 2022.

11 A. Hassanien, M. Amin, E. Aboutanios, and B. Himed. Dual-function radar communication systems: A solution to the spectrum congestion problem. *IEEE Signal Processing Magazine*, 36(5):115–126, 2019.

12 M. S. Herfeh and A. Chorti. Comparison of short blocklength Slepian-Wolf coding for key reconciliation. In *IEEE Statistical Signal Processing Workshop (SSP)*, 2021.

13 W. Jakes. *Microwave Mobile Communications*. Wiley-IEEE Press, 1994.

14 N. Kouiroukidis and G. Evangelidis. The effects of dimensionality curse in high dimensional kNN search. In *15th Panhellenic Conference on Informatics*, 2011.

15 F. Liu, C. Masouros, A. Petropulu, H. Griffiths, and L. Hanzo. Joint radar and communication design: Applications, state-of-the-art, and the road ahead. *IEEE Transactions on Communications*, 68(6):3834–3862, 2020.

16 L. Matti and L. Kari. Key drivers and research challenges for 6G ubiquitous wireless intelligence. University of Oulu, Oct 2019.

17 U. M. Maurer. Secret key agreement by public discussion from common information. *IEEE Transactions on Information Theory*, 39(3):733–742, 1993.

18 M. Mitev, A. Chorti, M. Reed, and L. Musavian. Authenticated secret key generation in delay-constrained wireless systems. *EURASIP Journal on Wireless Communications and Networking*, 2020:122, 2020.

19 M. Mitev, A. N. Barreto, T. M. Pham, M. Matthé, and G. Fettweis. Filterbank secret key generation rates in multipath channels. In *IEEE Global Communications Conference (GLOBECOM)*, 2022.

20 M. Mitev, A. N. Barreto, T. M. Pham, and G. Fettweis. Secret key generation rates over frequency selective channels. In *IEEE Conference on Vehicular Technology (VTC)*, 2022.

21 M. Mitev, A. Chorti, H. V. Poor, and G. Fettweis. What physical layer security can do for 6G security. *IEEE Open Journal of Vehicular Technology*, 4:375–388, 2023.

22 M. Mosca. Cybersecurity in an era with quantum computers: Will we be ready? *IEEE Security and Privacy*, 16(5):38–41, 2018.

23 NIST. Lightweight Cryptography Project, 2013. URL https://csrc.nist.gov/projects/lightweight-cryptography.

24 NIST. Post-Quantum Cryptography - Selected Algorithms, 2022. URL https://csrc.nist.gov/projects/post-quantum-cryptography/selected-algorithms-2022.

25 NIST. Selects ‘Lightweight Cryptography’ Algorithms to Protect Small Devices, 2023. URL https://www.nist.gov/news-events/news/2023/02/nist-selects-lightweight-cryptography-algorithms-protect-small-devices.

26 L. Reyzin. Some notions of entropy for cryptography. In *Information Theoretic Security*. Springer-Verlag, Berlin Heidelberg, 2011.

27 G. Smith. On the foundations of quantitative information flow. In *Foundations of Software Science and Computational Structures*. Springer-Verlag, Berlin Heidelberg, 2009.

28 N. Su, F. Liu, and C. Masouros. Sensing-assisted eavesdropper estimation: An ISAC breakthrough in physical layer security. *IEEE Transactions on Wireless Communications*, 23(4):3162–3174, 2022.

29 A. Teniou and B. Bensaber. Efficient and dynamic elliptic curve qu-vanstone implicit certificates distribution scheme for vehicular cloud networks. *Security and Privacy*, 1(1):e11, 2018.

30 T. Wild, V. Braun, and H. Viswanathan. Joint design of communication and sensing for beyond 5G and 6G systems. *IEEE Access*, 9:30845–30857, 2021.

31 J. Zhang, G. Li, A. Marshall, A. Hu, and Lajos Hanzo. A new frontier for IoT security emerging from three decades of key generation relying on wireless channels. *IEEE Access*, 8:138406–138446, 2020.

32 J. A. Zhang, Md. L. Rahman, K. Wu, X. Huang, Y. J. Guo, S. Chen, and J. Yuan. Enabling joint communication and radar sensing in mobile networks – A survey. *IEEE Communications Surveys & Tutorials*, 24(1):306–345, 2022.

33 M. Zoli, A. Barreto, S. Köpsell, P. Sen, and G. Fettweis. Physical-layer-security box: A concept for time-frequency channel-reciprocity key generation. *EURASIP Journal on Wireless Communications and Networking*, 2020:114, 2020.

34 M. Zoli, M. Mitev, A. Barreto, and G. Fettweis. Estimation of the secret key rate in wideband wireless physical-layer-security. In *ISWCS*, 2021.

35 M. Zoli, A. Barreto, and G. Fettweis. Investigating the eavesdropper attack in physical layer security wireless key generation: A simulation case study. In *IEEE Vehicular Technology Conference (VTC-Spring)*, 2021.

Part IV

Applications

11

Physical-Layer Authentication for 6G Systems

Stefano Tomasin[1], He Fang[2], and Xianbin Wang[3]

[1] *Department of Information Engineering, University of Padova, Padova, Italy*
[2] *School of Electronic and Information Engineering, Soochow University, Suzhou, China*
[3] *Department of Electrical and Computer Engineering, Western University, London, Ontario, Canada*

11.1 Authentication by Physical Parameters

The purpose of authentication is to ensure that a received *message* is truly coming from the declared sender rather than from an impersonating attacker. In a typical notation of security, agent Alice is the legitimate sender, agent Bob is the receiver, and agent Eve is the attacker. Authentication is the basis of secure communication, network access, and trusted cooperation in sixth-generation (6G) networks. To be more specific, organizations can keep their networks secure by permitting only authenticated users or processes to gain access to their protected resources, including computer systems, networks, databases, websites, and other network-based applications or services [12]. Moreover, only those uploaded data from authenticated devices/nodes can be trusted for further utilizations [1].

Physical-layer authentication (PLA) is a class of security mechanisms by which the authentication is based on the physical attributes of the *signal* received by Bob that operate as a unique signature of the transmitting device. A typical scenario provides that all agents are equipped with wireless transceivers and that an initial transmission from Alice to Bob has been authenticated by higher-layer mechanisms (e.g., based on cryptography and a previously shared key between Alice and Bob). Bob estimates the wireless channel (i.e., its impulse or frequency response) over which this initial authentic message passed and stores it as the reference channel. Upon reception of another message, Bob re-estimates the wireless channel over which this message passed and compares it with the reference channel. If the two estimates are close enough, the second message is deemed

Physical-Layer Security for 6G, First Edition. Edited by Parthajit Mohapatra, Nikolaos Pappas, Arsenia Chorti, and Stefano Tomasin.

authentic (i.e., coming from Alice), otherwise, it is discarded as fake (i.e., coming from Eve).

A few remarks are now in place. As with any other authentication mechanism, PLA also requires an initial step, where the identity is confirmed by some other trusted approach. A transmission by Eve will provide a different channel estimate since Eve is in a different position to Alice, thus, the signal propagation will be affected differently by surrounding objects and the relative position of Bob. The channels estimated on the two messages in general are different, even if Alice is transmitting both messages, as estimation errors are due to noise, interference, and reflections in the propagation environment that all change over time. Therefore, suitable techniques should be developed to mitigate the effects of such errors.

Formally, the authentication problem is a hypothesis testing (in Bayes formalism) or classification (in machine-learning formalism) problem, where Bob must decide between the two hypotheses (or must classify the second message in one of the two classes)

- $\mathcal{H}_0$: the message is coming from Alice,
- $\mathcal{H}_1$: the message is not coming from Alice.

As with any hypothesis testing problem, two errors may occur: a false alarm (FA), when Bob decides $\mathcal{H}_1$ for a message coming from Alice, and a missed detection (MD), when Bob decides $\mathcal{H}_0$ for a message coming from Eve.

Bob implements a test on the estimated channel to decide between the two hypotheses. Let $\boldsymbol{h}_{\text{ref}}$ be the vector representing the impulse response of the reference channel, and let $f(\boldsymbol{h}|\boldsymbol{h}_{\text{ref}}, \mathcal{H}_0)$ be the conditional probability density function (PDF) of the channel estimate $\boldsymbol{h}$ for the second message when Alice is transmitting (i.e., under hypothesis $\mathcal{H}_0$). Similarly, let $f(\boldsymbol{h}|\boldsymbol{h}_{\text{ref}}, \mathcal{H}_1)$ be the conditional PDF of the channel estimate $\boldsymbol{h}$, when Eve is transmitting (hypothesis $\mathcal{H}_1$). The optimal test by which Bob establishes the authenticity based on $\boldsymbol{h}$, i.e., the test providing the minimum MD probability for a given FA probability is the likelihood ratio test (LRT)

$$\mathcal{T} = \log \frac{f(\boldsymbol{h}|\boldsymbol{h}_{\text{ref}}, \mathcal{H}_0)}{f(\boldsymbol{h}|\boldsymbol{h}_{\text{ref}}, \mathcal{H}_1)} \underset{\mathcal{H}_1}{\overset{\mathcal{H}_0}{\gtrless}} \theta, \tag{11.1}$$

where θ is a parameter to ensure the desired FA probability $\pi_{\text{FA}} = \mathbb{P}[\mathcal{T} > \theta|\mathcal{H}_0]$. Typically, the statistics of the attack are unknown, as they depend on the position of the attacker and the propagation environment; in this case, we can resort to the generalized LRT (GLRT), i.e.,

$$\mathcal{T}' = \log f(\boldsymbol{h}|\boldsymbol{h}_{\text{ref}}, \mathcal{H}_0) \underset{\mathcal{H}_1}{\overset{\mathcal{H}_0}{\gtrless}} \theta. \tag{11.2}$$

For example, when the estimation error can be modeled as a Gaussian vector with zero-mean independent entries with variance σ^2, the GLRT simply becomes a test on the mean square error,

$$\mathcal{T}'' = ||\boldsymbol{h} - \boldsymbol{h}_{\text{ref}}||^2/\sigma^2 \underset{\mathcal{H}_1}{\overset{\mathcal{H}_0}{\gtrless}} \theta. \tag{11.3}$$

When the PDF of the estimation error is not available, machine-learning techniques can be adopted, to what becomes a classification problem. If a dataset of labeled channel estimates from both Alice and Eve is available, the model can be trained in a supervised manner to obtain a two-class classifier. A relevant example is given by a neural network trained to provide two outputs, corresponding to decisions for the two hypotheses $\mathcal{H}_0$ and $\mathcal{H}_1$. It has been shown [4] that when the neural network is complex enough and training converges to the optimum (using a minimum mean square error or cross-correlation loss function), such model performs as the optimal LRT. If a training set with only estimates of the channel from Alice is available, the model can be trained as a one-class classifier; also in this case, for a complex enough neural network and optimal convergence, the neural network performs as the GLRT [2].

The advantage of PLA is that channel features are difficult to forge or predict, while other basic higher-layer identifiers such as IP or MAC addresses can be easily intercepted and replicated. For more information on PLA, reference papers and surveys are [3, 10, 15, 27].

11.1.1 PLA and 6G Systems

Authentication is the basis of secure communication, network access, and trusted cooperation in 6G networks. To be more specific, organizations can keep their networks secure by permitting only authenticated users or processes to gain access to their protected resources, including computer systems, networks, databases, websites, and other network-based applications or services [12]. Moreover, only those uploaded data from authenticated devices/nodes can be trusted for further utilizations [1]. In the context of 6G systems, PLA is particularly suitable for IoT and device-to-device (D2D) communications, as a lightweight authentication procedure with low-power consumption. Moreover, operating at the lowest communication layer entails a significantly low latency since it reduces the computation overhead of the verification step. Indeed, verifying a digital signature in a vehicular network with a 400-MHz processor needs about 20 ms [16, 20], which is not compatible with the latency requirements of future networks. PLA can significantly reduce this time by operating directly at the physical layer.

Moreover, to handle a very large number of devices connected to the network and enable heterogeneous networks to cooperate, it is envisioned to have subnetworks connected to the main 6G networks. In this case, it is also expected that devices connecting to a subnetwork perform a local authentication, before (and when needed) authenticating to the main networks. In this context, having a simple and effective local authentication procedure becomes essential and PLA is well suited to handle the local authentication procedure [20]. PLA can also be integrated with other authentication mechanisms to strengthen security, especially when the number of connected devices grows significantly.

Among the technology of interest for 6G networks, distributed multiple-input–multiple-output (MIMO) systems and cell-free solutions are envisioned, with several local antennas spread in space to provide low-latency connection in a densely populated (by devices) area. In this scenario, an attacker can impersonate a local remote radio head to disrupt the device's functionalities. This is made easier by the presence of multiple remote radio heads that interfere with each other, in particular at the connection time, in what is known as the pilot contamination problem: when multiple devices send pilots to let corresponding terminals estimate the channel, and the signals interfere, the channel estimation is perturbed and degraded. It has been shown that cell-free systems are more vulnerable to this attack, since they are intrinsically more subject to pilot contamination [21].

PLA can be used to effectively avoid attacks from fake or rogue base stations [33]. These are base stations set up by an attacker to connect to legitimate devices and then send them malicious short messages (SMS) or steal personal data (such as positions and identifiers). Such threat mostly affects the old GSM networks, however, also new 6G devices should be able to withstand attacks based on the forced downgrading of their connection to GSM. Moreover, even 5G networks are vulnerable to rogue base stations forcing the user device to disclose its subscription permanent identifier (SUPI) [11].

11.2 Challenge-Response PLA for 6G

Although cryptographic techniques have been widely studied for authentication, they may fall short of the desired performance in many emerging scenarios of 6G wireless networks [5, 13, 18, 19]. The main reason is that appropriate key management procedures are necessary for conventional cryptographic techniques, including key pre-generation, distribution, update, destruction, and storage. These could increase latency, communication, and computation overheads to achieve better security performance, which is extremely undesirable for delay-sensitive communications and resource-constraint devices. Moreover, cooperation among multiple entities is required for both ring and group signature and group key agreement [9, 32], leading to excessive latency and significant communication overhead.

The new challenges posed by 6G networks and the new technologies that will be adopted both favor innovative solutions for PLA. Among those, the

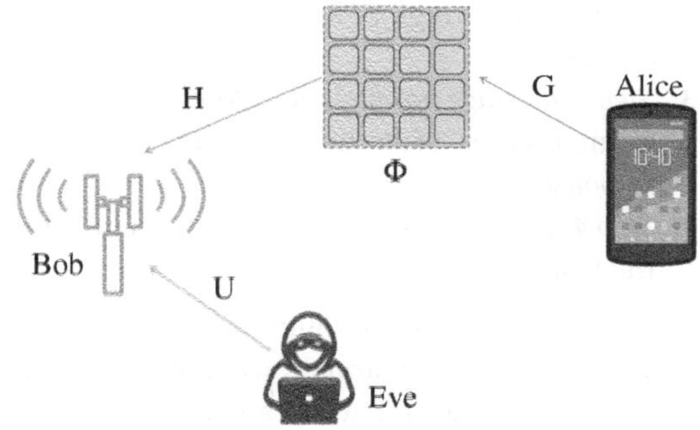

Figure 11.1 Model of the RIS-assisted MIMO communication system.

challenge-response (CR) approach (well known in cryptography) has been recently proposed for PLA [22]. In CR PLA, the verifier poses a challenge by *modifying the propagation environment* and the device under verification propagates its message over such a modified environment. Lastly, the verifier checks that the received message has propagated through the modified environment (by estimating the channel).

A relevant 6G example is shown in Figure 11.1, where the verifier Bob is a base station controlling a reflective intelligent surface (RIS) to authenticate a smartphone (Alice). A RIS is a panel of reflective elements, each acting as a receive and transmit antenna and adding a phase shift to the baseband reflected signal. The phase shift of each element can be chosen to obtain the desired reflection properties. By reconfiguring the RIS, Bob can control the propagation environment and verify that the signal he receives is coming from Alice rather than from Eve. With reference to Figure 11.1, for a RIS with M elements, the $M \times M$ diagonal matrix $\boldsymbol{\Phi}$ contains on its diagonal entry $\Phi_{m,m} = e^{j\phi_m}$, $m = 1, \ldots, M$, where ϕ_m is the phase shift of the mth element. The baseband-equivalent narrowband MIMO channels among devices are represented by matrices $\boldsymbol{H}$, $\boldsymbol{G}$, and $\boldsymbol{U}$, in particular, the RIS–Bob, Alice–RIS, and the Eve–Bob channels, respectively. For example, the overall Alice–Bob channel has an impulse response given by the matrix $\boldsymbol{H\Phi G}$.

These are the steps of the CR physical-layer security (PLS) authentication protocol:

1. *Step 1 channel state information (CSI)*: In this step, Alice sends Bob *several* pilot signals under *several RIS configurations* chosen by Bob; a higher-layer security mechanism is used to authenticate these transmissions; Bob estimates the channels using the pilot signals and stores them, together with the RIS configurations.
2. *Step 2, random configuration*: This is the challenge step, wherein Bob chooses a random RIS configuration: note that the RIS configuration may have already been explored in Step 1 or not. In case it was not explored before, Bob should

infer his resulting CSI to Alice with the new configuration, based on the results of Step 1.

3. *Step 3, message transmission*: Alice transmits the message over the RIS and Bob estimates the CSI as obtained from the received signal. The estimated CSI is the *response* from Alice;
4. *Step 4, channel check*: This is the authentication step, wherein Bob checks that the response matches the CSI predicted in Step 2 (expected response).

Note that the proposed solution (as conventional PLA) assumes a static scenario: mobility introduces channel variations that in general degrade the authentication procedure. Such degradation can be in part mitigated when it is possible to predict the channel variations.

CR PLA presents several advantages over conventional PLA. First, it is more robust to repeated attacks, since at each new transmission, the channel conditions are randomly modified, and an attacker not aware of the random RIS configuration does not know the expected response, thus he cannot forge an effective attack. This is particularly true if Eve transmits directly to Bob (as in Figure 11.1, not passing through the RIS) but it holds also when she passes through the RIS, since in this case, Eve must be able to forge the Alice-RIS channel so that it provides the expected Alice–Bob channel for *any* RIS configuration. When a single RIS configuration is used, instead, Eve may simply forge a channel that (passing through the fixed RIS) provides the expected Alice–Bob channel. Since typically RISs have many more elements (playing the role of antennas) than the base station, several Eve–RIS channels provide the same Alice–Bob channel, for a given RIS configuration.

A second advantage is that even when Eve has found a forged channel that breaks the PLA mechanism with *a given* RIS configuration, she is still not able to break the PLA mechanism with *other* RIS configurations, thus, the entire authentication system is not broken. For example, consider the case wherein the random RIS configuration $\boldsymbol{\Phi}$ is taken from a set of L-fixed configurations. Moreover, let Eve have a perfect estimate of the Alice–Bob for a specific configuration $\overline{\boldsymbol{\Phi}}$, thus, when this configuration is selected in the CR PLA protocol, the MD probability is 1. Instead, for any other RIS configuration $\boldsymbol{\Phi}' \neq \overline{\boldsymbol{\Phi}}$, let $\alpha < 1$ be the MD probability. This occurs for example when channel $\boldsymbol{U}$ is independent from both $\boldsymbol{H}$ and $\boldsymbol{G}$. In this case, even if Eve knows she has broken the PLA mechanism for configuration $\overline{\boldsymbol{\Phi}}$, she does not have any further information to improve its attack for other configurations. Then, assuming that Bob chooses uniformly at random one of the M configurations at each transmission, Eve will be able to break the CR PLA mechanism with probability

$$\epsilon = \frac{1}{M} + \alpha\frac{M-1}{M}, \tag{11.4}$$

where the first term is related to the case in which Eve correctly guesses the right RIS configuration, while the second term is related to a wrong choice still being accepted due to misdetection. Note that a random configuration of the RIS has also an impact on the communication performance (data rate), thus, a suitable trade-off between communication rate and security should be chosen. Moreover, in general, Eve may partially know the Alice–Bob channel, thus her attack may be more effective.

CR PLA can be implemented whenever the propagation environment is *partially controllable* by Bob. Indeed, such control enables Bob to face several challenges. Still, the control must be *partial*, so that CSI is still in part random and depending on the position of transmitting/receiving devices or, in general, their features: this is a key point for PLA.

There are other relevant possible applications of CR PLA in 6G networks. These include the use of authentication when drones are involved: in this case, the verifier can control the position of the drones (either the verifier or the verified drones, or both) to modify the propagation environment.

Preliminary results on the performance of CR PLA have been reported in [22], where it is shown that, while the conventional PLA can be easily broken after a few attacks (as the MD probability reaches 1), CR PLA instead keeps the MD probability below 10^{-3}.

11.3 Intelligent PLA Based on Machine Learning

PLA technique provides an effective approach to validating a device by exploiting the radio-frequency fingerprint (RFF), wireless communication link, device, and location-related attributes [24]. Some widely used physical-layer attributes include the carrier frequency offset (CFO) [8], channel impulse response (CIR)/channel frequency response (CFR) [25], received signal strength indicator (RSSI) [7]. These attributes are inherently related to the unique imperfection of communicating devices, physical location, and corresponding wireless communication environment, which can provide specific distinguishing features for identifying multiple devices, thus, providing natural dynamic protections for legitimate communications.

However, due to the dynamic communication environment of 6G networks encountered and the small difference in RFF between the two devices, the performance of PLA may suffer from the time-varying estimations of the channel parameters [6]. As shown in Figure 11.2, the conventional techniques are facing the following challenges:

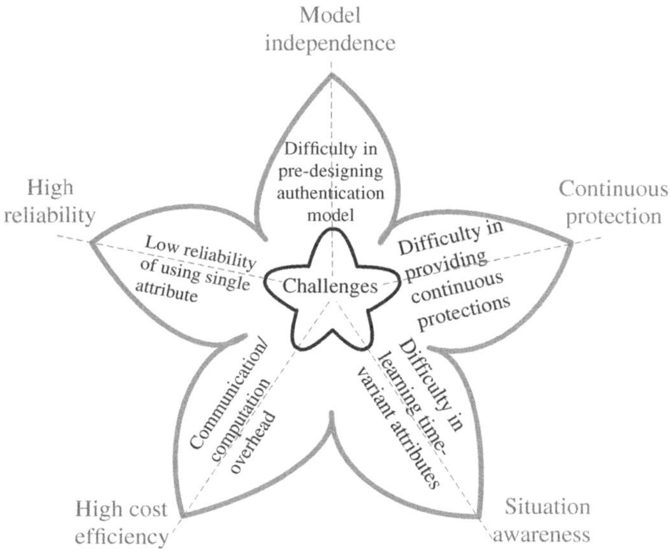

Figure 11.2 Challenges of traditional authentication methods.

Difficulty in predesigning a static authentication model: Because of the dynamic communication environment as well as the mobility of devices, it is difficult to predict the wireless communication channel. In this way, it is unavailable to predesign a static authentication model, especially in highly dynamic networks, e.g., Internet-of-Vehicle (IoV) [17]. Hence, the performance of the traditional statistic authentication model may be limited in the 6G systems.

Low authentication accuracy of using single attribute: Similarly, the performance of PLA using only one signal attribute will suffer from the complex dynamic environment as well as the limited information and diversity, leading to low reliability. Hence, utilizing multiple attributes could be a solution to improve the reliability of PLA.

High communication/computation overhead: The complex process of key/certification management leads to high communication/computation overhead. Moreover, to achieve better performance, high estimation and prediction accuracy of the PLA attribute are also required.

Difficulty in providing continuous protections: Traditional authentication methods usually perform one-time validation of the identity of the transmitter, which could not protect the system continuously, leading to several loopholes for Eve.

Difficulty in learning time-variant attributes: The conventional PLA technique compares the attribute estimation with the reference estimation, e.g., channel estimation. However, it may suffer from the dynamic wireless communication channel. Moreover, it may fail in accurately identifying the transmitter due to the channel de-correlation.

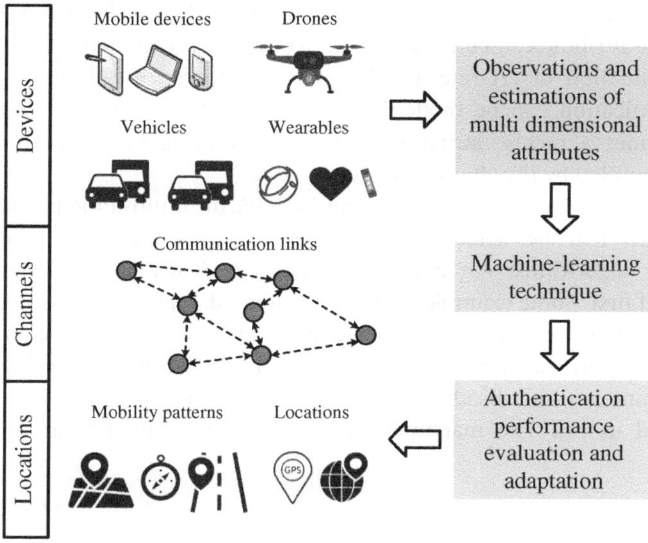

Figure 11.3 Intelligent PLA based on machine-learning algorithms.

In overcoming the above challenges, the machine-learning algorithms may provide a solution, since they provide powerful tools to learn the dynamic wireless communication and adversarial environment for self-optimization/organization of PLA [6, 26]. Moreover, they could help in the detection of time-varying PLA attributes and the adaptation of the PLA process. These will facilitate the PLA model design and the improvement of authentication intelligence and efficiency in the 6G networks. As shown in Figure 11.2, this chapter will design a machine-learning-based PLA approach to achieve efficient and continuous protection for 6G wireless communication systems.

11.3.1 Machine-Learning-Based PLA Approach

As shown in Figure 11.3, we utilize multiple attributes for designing a machine-learning-based PLA approach. The PLA approach design contains the following three steps:

- *Step I*: Utilizing multiple PLA attributes could not only improve the authentication accuracy but also increase the difficulty of being cracked by Eve. To be more specific, multiple PLA attributes provide multiple-dimensional information of channel, device, and physical environment, which obviously will enrich the diversity of identification of different devices. First, multiple PLA attributes will be estimated and collected in the complex dynamic 6G networks. According to the specific 6G application and communication environment, these PLA attributes can be analyzed for further authentication model design. To be more specific, each PLA attribute can be used for PLA to analyze the authentication accuracy, i.e., evaluation of FA rate and MD rate. Moreover, the design of intelligent authentication can be model-free relying on the estimation data of attributes without a precise structure of the time-varying physical-layer attributes as exemplified by the channel model.

- *Step II*: Based on the collected PLA attributes and analysis, the multi-dimensional attribute estimates can be selected and fused for PLA based on the machine-learning algorithms. An appropriate machine-learning algorithm should be selected first. Some examples may be considered: logistic regression, Bayesian learning, k-means clustering, principal component analysis (PCA), kernel learning, support vector machines (SVM), and reinforcement learning [14, 31]. The advantages and disadvantages of their application in PLA have been summarized in [6]. The machine-learning-based PLA model can be formulated as

$$\xi = F(\boldsymbol{H}[t]), \tag{11.5}$$

where $\boldsymbol{H}[t] = (H_1[t], H_2[t], \dots, H_N[t])$ are the estimations of N PLA attributes of Alice at time instant t, which can be seen as the inputs of the selected machine-learning algorithm. Then, its output is ξ, which means the transmitter is Alice or Eve (the message is from Alice or Eve). In this PLA model, selecting and developing an appropriate machine-learning algorithm as well as reducing the dimensionality of the authentication system also benefits the efficiency of the PLA process. Moreover, the observations of PLA attributes are required for the training in the machine-learning-based PLA model. This should be carefully considered in the design, since some 6G applications may not provide enough data for training, especially those for PLA attribute estimation of Eve.

- *Step III*: To provide adaptation of PLA and continuous protection for 6G systems, the design of a machine-learning-based PLA model relies on the sequential estimation of PLA attributes. Moreover, due to the dynamic network topology and wireless environment of 6G systems, the adaptive selection of PLA attributes is considered in the PLA model for better communication and security performance. The machine-learning algorithms can track the time-varying attributes by updating the inputs $\boldsymbol{H}[t] = (H_1[t], H_2[t], \dots, H_N[t])$, and output the accurate authentication results.

11.3.2 Performance Analysis

To validate and evaluate the presented intelligent PLA design, we use different machine-learning algorithms in this section, including the backpropagation

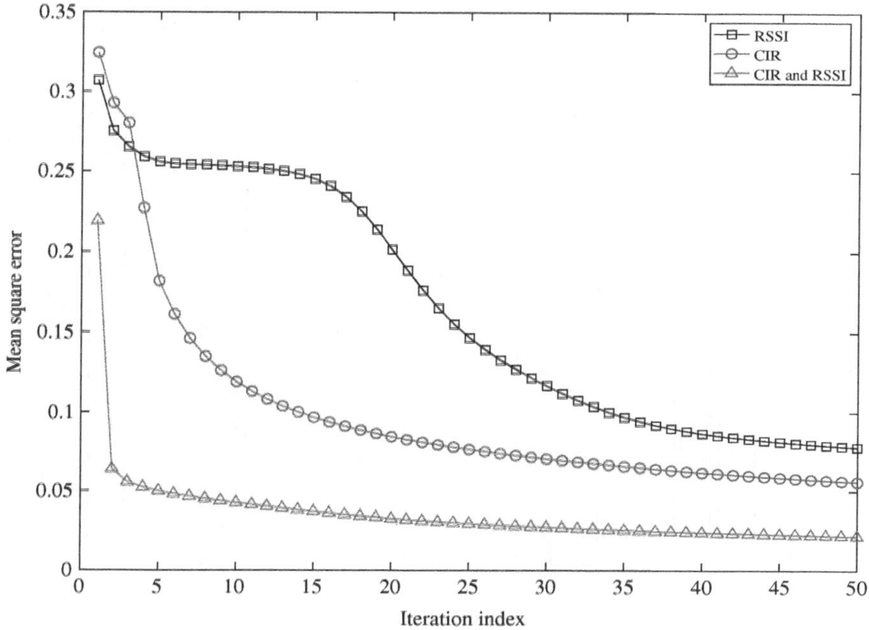

Figure 11.4 Training accuracy of intelligent PLA based on the BPNN algorithm.

neural network (BPNN) [28, 30] and SVM algorithms [23, 29]. In the BPNN algorithm, a three-layer BPNN is selected, and the sigmoid function is used as the activation function. Moreover, the RSSI and CIR are selected as the physical-layer attributes for authentication. Their estimates are normalized and used as the inputs of the selected machine-learning algorithms, and the authentication results are the outputs. To be more specific, the channel model in this section is built using the WINNER II module by MATLAB. The proposed scheme is studied in an urban scenario, where the center frequency is 5.25 GHz.

As shown in Figure 11.4, the CIR and RSSI are utilized as the inputs of the BPNN for PLA design. The learning rate is set to $\mu = 0.2$ and the initial parameters of BPNN are randomly generated. Given 500 CIR and RSSI samples of Alice, the mean square error of learning decreases significantly as the number of iterations increases from 0 to 50. It can be also observed that the intelligent authentication process using CIR and RSSI results in the best training performance compared to that of using CIR or RSSI separately. Moreover, the proposed scheme relying on RSSI has the slowest convergence speed and the highest steady-state value of convergence. The reason for this trend is that the CIR estimates are more reliable than the RSSI in the simulated static communication scenario. More attributes

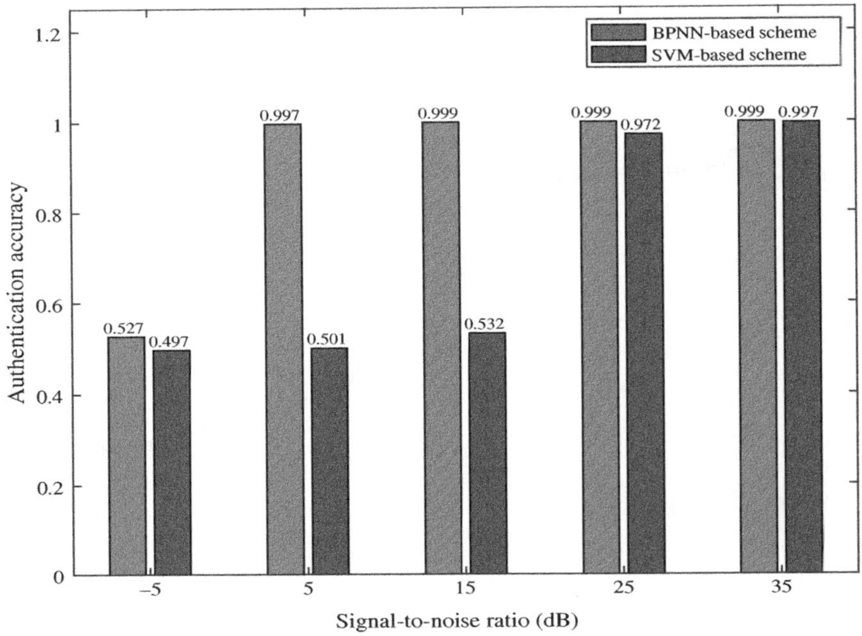

Figure 11.5 Authentication accuracy of intelligent PLA based on different machine-learning algorithms, i.e., BPNN algorithm and SVM algorithm.

used for authentication can provide more information and increase authentication reliability.

Figure 11.5 characterizes the authentication accuracy comparison between the BPNN-based scheme and the SVM-based scheme utilizing RSSI and CIR. The authentication accuracy of the BPNN-based scheme is higher than that of the SVM-based scheme in all cases. In particular, the authentication accuracy of the BPNN-based scheme achieves 99% in the case that the signal-to-noise ratio (SNR) is higher than 5 dB, while the SVM-based scheme cannot reach similar authentication accuracy until the SNR is higher than 25 dB. The reason is that the three-layer BPNN is used in the BPNN-based scheme, which provides better performance of training for the physical-layer authentication in this simulated communication scenario. When the SNR is −5 dB, the authentication accuracy values of both schemes are around 0.5. This means that both schemes are unsuitable for the physical-layer authentication using RSSI and CIR in this simulated communication scenario due to the terrible communication environment.

References

1 M. A. Al-Shareeda, M. Anbar, I. H. Hasbullah, and S. Manickam. Survey of authentication and privacy schemes in vehicular ad hoc networks. *IEEE Sensors Journal*, 21(2):2422–2433, 2021. doi: 10.1109/JSEN.2020.3021731.

2 F. Ardizzon and S. Tomasin. Generalized likelihood ratio test with one-class classifiers, 2022. URL https://arxiv.org/abs/2210.12494.

3 P. Baracca, N. Laurenti, and S. Tomasin. Physical layer authentication over MIMO fading wiretap channels. *IEEE Transactions on Wireless Communications*, 11(7):2564–2573, 2012.

4 A. Brighente, F. Formaggio, G. M. Di Nunzio, and S. Tomasin. Machine learning for in-region location verification in wireless networks. *IEEE Journal on Selected Areas in Communications*, 37(11):2490–2502, 2019.

5 D. Díaz-Sánchez, A. Marín-Lopez, F. A. Mendoza, P. A. Cabarcos, and R. S. Sherratt. TLS/PKI challenges and certificate pinning techniques for IoT and M2M secure communications. *IEEE Communications Surveys & Tutorials*, 21(4):3502–3531, 2019. doi: 10.1109/COMST.2019.2914453.

6 H. Fang, X. Wang, and S. Tomasin. Machine learning for intelligent authentication in 5G and beyond wireless networks. *IEEE Wireless Communications*, 26(5):55–61, 2019. doi: 10.1109/MWC.001.1900054.

7 H. Fang, X. Wang, and L. Xu. Fuzzy learning for multi-dimensional adaptive physical layer authentication: A compact and robust approach. *IEEE Transactions on Wireless Communications*, 19(8):5420–5432, 2020. doi: 10.1109/TWC.2020.2993175.

8 W. Hou, X. Wang, J.-Y. Chouinard, and A. Refaey. Physical layer authentication for mobile systems with time-varying carrier frequency offsets. *IEEE Transactions on Communications*, 62(5):1658–1667, 2014. doi: 10.1109/TCOMM.2014.032914.120921.

9 S. Jiang. Group key agreement with local connectivity. *IEEE Transactions on Dependable and Secure Computing*, 13(3):326–339, 2016. doi: 10.1109/TDSC.2015.2399299.

10 E. Jorswieck, S. Tomasin, and A. Sezgin. Broadcasting into the uncertainty: Authentication and confidentiality by physical-layer processing. *Proceedings of the IEEE*, 103(10):1702–1724, 2015.

11 R. P. Jover and V. Marojevic. Security and protocol exploit analysis of the 5G specifications. *IEEE Access*, 7:24956–24963, 2019. doi: 10.1109/ACCESS.2019.2899254.

12 R. Khan, P. Kumar, D. N. K. Jayakody, and M. Liyanage. A survey on security and privacy of 5G technologies: Potential solutions, recent advancements, and

future directions. *IEEE Communications Surveys & Tutorials*, 22(1):196–248, 2020. doi: 10.1109/COMST.2019.2933899.

13 M. Khodaei, H. Jin, and P. Papadimitratos. SECMACE: Scalable and robust identity and credential management infrastructure in vehicular communication systems. *IEEE Transactions on Intelligent Transportation Systems*, 19(5):1430–1444, 2018. doi: 10.1109/TITS.2017.2722688.

14 B. Mao, F. Tang, Y. Kawamoto, and N. Kato. Ai models for green communications towards 6G. *IEEE Communications Surveys & Tutorials*, 24(1):210–247, 2022. doi: 10.1109/COMST.2021.3130901.

15 U. M. Maurer. Authentication theory and hypothesis testing. *IEEE Transactions on Information Theory*, 46(4):1350–1356, 2000.

16 M. Mitev, A. Chorti, H. V. Poor, and G. Fettweis. What physical layer security can do for 6G security, 2022. URL https://arxiv.org/abs/2212.00427.

17 A. Nanda, D. Puthal, J. J. P. C. Rodrigues, and S. A. Kozlov. Internet of autonomous vehicles communications security: Overview, issues, and directions. *IEEE Wireless Communications*, 26(4):60–65, 2019. doi: 10.1109/MWC.2019.1800503.

18 K.-A. Shim. A survey of public-key cryptographic primitives in wireless sensor networks. *IEEE Communications Surveys & Tutorials*, 18(1):577–601, 2016. doi: 10.1109/COMST.2015.2459691.

19 H. Tan, M. Ma, H. Labiod, A. Boudguiga, J. Zhang, and P. H. J. Chong. A secure and authenticated key management protocol (SA-KMP) for vehicular networks. *IEEE Transactions on Vehicular Technology*, 65(12):9570–9584, 2016. doi: 10.1109/TVT.2016.2621354.

20 A. Teniou and B. A. Bensaber. Efficient and dynamic elliptic curve qu-vanstone implicit certificates distribution scheme for vehicular cloud networks. *Security and Privacy*, 1(1):e11, 2018.

21 S. Timilsina, D. Kudathanthirige, and G. Amarasuriya. Physical layer security in cell-free massive MIMO. In *2018 IEEE Global Communications Conference (GLOBECOM)*, pages 1–7, 2018. doi: 10.1109/GLOCOM.2018.8647876.

22 S. Tomasin, H. Zhang, A. Chorti, and H. V. Poor. Challenge-response physical layer authentication over partially controllable channels. *IEEE Communications Magazine*, 60(12):138–144, 2022. doi: 10.1109/MCOM.001.2200339.

23 L. Wang, Z. Liu, C. L. P. Chen, Y. Zhang, S. Lee, and X. Chen. Energy-efficient SVM learning control system for biped walking robots. *IEEE Transactions on Neural Networks and Learning Systems*, 24(5):831–837, 2013. doi: 10.1109/TNNLS.2013.2242486.

24 W. Wang, Z. Sun, S. Piao, B. Zhu, and K. Ren. Wireless physical-layer identification: Modeling and validation. *IEEE Transactions on Information Forensics and Security*, 11(9):2091–2106, 2016. doi: 10.1109/TIFS.2016.2552146.

25 L. Xiao, L. J. Greenstein, N. B. Mandayam, and W. Trappe. Using the physical layer for wireless authentication in time-variant channels. *IEEE Transactions on Wireless Communications*, 7(7):2571–2579, 2008. doi: 10.1109/TWC.2008.070194.

26 L. Xiao, X. Wan, X. Lu, Y. Zhang, and D. Wu. IoT security techniques based on machine learning: How do IoT devices use AI to enhance security? *IEEE Signal Processing Magazine*, 35(5):41–49, 2018. doi: 10.1109/MSP.2018.2825478.

27 N. Xie, Z. Li, and H. Tan. A survey of physical-layer authentication in wireless communications. *IEEE Communications Surveys & Tutorials*, 23(1):282–310, 2021.

28 Y. Xie, K. Wang, and H. Huan. BPNN based indoor fingerprinting localization algorithm against environmental fluctuations. *IEEE Sensors Journal*, 22(12):12002–12016, 2022. doi: 10.1109/JSEN.2022.3172860.

29 J. Xu, Y. Y. Tang, B. Zou, Z. Xu, L. Li, and Y. Lu. The generalization ability of online SVM classification based on Markov sampling. *IEEE Transactions on Neural Networks and Learning Systems*, 26(3):628–639, 2015. doi: 10.1109/TNNLS.2014.2361026.

30 M. Yang et al. Machine-learning-based scenario identification using channel characteristics in intelligent vehicular communications. *IEEE Transactions on Intelligent Transportation Systems*, 22(7):3961–3974, 2021. doi: 10.1109/TITS.2020.3001132.

31 A. Yang, J. Weng, K. Yang, C. Huang, and X. Shen. Delegating authentication to edge: A decentralized authentication architecture for vehicular networks. *IEEE Transactions on Intelligent Transportation Systems*, 23(2):1284–1298, 2022. doi: 10.1109/TITS.2020.3024000.

32 L. Zhang, Q. Wu, J. Domingo-Ferrer, B. Qin, and Z. Dong. Round-efficient and sender-unrestricted dynamic group key agreement protocol for secure group communications. *IEEE Transactions on Information Forensics and Security*, 10(11):2352–2364, 2015. doi: 10.1109/TIFS.2015.2447933.

33 Y. Zhang et al. Lies in the air: Characterizing fake-base-station spam ecosystem in China. In *Proceedings of the 2020 ACM SIGSAC Conference on Computer and Communications Security*, CCS '20, page 521–534, New York, NY, USA, 2020. Association for Computing Machinery. ISBN 9781450370899.

12

Securing the Future e-Health: Context-Aware Physical-Layer Security

Mehdi Letafati[1], Eduard Jorswieck[2], and Babak Khalaj[1]

[1] *EE Department, Sharif University of Technology, Tehran, Iran*
[2] *Institute for Communications Technology, TU Braunschweig, Brunswick, Germany*

12.1 Introduction

12.1.1 PHYSEC in 6G

Ongoing research in both industrial and academic circles is delving into the realm of the sixth generation (6G) of mobile communications, advancing along several key avenues. These include exploration into higher-frequency bands (sub-THz), integration of a larger number of antennas (extreme MIMO), and the proliferation of nodes (massive multiple access). Additionally, 6G aims to facilitate integrated sensing and communications, with learning capabilities dispersed across all network nodes, encompassing edge and mobile nodes alike. These advancements promise data exchange with elevated rates, reduced latency, and enhanced reliability.

On the one hand, within the large amounts of data generated by end users and by machines, sensitive information and private data need to be secured. On the other hand, the technologies mentioned enable new intelligent, adaptive, and context-aware attack mechanisms compromising security and safety. Scalability and resource efficiency are limited using cryptographic primitives. Physical-layer security (PHYSEC) schemes offer post-quantum security and safety, scalability, and resource efficiency. Therefore, PHYSEC is considered an important technology to secure the 6G communication links.

PHYSEC is becoming increasingly important for 6G wireless networks as a set of techniques to protect massive amounts of sensitive data, the evolving security threat landscape, the growing demand for privacy, and the need for

Physical-Layer Security for 6G, First Edition. Edited by Parthajit Mohapatra, Nikolaos Pappas, Arsenia Chorti, and Stefano Tomasin.

better spectrum utilization. PHYSEC provides an additional layer of protection for sensitive data, making it an essential component of secure 6G wireless communication systems.

There are many recent overview and survey papers that highlight the importance of PHYSEC for 6G networks [9, 11, 23, 49, 52]. The open security issues in 5G and the research challenges ahead of 6G are reviewed in [11]. Porambage et al. [52] provide a vision of 6G security and security key performance indicators (KPIs) with the tentative threat landscape based on the foreseen 6G network architecture. In [49], a systematic overview of security and privacy issues is provided based on future technologies for 6G in the physical, connection, and service layers, as well as through lessons learned from the failures of existing security architectures and state-of-the-art defenses. Finally, the paper [23] reviews a subset of promising PHYSEC techniques which can enable and support a range of crucial key value indicators (KVIs).

The fact that each PHYSEC solution needs to be tailored to the specific application and context is highlighted in the references above. In this chapter, our focus covers applications that require security in the context of e-health services and use cases.

Digital healthcare services, also known as e-health, seek to improve our quality of life through the utilization of wireless body area networks (WBANs) and Internet-of-Medical-Things (IoMT). Notably, 6G networks are regarded as the primary infrastructure linking various e-health services [46], facilitating the realization of personalized healthcare services. This requires gathering and exchange of sensitive information and vital data of many people from e-health sensors to different network nodes and medical cloud servers. These developments highlight new security and privacy challenges in future 6G e-health systems [52]. 6G-aided digital healthcare relies on the wireless communication among resource-constrained IoMT nodes, which highlights the need for lightweight and scalable mechanisms for communication security.

Communication security of the fifth-generation (5G) standard hinges on cryptographic keys, such as elliptic curve cryptography. However, the emergence of 6G-enabled e-health networks, with extensive peer-to-peer (P2P) "on-the-fly" communications, challenges the effectiveness of traditional approaches. Additionally, the arrival of quantum computers poses a threat to the confidentiality of asymmetric key encryption algorithms [11, 52]. Furthermore, due to the constrained resources, it might not be practical to implement "pre-deployed" keys for e-health nodes. In this regard, PHYSEC is envisioned to realize a paradigm shift from the traditional complexity-based approaches towards novel lightweight secrecy-assuring methods. It is noteworthy that 6G is anticipated to introduce device-level intelligence through the implementation of cutting-edge artificial intelligence (AI) and machine-learning (ML) algorithms

[11, 35]. Consequently, the capabilities of ML techniques to comprehend feature statistics of data-streams can also be integrated into PHYSEC protocols, thereby establishing innovative learning-based PHYSEC. In this chapter, we will also study learning-assisted PHYSEC schemes for e-health.

The chapter is organized as follows. First, we introduce PHYSEC for 6G networks in Section 12.1. Key-less and key-based PHYSEC under active and passive attacks are reviewed in Section 12.1.2. We then focus on PHYSEC key generation in Section 12.2, where a learning-assisted key generation scheme for wireless-powered e-health systems is proposed. A secure medical image transmission scheme based on key-less PHYSEC is studied in Section 12.3 as another use-case of incorporating PHYSEC into wireless e-health. Next, a proof-of-concept (PoC) testbed for group PHYSEC is studied in Section 12.4. Finally, Section 12.5 concludes this chapter by providing insightful future directions.

In this chapter, vectors are shown by bold lowercase letters. Matrices are denoted as bold uppercase symbols. The transpose and ℓ^2 norm of a vector are represented by $(\cdot)^{\mathrm{T}}$, and $||\cdot||$, respectively. The expected value and the entropy of random variable (RV) X are denoted by $\mathbb{E}[X]$ and $H(X)$, respectively. The mutual information of RVs X and Y is denoted by $I(X;Y)$.

12.1.2 Introduction to PHYSEC Solutions

Besides classical cryptography, PHYSEC [7] is a promising technique to enable secure data transmission. In contrast to cryptography, no shared key is required, but the physical properties of the wireless channel are exploited. Additional advantages of PHYSEC solutions compared to classical cryptography-based primitives are the lower complexity, better scalability, and the information-theoretic security achieved.

Post-quantum security refers to algorithms and techniques which are secure against a cryptanalysis attack by a quantum computer. Recently, it was shown that asymmetric key cryptography is weak against quantum computers which could *efficiently* solve the prime-factorization problem which forms the basis for many asymmetric public key algorithms [15]. Information theoretic coding and security schemes can provide post-quantum security [47]. And furthermore, they can scale well with the number of devices. Finally, resource efficiency in terms of energy and time is guaranteed.

12.1.2.1 General Model and Problem Formulations

The first step toward secure communications and networking is the description of the system model. Classically, this includes the attacker model and the description of the capabilities of the legitimate system. In PHYSEC, this includes a careful description of the signal model and the channel models. The description of the models ends in the development of suitable objective functions and constraints which operationally describe the communications and security requirements and goals.

The main objectives of the transmitter (Alice) during confidential communications are at least twofold: low error probability at the legitimate receiver (we usually call him Bob) while no information leakage to the eavesdropper (called Eve). Let us consider the general model in which Alice sends some channel input $\boldsymbol{x}$ based on the secure message m, Bob observes a signal $\boldsymbol{y}$, and Eve another signal $\boldsymbol{z}$. The two objectives are then formulated as

$$\max_{p_x} I(M;\boldsymbol{y}) \geq R_* \tag{12.1}$$

$$I(M;\boldsymbol{z}) \leq \varepsilon. \tag{12.2}$$

The mutual information between the random secure message M and the observation at Bob should be larger than a target rate while the information leakage, namely, the mutual information between the secure message and the observation at Eve should be below a arbitrarily small threshold ε. Note that (12.1) can be directly related to a lower bound on the error probability at Eve [6].

The advantage of the confidentiality objective in (12.2) compared to complexity-based cryptography measures is that it is post-quantum secure since no data processing at all, including quantum computers, can obtain useful information from observation $\boldsymbol{z}$ [9]. A disadvantage of the objective (12.2) is that it is sometimes difficult to obtain exact estimates or even bounds on this quantity since it depends on the properties of the eavesdropper such as its position, the propagation conditions, and its hardware and radio-frequency (RF) components. These are usually unknown, and the estimate of information leakage (12.2) is calculated based on assumptions and assertions.[1] Therefore, the context of the communication system model is important and needs to be considered when designing the PHYSEC mechanism.

As an alternative to wiretap coding which is a key-less approach, different secret key generation (SKG) techniques have been recently considered. Infrastructure-less key management systems support privacy, e.g., communicating in the presence of a possibly repressive government, avoiding industrial espionage, circumvention of malicious intrusion into processing steps in factories, or traffic communications of unmanned vehicles. SKG at the physical layer is one key enabler for resilient, safe, and scalable industry solutions, as well as consumer communications.

1 These assumptions are typically made when the system, signal, and channel models are defined.

12.1.2.2 Key-less Versus Key-Based Techniques

One approach to achieve the two goals in (12.1) and (12.2) is to apply wiretap coding [62], see also the corresponding chapter on wiretap coding in this book. For a more recent summary of wiretap coding schemes, refer to [8]. This approach is coined *keyless* since no explicit key exchange or generation is performed.

In order to apply the wiretap coding scheme, there must exist an advantage of the legitimate link compared to the eavesdropper link. Then, the basic idea is to include confusion messages for the eavesdropper in the construction of the wiretap codebook. This reduces the data rate of the secure messages for Bob and usually a secure data rate equal to

$$R_S = \max_{p_X} \left(I(\boldsymbol{x};\boldsymbol{y}) - I(\boldsymbol{x};\boldsymbol{z})\right) \tag{12.3}$$

can be achieved.

The alternative approach to achieve the two goals in (12.1) and (12.2) is to first obtain or generate a secret key at Alice and Bob and then apply it to create a confidential link. This can be achieved by standard cryptography primitives, e.g., the one-time pad in the simplest case. In this approach, the main focus lies on the SKG itself and its confidentiality and security constraints [7].

The are two main models for SKG: the channel and the source model. While the names might be misleading, the models are best explained from an information theoretic point of view [41]. In the source model, there is an underlying statistical model, where all entities, Alice, Bob, and Eve, have access to random observations X, Y, Z with an underlying joint probability distribution $p_{X,Y,Z}$. Based on their observations and using a public protocol F Alice and Bob decide on their random keys K_a and K_b, while Eve should not gain any information about these key bits based on the observations Z and the public protocol. The corresponding constraints for the SKG from the source model are formulated as

$$\mathbb{P}(K_a \neq K_b) \leq \varepsilon, \tag{12.4}$$

$$H(K_a) \geq \log |\mathcal{K}| - \varepsilon, \tag{12.5}$$

$$\mathbb{D}(P_{K_A,Z,F} || P_{K_a} P_Z P_F) \leq \varepsilon, \tag{12.6}$$

with the constraint (12.4) that guarantees that the keys at Alice and Bob agree, (12.5) that the generated secret key sequence is close to uniform distribution, and (12.6) that Eve cannot leak any information from the protocol and the source observations.

In practical applications, the source of the model has to be identified. One typical example is to use the random channel realizations in reciprocal wireless channels. In general, the properties of the wireless medium can be well exploited for security primitives [22, 38].

The second model for SKG is the channel model, where the secret key is generated during the transmission over the channel. In scenarios, where the source model is not applicable because there is no suitable random source available, e.g., in static fading scenarios, the alternative is that Alice transmits a random sequence as source of randomness (SoR), while Bob and Eve observe differently distorted versions of the sequence. Based on this difference, Bob and Alice can agree on a secret key, by public discussion. The scheme was independently introduced 30 years ago in [1, 45].

12.1.2.3 Active and Passive Attacks

The attacker model is an important part of the system description and needs to be carefully described. It is distinguished between active and passive attacks. For example, eavesdropping and jamming are two of the well-known passive and active attacks, respectively. These two attackers could operate separately or they could collaborate as in the case of the man-in-the-middle (MiM) attack. There, the attacker places itself between transmitter and receiver and tries to deceive legitimate entities by injecting fake signals.

For passive attacks, eavesdropper scenarios are usually distinguished between separate eavesdroppers or colluding eavesdroppers, where the second case basically correspond to a distributed antenna array scenario [16]. Another distinction is made between eavesdroppers who are interested in the secure data or wardens who are interested in the existence of a communication link between two network nodes. The latter is coined as covert or stealth communication [3].

In addition to passive eavesdroppers also, active attacks can simultaneously be performed. There are plenty of active attacks. In the following, we focus only on attacks related to the physical and medium access control layer. The arbitrarily varying wiretap channel (AVWC) addresses this issue and provides information-theoretic guarantees on how much information can be transmitted reliably and securely, even in the presence of active and passive adversaries [51, 60]. Arbitrarily varying means that the channel state can change with every symbol of the transmitted codeword. This allows the attacker, i.e., the jammer, to disturb the communication to Bob up to the point of zero data rate by symmetrization of the channel [10]. In [20], the AVWC with noncausal side information at the jammer is considered. Noncausal side information here means that codewords are known at an active adversary before transmission.

The MiM attacks are active attacks where the smart attacker is also able to eavesdrop part of the communications but at the same time can participate in the communications by actively transmitting signals. Depending on the PHYSEC technique applied, there are different ways to deceive Alice and Bob. In the case of SKG, the attacker can send signals to disturb the SKG process [31, 37]. In case of the source model, e.g., from channel estimation in reciprocal channels, the attacker tries to disturb the channel estimates in the two directions differently. This differs from traditional jamming attacks where the goal is to destroy the communication between Alice and Bob [30, 33]. Technically speaking, for the case of MiM attack on PHYSEC key generation schemes, the MiM (denoted by $\mathcal{M}$) first listens to the exchanged packets between Alice and Bob to obtain the CSI of his communication links to the legitimate nodes. Gaining the knowledge of wireless channels, he then tries to inject "poisoned packets," i.e., adversarially precoded malicious signals, to cause a fake SoR at legitimate endpoints. The adversarial precoder is designed in a way that the same fake signal is "observed" at both legitimate parties, making Alice and Bob "believe" that the SoR is what $\mathcal{M}$ has sent. It is shown in [34] that the kernel matrix (a.k.a. null space) of the difference channel vectors between $\mathcal{M}$-to-Alice and $\mathcal{M}$-to-Bob can be exploited for the adversarial precoder of MiM attacks on PHYSEC key generation.

12.2 PHYSEC Key Generation

The PHYSEC key generation process generally contains four stages, namely, channel probing, quantization, information reconciliation, and privacy amplification. At the beginning, Alice and Bob alternately transmit pilot signals to obtain correlated channel measurements. Then, they quantize these analog channel measurements into digital bits. Although the uplink and downlink channels are reciprocal, due to the calibration errors in uplink/downlink RF chains, and the temporal variation of the channel and the noise, the measurements of the uplink and downlink channel are not identical but highly correlated. Next, information reconciliation is used to enable Alice and Bob to agree on the same key through error detection protocols or error correction codes [19]. Finally, privacy amplification eliminates any potential information leakage to eavesdroppers.

Based on channel reciprocity, the channel probing stage shares the common random sources between legitimate users to generate the secret keys. Most PHY key generation (PKG) implementations are realized in the time division duplex (TDD) mode in order to utilize the channel reciprocity. Specifically, Alice and Bob alternately transmit probing signals to estimate the CSI, where the sampling time difference between them should be smaller than the channel coherence time, indicating that the coherence time limits the pilot overhead.

A general assumption in SKG is that Eve is located at least half wavelength away from base station (BS) and all of the user terminals (UTs), because half wavelength is very close, e.g., 6.25 cm for 2.4 GHz and Eve might get detected within a distance below that. Therefore, Eve's channel observations are often assumed to be independent of those between the BS and UTs. Of course, the wireless decorrelation property of the channel depends heavily on the communications scenario.

Just imagine a scenario in which there are only line-of-sight paths between all nodes in the network. In this case, the flat fading channel realizations might be highly correlated and also do not provide sufficient randomness for generating secret keys. Therefore, it is important to keep the context of the PHYSEC key generation process in mind.

12.2.1 Learning-Aided PHYSEC for e-Health

In this part, we study a lightweight ML-aided scheme for PHYSEC key generation in a wireless-powered cooperative communication scenario. Exploiting the idea of reservoir computing [5] (i.e., utilizing a pool of recurrent neurons for computation and feature extraction), echo state networks (ESNs) are employed for the "randomness distillation" phase of PHYSEC key generation. To this end, one of the legitimate parties employs ML-based sequence-to-sequence (Seq2Seq) regression to predict the data sequence of the other side, with the goal of compensating for potential mismatches among legitimate parties. ESNs exploit a reservoir of neurons with short-term memory and sparse connections. This will be elaborated in more detail in Section 12.2.1.1.

According to Figure 12.1, assume a patient who is monitored via an e-health-enabled platform. Sensors are implanted on/in the body to measure personal medical information, such as heart rate, body temperature, and blood pressure. The data is gathered by a medical device, called Bob (B), and it is routed to an edge device, called Alice (A). As illustrated in Figure 12.1, Alice and Bob do not have a direct communication link because of heavy shadowing;

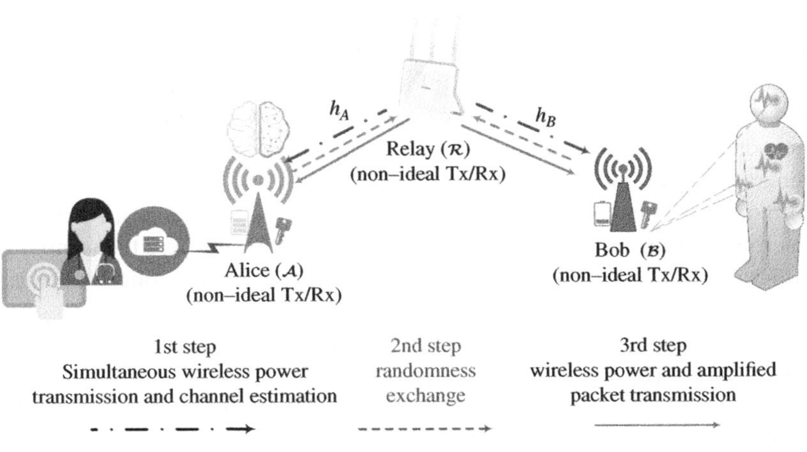

Figure 12.1 Learning-assisted PHYSEC key generation for wireless-powered cooperative e-health networks. Source: Adapted from [36].

or the direct link is too weak, such that an intermediate node, R, relays the information-bearing signals. The relay is programmed to amplify and forward the packets, without tampering with the contents of the packets [2].

The helper relay is assumed to also serve as an energy supplier for Alice and Bob via energy harvesting (EH). That is, the medical devices of Alice and Bob harvest the radio-frequency (RF) energy provided by R through *simultaneous wireless power and packet exchange* [58].

The concept of *reservoir learning* is exploited at the edge device of Alice to have a lightweight intelligent security solution for the system. By implementing an ESN, Alice learns the observation sequence of Bob, which is then utilized for SKG. The learning model is implemented at Alice's side, because it is connected to the medical cloud server, and it can support for computation resources of the algorithm.[2] The data sequence of observations are learned, which can be utilized as the source for wireless key generation.

Considering multi-carrier communication over N_c parallel block-fading channels [2], the wireless channel of A-to-R and B-to-R links at a given block $1 \leq i \leq N_c$ are denoted, respectively, by $\boldsymbol{h}_A^{(i)} \in \mathbb{C}^{n_R \times 1}$ and $\boldsymbol{h}_B^{(i)} \in \mathbb{C}^{n_R \times 1}$. We consider the channel model of [58] for our scheme.

For PHY key generation, a common SoR is first obtained from the PHY attributes. To this end, A and B start exchanging pilot packets with the help of R as illustrated in Figure 12.1. The local observations at legitimate endpoints can be modeled as given below:

$$y_A = \underbrace{\sqrt{1-\rho}\boldsymbol{h}_A^T \boldsymbol{G} \boldsymbol{h}_B x_B x_A}_{\text{Common Randomness}} + \underbrace{\varrho_A x_A}_{\text{Residual Terms}}, \tag{12.7}$$

$$y_B = \underbrace{\sqrt{1-\rho}\boldsymbol{h}_B^T \boldsymbol{G} \boldsymbol{h}_A x_A x_B}_{\text{Common Randomness}} + \underbrace{\varrho_B x_B}_{\text{Residual Terms}}, \tag{12.8}$$

where we have omitted the superscript (i) for notation brevity, i.e., the signals are expressed for each parallel sub-channel. In (12.7) and (12.8), x_A and x_B denote random pilot packets exchange by Alice and Bob, respectively.[3] Moreover, ρ portion of the power of the received signals is exploited for energy supply. Also note that $\boldsymbol{G}$ denotes the relaying matrix based on the amplification gain over each antenna of the relay. Moreover, ϱ_N for $N \in \{A, B\}$ formulates the residual uncertainties within the observations of legitimate parties, due to random noise,

2 Security of cloud connection is handled by the available networking standards, and it is not the focus of this chapter.

3 Pilots are drawn from QPSK constellation symbols with i.i.d. discrete uniform phase distribution $U\left(\left\{\pm\frac{\pi}{4}, \pm\frac{3\pi}{4}\right\}\right)$. It has been shown in [31, 34, 37] that it is an effective strategy against active attacks, while also enhancing the randomness richness.

channel estimation, and hardware mismatches [36]. We mention that (12.7) and (12.8) are obtained after canceling out the self interference signals from the received version of the exchanged pilots, and then post-multiplying the local data with the corresponding pilot signals x_A and x_B at Alice and Bob, respectively [36].

12.2.1.1 Neural Network Implementation

Invoking (12.7) and (12.8), although a common SoR can be observed by legitimate end-points, there also exist unbalanced residual terms within the observations of Alice and Bob. To solve this issue, we are inspired by the notion of *reservoir computing* [5] and employ an ESN that compensates the observation mismatches. Alice aims to predict the data sequence of Bob, via ESN-based sequence-to-sequence (Seq2Seq) prediction.

The idea of ESN is to employ utilize a relatively large reservoir of neurons with sparse connections and short-term memory, as depicted in Figure 12.2. At a given time-slot n, the input $\boldsymbol{u}(n)$ is the sequence of Alice's observations, and the ESN outputs a vector $\boldsymbol{y}_o(n)$ as the prediction about Bob's data. ESNs maintain a very simple training process. This is because the connections in the reservoir pool are assigned randomly, and only the output layer is trained during the training phase. The training process is carried out with the help of an offline training set $\mathcal{T} \triangleq \{(\boldsymbol{u}(i), \boldsymbol{y}_B(i))\}_{i\in[N_\mathcal{T}]}$, which is generated based on (12.7) and (12.8) for $N_\mathcal{T}$ (training size) realizations of communication channels. The loss function is to minimize the difference between Bob's observations (i.e., the desired output vector $\boldsymbol{y}_B$), and the predicted sequence at the output of the ESN, $\boldsymbol{y}_o(n)$. As a widely adopted

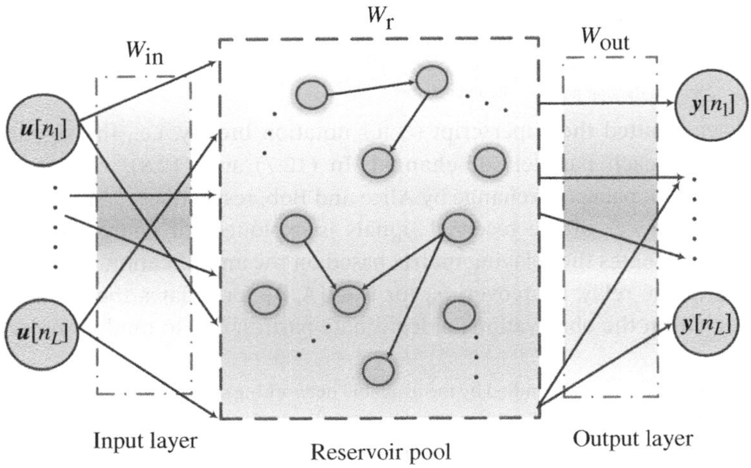

Figure 12.2 Echo state network neural architecture. Source: [36]/IEEE/CC BY 4.0.

measure, we employ mean-square error (MSE) as follows:

$$\mathsf{MSE} = \frac{1}{\mathsf{N}_{\mathcal{T}}} \sum_{i=1}^{\mathsf{N}_{\mathcal{T}}} ||\mathbf{y}_{\mathsf{o}}(i) - \mathbf{y}_{\mathsf{B}}(i)||^2. \tag{12.9}$$

After that the training is finished, for each round of relay-aided packet exchange between A and B, Alice passes her observation sequence to the ESN for ML-based prediction about Bob's data. This only requires a forward propagation from the input to the output layer of the trained ESN.

Figure 12.3 shows the observations mismatch between Alice and Bob. The figure indicates that the proposed approach showcases a generalized performance over a wide range of transmit powers while it has been trained for a fixed power $P_{\mathsf{R}} = 20\,\text{dBm}$. Figure 12.3 demonstrates that more than 32% improvement in terms of observation mismatch is obtained compared with a PHY-only SKG scheme [2, 30]. The PHY-only benchmark directly utilizes the PHY-extracted observations for the key generation process (quantization, information reconciliation, and hashing blocks), rather than employing any neural network. The ESN, which contains both a sparse version of the affine connections and the recurrent short-term memory of the previous states, outperforms the fully-connected neural network [17, 21] by about 15% gain. Moreover, Figure 12.4 studies the required

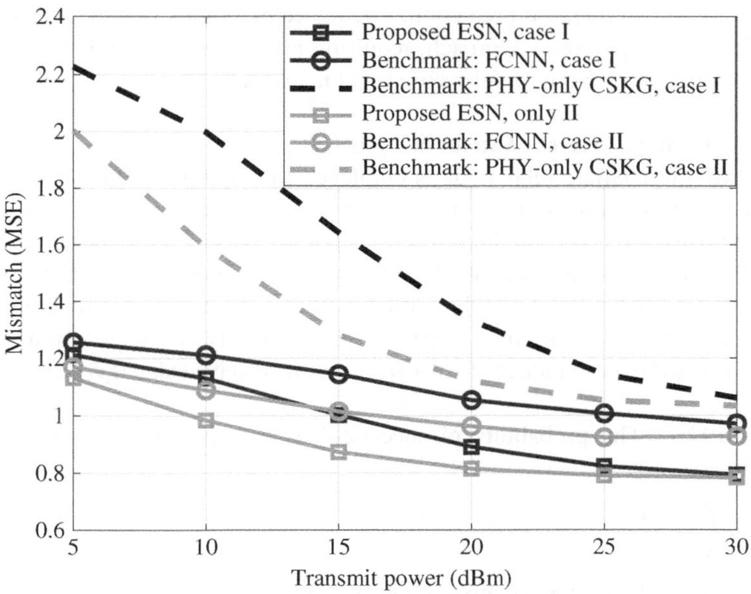

Figure 12.3 Observation mismatches for learning-aided versus conventional PHYSEC key generation. Source: Adapted from [36].

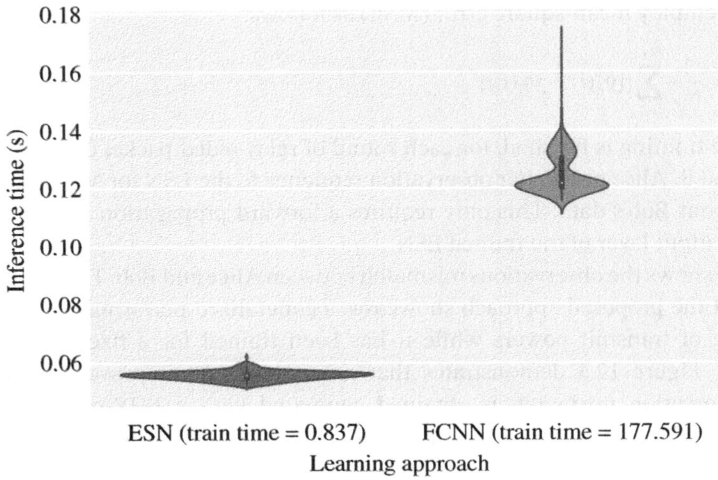

Figure 12.4 Inference time for ESN versus fully-connected network. Source: Adapted from [36].

training time and the inference time over 5000 time stamps. The distribution of inference time is plotted as a violin plot, which is empirically obtained from 450 runs of ML-based Seq2Seq prediction. The figure highlights the lightweight computations of the proposed approach, requiring much less computation time (and complexity) compared with fully-connected networks.

12.2.1.2 Information-Theoretic Secrecy Analysis

Although the intermediate relay node is employed to amplify and forward the packets, it might be curious about the content of the signals received. This is known as untrusted relay in the literature [30, 37]. We provide an upper bound for the probability of successful eavesdropping by R. Considering, without loss of generality, that A and B agree on the bit sequence of Alice ($\boldsymbol{q}_{\mathrm{A}}$) after reconciliation phase, we characterize this secrecy metric as the probability of the event that R correctly extracts the bit sequence of Alice $\boldsymbol{q}_{\mathrm{A}}$, given its received signal $\boldsymbol{y}_{\mathrm{R}}$ [2].

Proposition 12.1 The probability of successful eavesdropping, $\mathcal{P}^{\mathrm{E}}$, is upper bounded by

$$\mathcal{P}^{\mathrm{E}} \leq \left(1 - \frac{H(\boldsymbol{q}_{\mathrm{A}}) - I(\boldsymbol{y}_{\mathrm{A}};\boldsymbol{y}_{\mathrm{R}}) - 1}{\mathrm{N_q}}\right)^{N_c}, \tag{12.10}$$

where $\mathrm{N_q}$ denotes the number of bits used for mapping the observations into the corresponding bit sequence. The proof can be found in [36], which is obtained using Fano's inequality and data-processing inequality (DPI) [12]. Also note that obtaining a closed-form expression for the secret key rate and information leakage rate is not tractable due to the non-Gaussian behavior of the data sequences of Alice, Bob, and relay. We empirically study these metrics in Section 12.4 for a real-world setup.

In the following subsections, other variants of PHY-based SKG are studied, which can provide useful insights on how to reinforce the e-health wireless systems. Specifically, covert and MiM-resistant SKG schemes are addressed. Besides, a multiuser SKG scheme for MIMO systems is studied to provide a generic key-based PHYSEC solution that can be used for future e-health systems.

12.2.2 Covert or Stealthy SKG

Private eHealth data should be communicated anonymously in order to avoid attackers to map data to transmitters or receivers [26]. Furthermore, stealthy authentication is mentioned in eHealth applications and general 6G scenarios in [25]. Covert or stealthy SKG refers to the process of generating a secret key between two communication devices in a manner that is hidden from an eavesdropper. In covert key generation, the two devices may use various techniques to conceal the fact that they are generating a key. For example, they might use the cover traffic to generate the key, or they might embed the key generation process into seemingly harmless data. The key is generated using randomness that is hard for an eavesdropper to predict, such as the timing of packet transmissions, the measurement of fading in the wireless channel, or the thermal noise in the communication system.

Covert key generation is important in situations where the mere existence of a secret key must be kept confidential, such as in military or intelligence operations. However, covert communications also provide private and anonymous links for sensible eHealth data. The success of covert key generation depends on the ability of the two devices to generate a key that is both secure and hidden from an eavesdropper.

Consider the following motivating example, which is a simplified model specialized from [42]. Two agents, Alice and Bob, want to establish a communication that does not raise the curiosity of a warden Willie, whose duty is to monitor if there is any suspicious activity and also decrypts the data. The additional constraint against the warden Willie is formulated by

$$\mathbb{D}(P_Z||Q_Z) \leq \varepsilon, \tag{12.11}$$

where P_Z and Q_Z are the output distributions at Willie induced by meaningful and meaningless signals, respectively. The operational meaning of the constraint in (12.11) can be well explained by hypothesis testing [18]. By this viewpoint, if the constraint is fulfilled, the adversary's best strategy is to blindly guess whether the current transmitted signal is meaningful or meaningless.

The main results in [42] are summarized as follows.

Theorem 12.1 If (X, Y, Z) are drawn from the common random source described by the quadruple $(\mathcal{X}, \mathcal{Y}, \mathcal{Z}, P_{XYZ})$, then the stealthy strong secret key capacity C_{SK} of the source model with the stealth constraint is bounded by

$$\max\{I(X;Y) - I(X;Z), I(Y;X) - I(Y;Z)\} \leq C_{\text{SK}} \\ \leq \min\{I(X;Y|Z), I(X;Y)\}. \tag{12.12}$$

Furthermore, if (X, Y, Z) forms a Markov chain $X - Y - Z$, the stealthy strong secret key capacity is

$$C_{\text{SK}} = I(X;Y) - I(X;Z). \tag{12.13}$$

A sufficient condition to obtain the secret key capacity in (12.13) is derived in the next result.

Theorem 12.2 If a common random source $(\mathcal{X}, \tilde{\mathcal{Y}}, \tilde{\mathcal{Z}}, P_{X\tilde{Y}\tilde{Z}})$ is stochastically degraded such that $P_{\tilde{Y}|X} = P_{Y|X}$ and $P_{\tilde{Z}|X} = P_{Z|X}$, where $X - Y - Z$, then $C_{\text{SK}} = I(X;Y) - I(X;Z)$.

One possible application of this characterization is the fast fading Gaussian Maurer's (satellite) model [48]. The three observations $X = A_X S_0 + N_X$, $Y = A_Y S_0 + N_Y$, and $Z = A_Z S_0 + N_Z$ where $N_X, N_Y, \text{and} N_Z$ are independent AWGNs at Alice, Bob, and Willie, respectively, while both are with zero mean and unit variance. For A_X, A_Y, and A_Z fading distributions could be chosen. Taking advantage of the same marginal property [13], we can construct an equivalent channel, where we can identify the fading channels that can be reordered in the equivalent channel to keep the trichotomy order fixed [43].

Extension to the case where the warden can also actively influence the channel state is considered in [55]. It investigates the problem of covert and SKG over a state-dependent discrete memory-less channel with one-way public discussion in which an adversary, the warden, may arbitrarily choose the channel state.

12.2.3 SKG in Multiuser Massive MIMO

Massive MIMO technology is applied in the fifth generation of mobile communications as well as is envisioned for the sixth generation. SKG is a basic requirement for many cryptographic protocols and thereby relevant and important for private eHealth communications, too. For the SKG from the source model by channel reciprocity, the additional spatial degrees of freedom provide additional sources of randomness. Li et al. [39] address SKG with multiusers simultaneously in a narrow band massive MIMO system.

A BS simultaneously generates secret keys $\kappa = \{\kappa_1, \kappa_2, \ldots, \kappa_K\}$ with K UTs. The BS is equipped with M antennas and the kth UT is equipped with N_k antennas. Under the TDD operation, based on the reciprocal uplink and downlink channels, the BS generates the pairwise key κ_k with the kth UT.

As mentioned before, the context of the system including the channel model has an important impact on the achievable secret key rates. A geometric channel model with N_P paths is considered. Then, the $N_k \times M$ physical MIMO channel matrix in the downlink associated with the pth path of the kth UT can be expressed as [56].

$$\boldsymbol{H}_{k,p}^{\mathrm{DL}} = \alpha_{k,p}\mathbf{a}_{\mathrm{UT},k}(\theta_{k,p})\mathbf{a}_{\mathrm{BS}}^{H}(\varphi_{k,p}), \tag{12.14}$$

where $\alpha_{k,p}$ is the complex channel gain of the pth path, $\mathbf{a}_{\mathrm{UT},k}(\theta_{k,p})$ is the UT antenna array response vector with the AoA $\theta_{k,p}$, and $\mathbf{a}_{\mathrm{BS}}(\varphi_{k,p})$ is the BS antenna array response vector corresponding to AoD $\varphi_{k,p}$. For uniform linear array (ULA), these vectors are given by

$$\mathbf{a}_{\mathrm{UT},k}(\theta_{k,p}) = \frac{1}{\sqrt{N_k}}\left[1, e^{-j\frac{2\pi}{\lambda}d\sin(\theta_{k,p})}, \ldots, e^{-j(N_k-1)\frac{2\pi}{\lambda}d\sin(\theta_{k,p})}\right]^T$$
$$\mathbf{a}_{\mathrm{BS}}(\varphi_{k,p}) = \frac{1}{\sqrt{M}}\left[1, e^{-j\frac{2\pi}{\lambda}d\sin(\varphi_{k,p})}, \ldots, e^{-j(M-1)\frac{2\pi}{\lambda}d\sin(\varphi_{k,p})}\right]^T,$$

with wavelength λ and distance d between the adjacent antennas. For a narrow-band channel model, the channel coefficient of the kth UT is given by

$$\boldsymbol{H}_{k}^{\mathrm{DL}} = \sum_{p=1}^{N_P}\boldsymbol{H}_{k,p}^{\mathrm{DL}}. \tag{12.15}$$

The channel covariance matrices at the BS and the UT sides are $\boldsymbol{R}_{\mathrm{BS},k}$ and $\boldsymbol{R}_{\mathrm{UT},k}$. The uplink and downlink channels are reciprocal, i.e., $\boldsymbol{H}_k^{\mathrm{UL}} = (\boldsymbol{H}_k^{\mathrm{DL}})^T$, but the channel estimations are affected by noise. Beam domain transform samples the original physical channel by two series of uniformly distributed beams/angles over $[0, 2\pi]$, i.e., transmitting and receiving beams/angles. According to [57], the beam domain channel is

$$\tilde{\boldsymbol{H}}_k^{\mathrm{DL}} = \boldsymbol{A}_{\mathrm{UT},k}^{H}\boldsymbol{H}_k^{\mathrm{DL}}\boldsymbol{A}_{\mathrm{BS}}, \tag{12.16}$$

where $\boldsymbol{A}_{\mathrm{UT},k} = \left[\mathbf{a}_{\mathrm{UT},k}(\theta_1), \mathbf{a}_{\mathrm{UT},k}(\theta_2), \ldots, \mathbf{a}_{\mathrm{UT},k}(\theta_{N_k})\right] \in \mathbb{C}^{N_k \times N_k}$ and $\boldsymbol{A}_{\mathrm{BS}} = [\mathbf{a}_{\mathrm{BS}}(\varphi_1), \mathbf{a}_{\mathrm{BS}}(\varphi_2), \ldots, \mathbf{a}_{\mathrm{BS}}(\varphi_M)] \in \mathbb{C}^{M \times M}$ are the sampling matrices at the kth UT and the BS, respectively. They satisfy $\boldsymbol{A}_{\mathrm{UT},k}^{H}\boldsymbol{A}_{\mathrm{UT},k} = \boldsymbol{I}$, $\boldsymbol{A}_{\mathrm{BS}}^{H}\boldsymbol{A}_{\mathrm{BS}} = \boldsymbol{I}$. The (n, m)th element of $\tilde{\boldsymbol{H}}_k^{\mathrm{DL}}$ represents the channel gains from AoD φ_m to AoA θ_n, where φ_m and θ_n are the mth and nth sample angles, which satisfy $\sin(\varphi_m) = 2m/M - 1$ and $\sin(\theta_n) = 2n/N_k - 1$ [53]. When the antenna spacing is half the wavelength, that is, $d = \lambda/2$, the matrices $\boldsymbol{A}_{\mathrm{UT}}$ and $\boldsymbol{A}_{\mathrm{BS}}$ become the unitary discrete Fourier transform (DFT) matrix, defined as [64].

Theorem 12.3 When the number of antennas grows to infinity, the beam domain channel $\tilde{\boldsymbol{H}}_k^{\mathrm{DL}}$ tends to $\boldsymbol{G}_k^{\mathrm{DL}}$, i.e., for arbitrary n and m,

$$\lim_{M,N_k \to \infty} [\tilde{\boldsymbol{H}}_k^{\mathrm{DL}} - \boldsymbol{G}_k^{\mathrm{DL}}]_{n,m} = 0, \tag{12.17}$$

where $\boldsymbol{G}_k^{\mathrm{DL}} \in \mathbb{C}^{N_k \times M}$ is given by

$$\begin{aligned}[\boldsymbol{G}_k^{\mathrm{DL}}]_{n,m} = &\sum_{p=1}^{N_{\mathrm{P}}} \alpha_{k,p} \delta(\theta_{k,p} - \arcsin(2n/N_k - 1)) \\ &\times \delta(\varphi_{k,p} - \arcsin(2m/M - 1)).\end{aligned} \tag{12.18}$$

The proof is found in the appendix in [39].

In the beam domain, the channel is sparse. A few elements contain most of the channel information. The UT and the BS transmit sounding signals to acquire the statistical CSI. The BS and UTs design precoding and receiving matrices based on the statistical CSI, in order to reduce the dimension of channel estimation. Then, the effective channel parameters are estimated efficiently. After quantization, the information reconciliation and privacy amplification procedures are used to generate consistent and private secret keys, similarly with the point-to-point SKG. Therefore, the information reconciliation and privacy amplification phases are not highlighted in the following.

Let $\boldsymbol{S}_k^{\mathrm{DL}} \in \mathbb{C}^{M_e \times T_{\mathrm{D}}}$ denote the downlink pilot from BS to UT k within T_{D} time slots, which satisfies $\boldsymbol{S}_k^{\mathrm{DL}}(\boldsymbol{S}_k^{\mathrm{DL}})^H = \boldsymbol{I}$. By the least square (LS) estimation, UT k estimates the downlink CSI as

$$\begin{aligned}\boldsymbol{Z}_k^{\mathrm{DL}} &= \boldsymbol{Y}_k^{\mathrm{DL}}(\boldsymbol{S}_k^{\mathrm{DL}})^H = \boldsymbol{C}_k^H \boldsymbol{H}_k^{\mathrm{DL}} \boldsymbol{P}_k \\ &+ \boldsymbol{C}_k^H \boldsymbol{H}_k^{\mathrm{DL}} \sum_{k' \neq k} \boldsymbol{P}_{k'} \boldsymbol{S}_{k'}^{\mathrm{DL}} (\boldsymbol{S}_k^{\mathrm{DL}})^H + \boldsymbol{C}_k^H \boldsymbol{N}_k (\boldsymbol{S}_k^{\mathrm{DL}})^H,\end{aligned} \tag{12.19}$$

where $\boldsymbol{P}_k$ and $\boldsymbol{C}_k$ are the pre- and decoding matrices of link k.

In the uplink transmission, let $\boldsymbol{S}_k^{\mathrm{UL}} \in \mathbb{C}^{N_e \times T_{\mathrm{U}}}$ denote the pilot transmitted by UT k within T_{U} time slots, satisfying $\boldsymbol{S}_k^{\mathrm{UL}}(\boldsymbol{S}_k^{\mathrm{UL}})^H = \boldsymbol{I}$. By employing the LS estimation, the estimated effective channel of UT k can be expressed as

$$\begin{aligned}\boldsymbol{Z}_k^{\mathrm{UL}} &= \boldsymbol{Y}^{\mathrm{UL}}(\boldsymbol{S}_k^{\mathrm{UL}})^H = \boldsymbol{P}_k^T \boldsymbol{H}_k^{\mathrm{UL}} \boldsymbol{C}_k^* \\ &+ \boldsymbol{P}_k^T \sum_{k' \neq k} \boldsymbol{H}_{k'}^{\mathrm{UL}} \boldsymbol{C}_{k'}^* \boldsymbol{S}_{k'}^{\mathrm{UL}} (\boldsymbol{S}_k^{\mathrm{UL}})^H + \boldsymbol{P}_k^T \boldsymbol{N} (\boldsymbol{S}_k^{\mathrm{UL}})^H.\end{aligned} \tag{12.20}$$

In the uplink and downlink transmissions, the BS and UTs vectorize the estimated effective channel matrices $\boldsymbol{Z}_k^{\text{UL}}$ and $\boldsymbol{Z}_k^{\text{DL}}$ and employ them to generate the secret key, where the reciprocal component between the BS and UT k is $\boldsymbol{C}_k^H \boldsymbol{H}_k^{\text{DL}} \boldsymbol{P}_k$ with a small dimension of $N_{\text{e}} \times M_{\text{e}}$. The number of secure bits for the link from the BS to UT k in the mutual information can be expressed as [41]

$$I_k = \min_{i \neq k} I(\mathbf{z}_k^{\text{DL}}; \mathbf{z}_k^{\text{UL}} | \mathbf{z}_i^{\text{DL}}), \tag{12.21}$$

where $\mathbf{z}_k$ is the realization of the random variable Z_k. Denote the precoding and receiving matrices in the beam domain as $\tilde{\boldsymbol{P}}_k = \boldsymbol{A}_{\text{BS}}^H \boldsymbol{P}_k$ and $\tilde{\boldsymbol{C}}_k = \boldsymbol{A}_{\text{UT},k}^H \boldsymbol{C}_k$, respectively. Let $\boldsymbol{V}_k = \boldsymbol{\Lambda}_k^{1/2} \left(\sum_{k'} (\tilde{\boldsymbol{P}}_{k'} \boldsymbol{S}_{k'k}^{\text{DL}})^T \otimes \tilde{\boldsymbol{C}}_k^H \right)^H$ and $\boldsymbol{V}_{kk'} = \boldsymbol{\Lambda}_{k'}^{1/2} \left(\tilde{\boldsymbol{P}}_k^T \otimes \boldsymbol{S}_{k'k}^{\text{UL}} \tilde{\boldsymbol{C}}_{k'}^H \right)^H$, where $\boldsymbol{\Lambda}_k = \mathbb{E}\{\text{vec}(\tilde{\boldsymbol{H}}_k)\text{vec}(\tilde{\boldsymbol{H}}_k)^H\}$ is the full correlation of the beam domain channel.

Theorem 12.4 When the channels of different UTs become independent, the secret key rate of the kth UT is given by

$$\begin{aligned} I(\mathbf{z}_k^{\text{DL}}; \mathbf{z}_k^{\text{UL}}) = -\log\det \Bigg(& \boldsymbol{I} - \boldsymbol{V}_{kk} \left(\sum_{k'} \boldsymbol{V}_{kk'}^H \boldsymbol{V}_{kk'} + \left(\boldsymbol{P}_k^T \boldsymbol{P}_k^* \otimes \boldsymbol{I}_{T_{\text{U}}} \right) \right)^{-1} \boldsymbol{V}_{kk}^H \\ & \times \boldsymbol{V}_k \left(\boldsymbol{V}_k^H \boldsymbol{V}_k + \boldsymbol{I}_{T_{\text{D}}} \otimes \boldsymbol{C}_k^H \boldsymbol{C}_k \right)^{-1} \boldsymbol{V}_k^H \Bigg). \end{aligned} \tag{12.22}$$

The proof of the statement can be found in the appendix in [39].

We conclude this section with a numerical evaluation of the achievable secret key rate in a multiuser massive MIMO system. Figure 12.5 compares the secret key rate for different number of users. From the results, we can find that as the number of UTs increases, the secret key rate grows up approximately linearly. This could be explained by the spatial degrees of freedom which result in the corresponding multiplexing gain. For example, the key rate of 4 UTs approaches twice than that of 2 UTs. However, as the number of UTs continues increasing, the key rate of each UT becomes lower, due to the interference or information leakage among UTs.

12.2.4 Robust MiM Attack-Resistant SKG for Multi-carrier MIMO Systems

In this section, we will take a closer look at the challenges for SKG which were kept outside the presentations in the last two sections. In [40], the bottlenecks of imperfect channel reciprocity, nearby attack, and high temporal auto-correlation on the performance of PHYSEC SKG are discussed and a new method called Two-Band Multiple-Antenna Loop-bAck key generation (TB-MALA) derived.

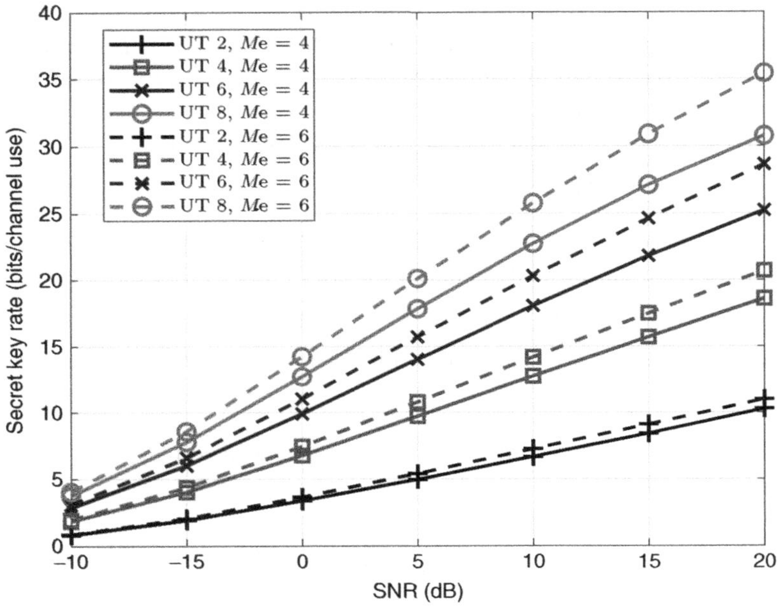

Figure 12.5 Secret key rate comparison for different number of UTs. Source: [39]/with permission of IEEE.

The following practical challenges occur in channel-based SKG [40]: (i) Imperfect channel reciprocity: Channel reciprocity is a primal condition for the PKG to succeed. The two-way uplink and downlink channels are in fact estimated involving the impacts of transceiver impairments. It leads to imperfect reciprocity with mismatches between transmit and receive modes [44]. The imperfect reciprocity causes significant differences between keys and results in decryption failure [65]. (ii) Risk of nearby eavesdroppers: It is usually believed that wireless channels are independent by a minimum distance of multiple wavelengths. When eavesdroppers are close to either of the legitimate nodes, their estimates of wiretap channel may be highly correlated with the legitimate channel. (iii) High temporal auto-correlation: Temporal variation is an essential concern in the design as it contributes to sufficient random keys during a period of time. When the user mobility is limited, the channel variation is mostly provided by the movement of nearby scatters. In this case, the channel fluctuation is usually insufficient and the channel measurement shows high auto-correlation, which lead to low-entropy keys.

Most of the challenges, at least one and two, can be approached by a proper design of privacy amplification [4]. However, if the advantage of the legitimate link over the wiretapper is too small, there might be no room for privacy amplification.

Therefore, we will tackle the challenges on the physical layer with smart signal processing methods.

The typical three node setup for SKG is considered. A MIMO link between Alice and Bob is assumed which secures the communication by using physical-layer secret keys. Alice and Bob dynamically generate symmetric keys by exploiting the randomness of their corresponding channels with reciprocity. Meanwhile, there exists a passive eavesdropper, namely, Eve who intends to infer the secret key based on its own channel observations and the information transmitted over public channels.

Denote the channel from user i to user j by $\boldsymbol{P}_{ij} \in \mathbb{C}^{N_j \times N_i}$ where $i, j \in \{A, B, E\}$. Let us assume that perfect reciprocity of the physical channel holds, i.e., $\mathbf{P}_{ij}(t+\tau) = \mathbf{P}_{ji}^T(t)$, when τ is a time shift no larger than the coherence time. Let us also assume that we have a rich scattering environment and all channels are full rank.

With the consideration of RF front ends, an effective channel model is established. In general, the effective channel comprises a cascade of transmitter gain, physical propagation channels, and receiver gain. The effective channels of Alice and Bob, i.e., $\boldsymbol{H}_{\mathrm{A}}$ and $\boldsymbol{H}_{\mathrm{B}}$, are written as $\mathbf{H}_{\mathrm{A}} = \mathbf{R}_{\mathrm{A}}\mathbf{P}_{\mathrm{BA}}\mathbf{T}_{\mathrm{B}}$, and $\mathbf{H}_{\mathrm{B}} = \mathbf{R}_{\mathrm{B}}\mathbf{P}_{\mathrm{AB}}\mathbf{T}_{\mathrm{A}}$ where diagonal matrices $\boldsymbol{T}_{\mathrm{A}}$ and $\boldsymbol{R}_{\mathrm{A}}$, respectively, represent the transmit and receive gains at Alice, $\boldsymbol{T}_{\mathrm{B}}$ and $\boldsymbol{R}_{\mathrm{B}}$, respectively, represent the transmit and receive gains at Bob. Similarly, the effective channels of Eve are denoted by $\boldsymbol{H}_{\mathrm{E1}} = \boldsymbol{R}_{\mathrm{E}}\boldsymbol{P}_{\mathrm{AE}}\boldsymbol{T}_{\mathrm{A}}$ and $\boldsymbol{H}_{\mathrm{E2}} = \boldsymbol{R}_{\mathrm{E}}\boldsymbol{P}_{\mathrm{BE}}\boldsymbol{T}_{\mathrm{B}}$.

The main idea of the proposed new SKG scheme is to exploit the frequency diversity of the channel. In TDD systems with multi-carrier modulations, multiband (MB)-PHY can be implemented by selecting different subcarriers for each transmission. Since the channel estimates are also necessary for data demodulation, it does not impose extra complexity for channel estimation. Band f_1 and band f_2 are two distinct bands with a band separation $\Delta f = |f_1 - f_2|$ greater than the coherence bandwidth of the physical channel; therefore the physical channels over band f_1 and band f_2 are assumed to be independent.

Theorem 12.5 For the TB-MALA, Alice and Bob achieve in probability the same raw key that

$$\mathbf{k}_{\mathrm{A}}' = \mathbf{k}_{\mathrm{B}}' = \mathbf{Q}\left(\mathrm{vec}\left((\mathbf{H}_{\mathrm{B}}^{f_1})^T \odot \mathbf{H}_{\mathrm{A}}^{f_2} \odot \mathbf{G}_{\mathrm{A}} \odot \mathbf{G}_{\mathrm{B}}^T\right)\right), \tag{12.23}$$

for asymptotic high signal-to-noise ratios (SNRs).

The proof is provided in [40].

Furthermore, the proposed SKG scheme is secure against the multiply-divide (MD) attack, the nearby attack, and it can reduce the temporal auto-correlation [[40], Section V].

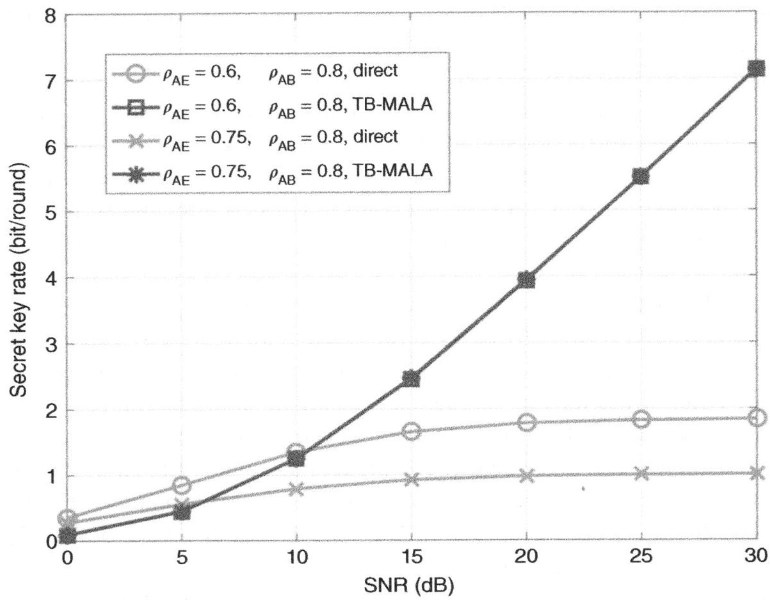

Figure 12.6 Secret key rate comparison of the direct and TB-MALA schemes Source: [40]/with permission of IEEE.

In the following numerical assessment, it is illustrated that the correlation between the channels to the eavesdropper and those of Alice and Bob has a significant impact on the performance of the standards SKG scheme, while TB-MALA is robust against this nearby-attack.

Figure 12.6 compares the secret key rate of the direct scheme and the TB-MALA scheme, where the correlation coefficient between Alice and Bob is 0.8, and the correlation between Alice and Eve is 0.6 and between Bob and Eve 0.75. From the numerical results, we can find that the secret key rate of the TB-MALA increases with the SNR and it keeps constant with the increase of the correlation between Alice and Eve ρ_{AE}.

For further numerical assessments and discussions, we refer the interested reader to [40].

12.3 Key-less PHYSEC for Medical Image Transmission

In this section, a content-aware key-less secure scheme is studied for medical image transmission in wireless healthcare. We take into account the fact that different regions of an image maintain different importance levels from the security perspective. More specifically, the region of interest (RoI) segments mostly contain the diagnostic information and require more stringent reliability and confidentiality requirements compared with the background (BG) segment. We show that key-less PHYSEC can provide a reliable medical image transmission, while taking confidentiality and delay limits into account.

Conventionally, the PHYSEC methods consider secrecy capacity (SC) as the performance metric to provide perfect secrecy [30, 33]. This approach for secure transmission requires a much lower data rate, which can result in transmission delay. On the contrary, the fountain coding (FC) scheme has been proposed as a low-cost alternative for PHYSEC [54]. This approach is capable of significantly increasing the transmission rate (bounded by Shannon capacity rather than SC) between the legitimate nodes [54]. We will exploit the FC scheme in this section for key-less PHYSEC.

12.3.1 Content- and Delay-Aware Design

Consider a multiple-input single-output (MISO) wireless image transmission system, as depicted in Figure 12.7. A medical device with n_T antennas is assumed as the source node S, aiming to send medical images to a legitimate destination node, denoted by D, which can be the a single-antenna medical device of a doctor. Randomly located non-colluding single-antenna Eves are considered to overhear the communication, and it is assumed that they are located randomly on a two-dimensional (2D) plane according to the homogeneous Poisson point process Φ_E with density λ_E. That is, the ith Eve is located at a random distance r_i from S, where $i \in \Phi_E$.

The source image is split into N source packets each representing their corresponding content with b bits. At each transmission slot (TS), a subset of packets are selected and encoded via FC method, to construct the transmitting frame. The duration of each TS is denoted by T, and the available bandwidth is denoted by B. The frame is then sent over the wireless link via employing prevalent beamforming techniques. The legitimate destination node D and also Eves aim to decode a sufficient number of packets to be able to recover the original data corresponding to the medical image. A feedback link from D to S exist which is used for acknowledging the packets received at D successfully. Transmission stops once D successfully decodes N independent fountain-coded source packets, or when the delay limit arrives. Confidentiality requirement is satisfied as long as the number of independent fountain-coded packets obtained by Eves is less than N. The operational meaning of this concept is formulated in [35].

The channel vector from S to any node j is denoted by $\mathbf{h}_{sj} = G\boldsymbol{h}_{sj} r_j^{-\frac{\eta}{2}}$, where r_j shows the corresponding distance between the nodes, G denotes the path-loss constant which depends on the operating frequency, and $\boldsymbol{h}_{sj} \in \mathbb{C}^{n_T \times 1}$ denotes

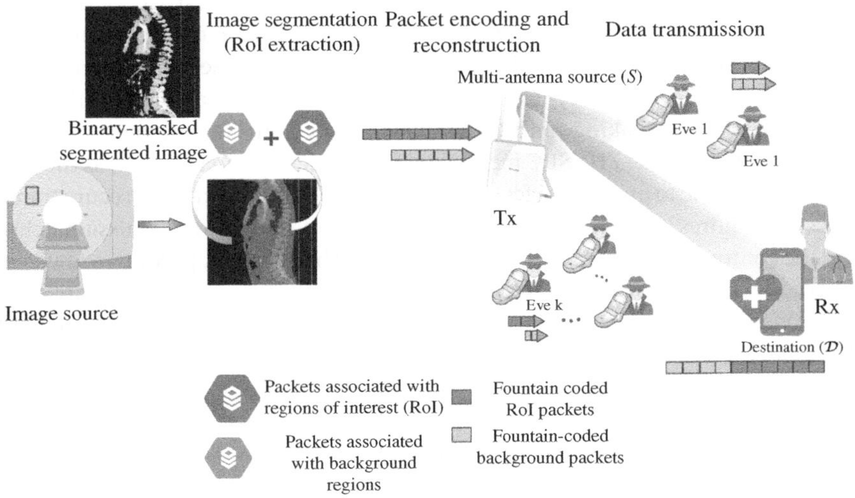

Figure 12.7 Secure medical image transmission. Source: [32]/with permission of IEEE.

the small-scale fading vector, with i.i.d. elements $\boldsymbol{h}_{sj,i} \sim \mathcal{CN}(0,1), 1 \leq i \leq n_{\mathrm{T}}$. Accordingly, the small-scale channel gain is denoted by $g_{sd} \triangleq ||\boldsymbol{h}_{sd}||^2$.

For the image transmission, the original image is first segmented into two sets of RoI packets Π_{roi} and BG packets Π_{bg}, where $\Pi_{\mathrm{roi}} \cap \Pi_{\mathrm{bg}} = \emptyset$, $|\Pi_{\mathrm{roi}}| = N_{\mathrm{roi}}$, and $|\Pi_{\mathrm{bg}}| = N_{\mathrm{bg}} = N - N_{\mathrm{roi}}$. As a content-aware transmission, source packets have different priorities, according to their content, for being transmitted. To elaborate, RoI-based pixels corresponding to salient segments and pixels containing edges are more likely to possess greater significance for an observer. Hence, a utility function is formulated as given below, to assess RoI and BG packets:

$$u(\pi) = w\varphi_{\mathrm{edge}}(\pi) + \varphi_{\mathrm{sal}}(\pi), \pi \in \{\Pi_{\mathrm{roi}}, \Pi_{\mathrm{bg}}\}, \tag{12.24}$$

with $w \geq 0$ denoting the weightening, and φ_{edge} and φ_{sal} denoting the average edge value and the average saliency value of each pixel, respectively. Then, the RoI and BG packets can be sorted from the highest to the lowest priority $\Pi_{\mathrm{roi}} = \{\pi_1^{\mathrm{roi}}, \pi_2^{\mathrm{roi}}, \dots, \pi_{N_{\mathrm{roi}}}^{\mathrm{roi}}\}$ and $\Pi_{\mathrm{bg}} = \{\pi_1^{\mathrm{bg}}, \pi_2^{\mathrm{bg}}, \dots, \pi_{N_{\mathrm{bg}}}^{\mathrm{bg}}\}$.

Without loss of generality, consider the tth TS. The source node S calls the most recently recovered source packet, and XORs it with $L(t)$ unrecovered packets resulting in $L(t)$ coded packets, $c_l^{(t)}, 1 \leq l \leq L(t)$, where the choice of $L(t)$ is provided next. Mathematically, we have

$$c_l^{(t)} = \pi_r^k \oplus \pi_l^k, \quad \text{for } 1 \leq l \leq L(t), k \in \{\mathrm{roi}, \mathrm{bg}\}, \tag{12.25}$$

where the subscript r shows that the packet π_r^k has been successfully recovered previously. A capacity-achieving code at the physical layer (PHY) is then applied to the fountain-coded packets $\mathbf{c}^{(t)} = [c_0^{(t)}, \dots, c_{L(t)}^{(t)}]$ to construct the transmitting frame. In addition, in order to realize the superiority for the legitimate link, compared to the eavesdropping links, maximum ratio transmission (MRT) beamforming technique is employed. Then, $L(t)$ can be formulated as follows:

$$L(t) = \left\lfloor \frac{BT}{b} \log_2(1 + \gamma_D) \right\rfloor, \tag{12.26}$$

where $\lfloor \cdot \rfloor$ denotes the floor operator. Moreover, $\gamma_D = \rho G^2 r_d^{-\eta} ||\boldsymbol{h}_{sd}||^2$, which formulates the instantaneous signal to noise ratio (SNR) at the destination, and ρ denotes the transmit SNR. We note that this choice for $L(t)$ is to make sure that the transmitted packets can be correctly decoded by D when capacity-achieving codes are employed.

Defining the instantaneous channel gain as $g_{sd} \triangleq ||\boldsymbol{h}_{sd}||^2$, during each TS, the prioritized RoI packets $\left\{ \pi_1^{\text{roi}}, \dots, \pi_{L(t)}^{\text{roi}} \right\}$ are XORed with a most recently recovered RoI packet π_r^{roi}, if g_{sd} is greater than the threshold $\tau(t) = \frac{\nu}{\exp(\theta)}$. Otherwise, $L(t)$ BG packets are used for coding and transmission. Mathematically speaking, we have the following two events:

$$\begin{cases} \Psi_1, & \text{if } g_{sd} > \frac{\nu}{\exp(\theta)}, \\ \Psi_0, & \text{if } g_{sd} \le \frac{\nu}{\exp(\theta)}, \end{cases} \tag{12.27}$$

where $\theta \triangleq t/T_{\text{lim}}$ and T_{lim} is the delay limit. Notably, $\exp(\theta)$ shows the urgency of completing image transmission given the delay limit T_{lim}. Moreover, $\nu > 0$ is a constant. The reason behind considering a time-varying threshold can be explained as follows. We aim to be compatible with the delay limits of system. Accordingly, when the number of passed TSs increases, the threshold decreases, meaning that we tend to send RoI packets with a less stringent limit on the quality of the legitimate link, while only less important packets (BG and RoI) remain.

12.3.1.1 Security Level Adjustment

Considering the fact that RoI packets contain important diagnostic information about the medical images, we further aim to limit the intercept probability (IP) at the distributed Eves when RoI packets are transmitted. By defining the instantaneous SNR at passive Eves $\gamma_{\text{E}_i} = \rho G^2 r_i^{-\eta} |\boldsymbol{w}^{\text{T}} \boldsymbol{h}_{se_i}|^2$ (with $\boldsymbol{w}$ showing the beamforming vector), and $\gamma_{\text{E}} \triangleq \max_{i \in \Phi_{\text{E}}} \gamma_{\text{E}_i}$ as the equivalent wiretap channel of randomly located Eves, the IP at a given TS should satisfy

$$\mathsf{IP}(t) \triangleq \mathsf{Pr} \left\{ \frac{BT}{b} \log_2(1 + \gamma_{\text{E}}) > L(t) \middle| \Psi_1 \right\} \le \epsilon_{\mathsf{IP}}. \tag{12.28}$$

One can adjust ϵ_{IP} to set different security levels. Moreover, the appropriate threshold value ν in (12.27) can be found by satisfying (12.28).

12.3.1.2 Evaluations

For the numerical evaluations, we consider a 512-by-512 chest computed tomography (CT) in the format of uint-16 as the original medical image to be transmitted, Figure 12.8 shows the total TSs required for the legitimate receiver to successfully decode all data packets corresponding to the medical image. The total delay limit is set to $T_{\text{lim}} = 3000$ TSs. The destination node is placed at location [20 m, −20 m] on a 2D plane, the source is placed at the origin, while Eves are randomly located according to Poisson point process (PPP) with density $\lambda_{\text{E}} = 0.0002$. according to the figure, by relaxing the security requirement, i.e., increasing the IP, the required time for image recovery decreases. The figure further indicates an important trade-off between security and quality of service (QoS) requirements. This trade-off highlights that the strict constraint of perfect secrecy could be relaxed by accepting a limited level of interception, in exchange for significantly improving the QoS-related performance metrics. For example, with 10 dBm transmit power, an IP of 0.15 is a good choice to ensure image content secrecy, while also improving the image recovery time. This sheds light on new methods towards application-centric security design.

To visualize the experimental result, Figure 12.9 is provided, highlighting that all of the image packets are successfully delivered and decoded by $\mathcal{D}$ without any

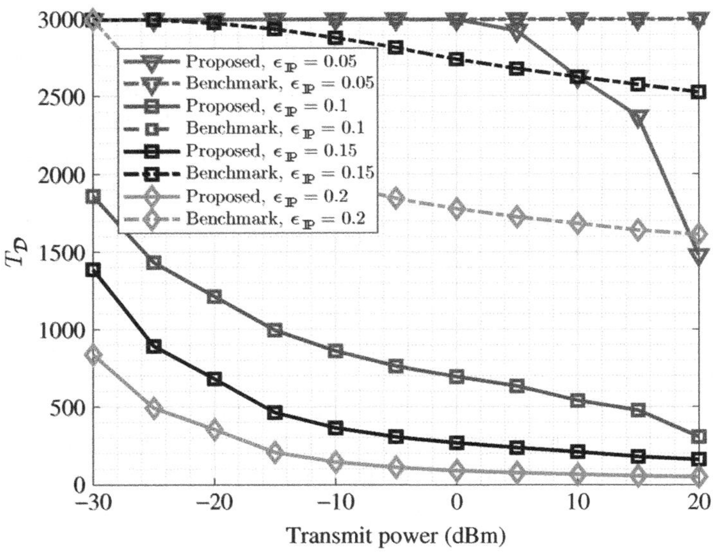

Figure 12.8 Required time for $\mathcal{D}$ to complete recovering the entire file versus transmit power (in dBm) for different values of ϵ_{IP} [32]/with permission of IEEE.

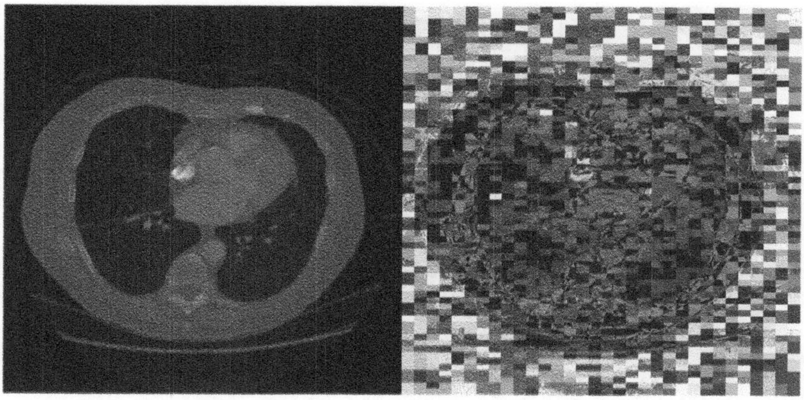

Figure 12.9 Recovered medical image at $\mathcal{D}$ versus the recovered one at Eves . Source: [32]/with permission of IEEE.

distortion, while no useful information can be obtained by Eves due to their heavily distorted image.

12.4 Proof-of-Concept Study

In this section, we provide a PoC testbed for group SKG and agreement protocol based on PHYSEC to validate the feasibility of PHYSEC protocols for a multiuser wireless system.[4] A cluster of ESP-32 smart modules[5] are employed for this purpose as shown in Figure 12.10. Secrecy of the studied PoC is experimentally evaluated in terms of the mutual information metric and secret key rate (SKR).

We consider a cluster of smart modules including a cluster head (CH), which can exchange data with the intra-cluster clients and the CH of other subnetworks. In order to secure the intra-cluster over-the-air connections, we propose to implement a Group SKG protocol on-device, where the smart modules collaboratively extract groupwise secret keys over time and in a fully-decentralized manner. The inter-cluster communications between different sub-networks are realized through the communication of CHs, where their corresponding wireless security can be safeguarded by the prevalent point-to-point (P2P) SKG schemes studied in the previous sections. For the G-SKG protocol, a source of "common randomness"

4 The authors would like to thank Mohammad Fahim for his help in implementing the algorithms.
5 https://docs.espressif.com/projects/esp-idf/en/latest/esp32.

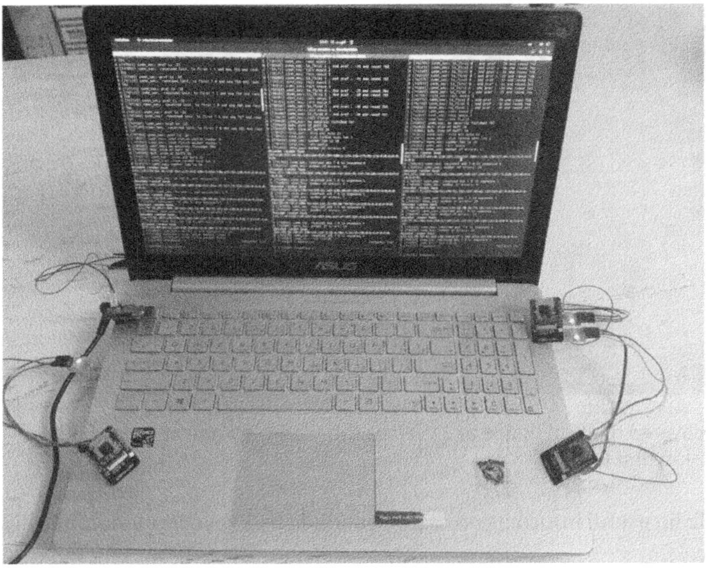

Figure 12.10 Implemented testbed for group SKG.

is required as the raw data to be distilled for extracting the shared secret key. To this end, the received signal strength indicator (RSSI) values are taken into account as a promising PHY attribute for PHYSEC key generation algorithms. During each round of PHY attribute distillation, the CH sends a pilot packet to the clients within the cluster. The first node receiving the pilot is considered as the reference node and introduces itself to the CH by sending back a response, upon which the corresponding RSSI is calculated at the CH as the reference PHY attribute $\texttt{RSSI}_{\text{ref}}$. The rest of the clients then take turns exchanging pilot packets (and obtaining their corresponding RSSIs) with the CH, respectively. By broadcasting the difference between $\texttt{RSSI}_{\text{ref}}$ and the individual RSSIs obtained at the CH, regular clients can find an imperfect copy of reference PHY attributes, using their local RSSIs.

Figure 12.11 shows the obtained RSSI reference attributes at the intra-cluster nodes (ESP-32 modules), which indicates that the legitimate nodes have been successful in obtaining a common SoR. Moreover, the secrecy of our PoC solution in terms of mutual information (MI) and SKR are illustrated in Figures 12.12 and 12.13, respectively. The empirical mutual information values are calculated numerically [63]. We employed an esp-32 module as an external node who listens to the exchanged packets, playing the role of an eavesdropper. Remarkably, one can see that the information leakage to the eavesdropper in a practical testbed

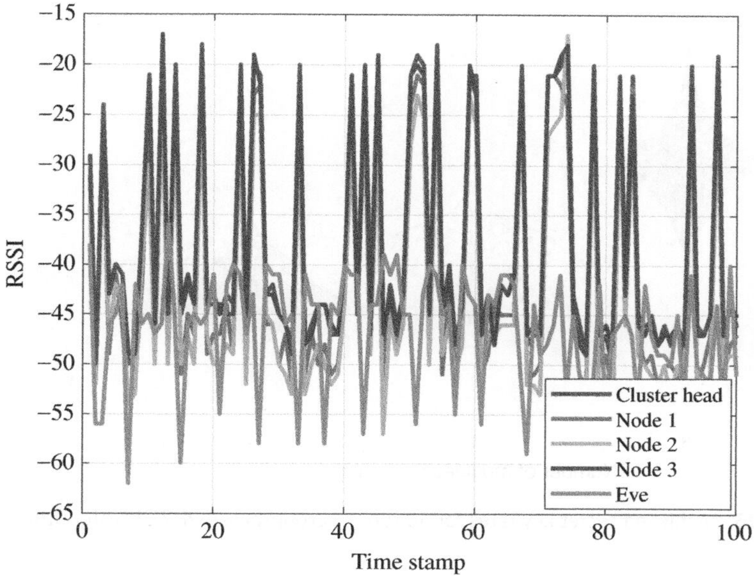

Figure 12.11 Raw RSSI values obtained from the testbed.

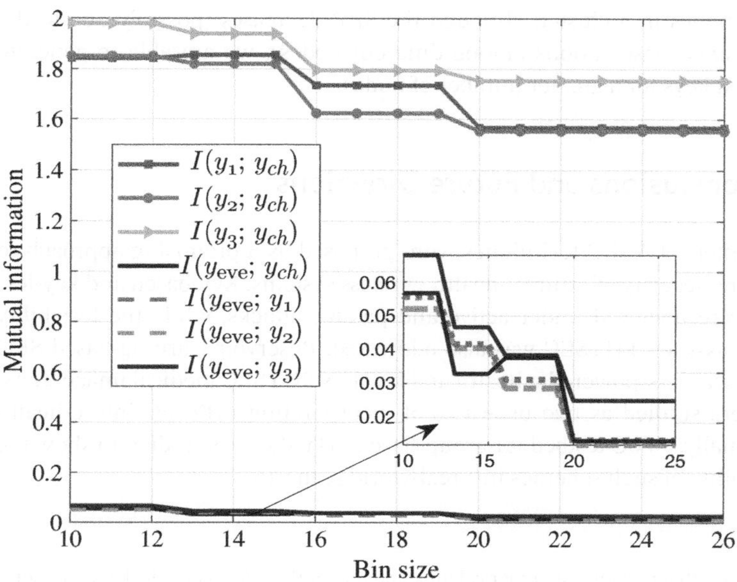

Figure 12.12 Secrecy performance in terms of mutual information metric.

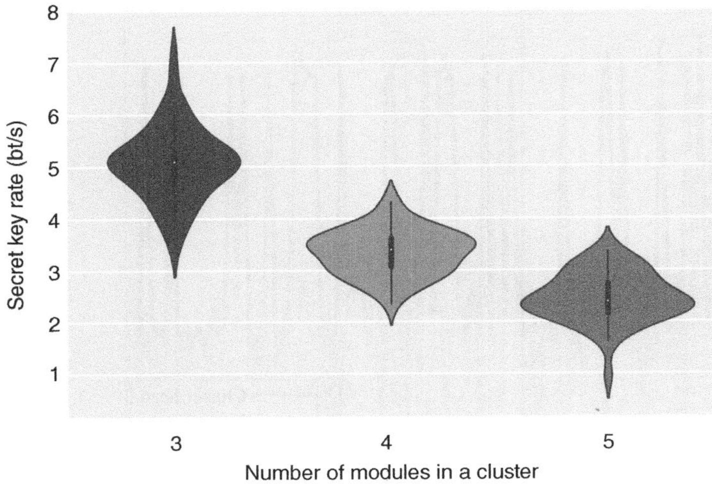

Figure 12.13 Achieved secret key rates for 50 runs of group-based PHY SKG algorithm.

is negligible, which clearly verifies the theoretical models and assumptions.[6] The SKRs achieved over 50 rounds of key agreement, are plotted in Figure 12.13 as a distribution violin plot. A trade-off can be observed that by increasing the number of nodes in a cluster, although the SKR decreases (possibly due to the more diversified observations among different nodes), we are able to generate secret shared keys for a greater number of nodes.

12.5 Conclusions and Future Directions

In this chapter, PHYSEC techniques were proposed as a promising approach to safeguard the security of future e-health wireless systems. Key-based and key-less PHYSEC were reviewed under active and passive attacks, while the feasibility of learning-assisted PHYSEC was also addressed. Reservoir learning-based SKG scheme for wireless-powered e-health and key-less PHYSEC medical image transmission were studied as two use-cases of incorporating PHYSEC into e-health systems. Finally, a PoC testbed for group-based PHYSEC was studied to show the generalizability of such schemes in a real-world scenario.

---

6 We emphasize that providing a numerical upper bound for the information leakage will not be accurate and generic in this case, since we are calculating approximations of the mutual information numerically, and such algorithms have different estimation errors depending on the context and scenarios.

The following items can be suggested for the future research directions at the intersection of PHYSEC and e-health:

(1) Very recently, thanks to reconfigurable intelligent surfaces (RISs) with their excellent ability on electromagnetic wave control, the wireless channel environment can be customized [61]. The survey paper [14] overviews the RIS-aided PHY-SKG in a static indoor environment, including its potential application scenarios, channel model, and hardware architectures.
(2) In terms of learning-aided e-health platforms, scalability and privacy awareness are two important factors required for the learning algorithms used for healthcare wireless systems. In this regard, the concept of federated learning (FL) has gained significant attention within the recent years [50]. The general idea of FL is to share the local model updates of the learning-based e-health nodes. Hence, securing the over-the-air communications of such nodes with the aid of PHYSEC techniques needs careful investigation in terms of the *secrecy-utility trade-off* for FL systems. In addition, privacy and security challenges of emerging digital healthcare technologies, such as Metaverse healthcare, are of pivotal importance, where PHYSEC techniques could be leveraged to reinforce the security [28, 29].
(3) The challenging aspect of utilizing FL-aided platforms for digital healthcare systems is that FL relies upon a coordinator node (i.e., the aggregator server), that might be honest-but-curious from the security perspective [27]. Then, the idea of swarm AI is also proposed as a fully-decentralized platform for healthcare applications [59]. However, the extension of swarm learning ideas to be deployed over wireless healthcare systems can raise critical challenges in terms of ensuring security of the shared models. In this regard, the PHYSEC approaches can be leveraged to help improve the secrecy of over-the-air swarm learning algorithms.
(4) For key-less PHYSEC over e-health networks, an important research direction would be to study learning-aided key-less schemes that consider joint source and channel coding via deep learning [24]. This way, a model-independent end-to-end transmission scheme can be realized without being restricted to any specific transmission model or modulation scheme. This concept, which is already known as DeepJSCC, should be further studied in order to realize a secure version of it when we aim to apply DeepJSCC to e-health systems that deal with vital data of many people.

References

1 R. Ahlswede and I. Csiszár. Common randomness in information theory and cryptography - I: Secret sharing. *IEEE Transactions on Information Theory*, 39(4):1121–1132, 1993.

2 N. Aldaghri and H. Mahdavifar. Physical layer secret key generation in static environments. *IEEE Transactions on Information Forensics and Security*, 15:2692–2705, 2020. doi: 10.1109/TIFS.2020.2974621.

3 B. A. Bash, D. Goeckel, D. Towsley, and S. Guha. Hiding information in noise: Fundamental limits of covert wireless communication. *IEEE Communications Magazine*, 53(12):26–31, 2015. doi: 10.1109/MCOM.2015.7355562.

4 C. H. Bennett, G. Brassard, C. Crepeau, and U. M. Maurer. Generalized privacy amplification. *IEEE Transactions on Information Theory*, 41(6):1915–1923, 1995. doi: 10.1109/18.476316.

5 F. M. Bianchi, S. Scardapane, S. Løkse, and R. Jenssen. Reservoir computing approaches for representation and classification of multivariate time series. *IEEE Transactions on Neural Networks and Learning Systems*, 32(5):2169–2179, 2021. doi: 10.1109/TNNLS.2020.3001377.

6 I. Bjelaković, H. Boche, and J. Sommerfeld. Secrecy results for compound wiretap channels. *Problems of Information Transmission*, 49(1):73–98, Jan 2013. ISSN 1608-3253. doi: 10.1134/S0032946013010079.

7 M. Bloch and J. Barros. *Physical-Layer Security*. Cambridge University Press, 2011. doi: 10.1017/CBO9780511977985.

8 M. Bloch, M. Hayashi, and A. Thangaraj. Error-control coding for physical-layer secrecy. *Proceedings of the IEEE*, 103(10):1725–1746, 2015. doi: 10.1109/JPROC.2015.2463678.

9 M. Bloch, O. Günlü, A. Yener, F. Oggier, H. V. Poor, L. Sankar, and R. F. Schaefer. An overview of information-theoretic security and privacy: Metrics, limits and applications. *IEEE Journal on Selected Areas in Information Theory*, 2(1):5–22, 2021. doi: 10.1109/JSAIT.2021.3062755.

10 H. Boche and R. F. Wyrembelski. Comparison of different attack classes in arbitrarily varying wiretap channels. In *2012 IEEE International Workshop on Information Forensics and Security (WIFS)*, pages 270–275, 2012. doi: 10.1109/WIFS.2012.6412661.

11 A. Chorti, A. N. Barreto, S. Köpsell, M. Zoli, M. Chafii, P. Sehier, G. Fettweis, and H. . Poor. Context-aware security for 6G wireless: The role of physical layer security. *IEEE Communications Standards Magazine*, 6(1):102–108, 2022. doi: 10.1109/MCOMSTD.0001.2000082.

12 T. M. Cover and J. A. Thomas. *Elements of Information Theory (Wiley Series in Telecommunications and Signal Processing)*. Wiley-Interscience, USA, 2006. ISBN 0471241954.

13 A. El Gamal and Y.-H. Kim. *Network Information Theory*. Cambridge University Press, 2011.

14 N. Gao, Y. Han, N. Li, S. Jin, and M. Matthaiou. When physical layer key generation meets RIS: Opportunities, challenges, and road ahead, 2022. URL https://arxiv.org/abs/2210.02337.

15 C. Gidney and M. Ekera. How to factor 2048 bit RSA integers in 8 hours using 20 million noisy qubits. *Quantum*, 5:433, Apr 2021. ISSN 2521-327X. doi: 10.22331/q-2021-04-15-433.

16 S. Goel and R. Negi. Secret communication in presence of colluding eavesdroppers. In *MILCOM 2005 - 2005 IEEE Military Communications Conference*, pages 1501–1506 Vol. 3, 2005. doi: 10.1109/MILCOM.2005.1605889.

17 X. Guan, N. Ding, Y. Cai, and W. Yang. Wireless key generation from imperfect channel state information: Performance analysis and improvements. In *2019 IEEE International Conference on Communications Workshops (ICC Workshops)*, pages 1–6, 2019. doi: 10.1109/ICCW.2019.8756656.

18 J. Hou and G. Kramer. Effective secrecy: Reliability, confusion and stealth. In *2014 IEEE International Symposium on Information Theory*, pages 601–605. IEEE, 2014.

19 C. Huth, R. Guillaume, T. Strohm, P. Duplys, I. A. Samuel, and T. Güneysu. Information reconciliation schemes in physical-layer security: A survey. *Computer Networks*, 109:84–104, 2016. ISSN 1389-1286. doi: 10.1016/j.comnet.2016.06.014. URL https://www.sciencedirect.com/science/article/pii/S1389128616301864. Special issue on Recent Advances in Physical-Layer Security.

20 C. R. Janda, M. Wiese, E. A. Jorswieck, and H. Boche. Arbitrarily varying wiretap channels with non-causal side information at the jammer. *IEEE Transactions on Information Theory*, 69(4):2635–2663, 2023. doi: 10.1109/TIT.2023.3245722.

21 L. Jiao, G. Sun, J. Le, and K. Zeng. Machine learning-assisted wireless PHY key generation with reconfigurable intelligent surfaces. In *Proceedings of the 3rd ACM Workshop on Wireless Security and Machine Learning*, WiseML '21, page 61–66, New York, NY, USA, 2021. Association for Computing Machinery. ISBN 9781450385619. doi: 10.1145/3468218.3469042.

22 E. Jorswieck, S. Tomasin, and A. Sezgin. Broadcasting into the uncertainty: Authentication and confidentiality by physical-layer processing. *Proceedings of the IEEE*, 103(10):1702–1724, 2015. doi: 10.1109/JPROC.2015.2469602.

23 E. Jorswieck, P.-H. Lin, and C. Janda. Physical layer security based enabling technologies for 6G communications values. In *IEEE Future Networks World Forum*. IEEE, Oct 2022.

24 S. A. A. Kalkhoran, M. Letafati, E. Erdemir, B. H. Khalaj, H. Behroozi, and D. Gündüz. Secure deep-JSCC against multiple eavesdroppers. In *GLOBECOM 2023 - 2023 IEEE Global Communications Conference*, pages 3433–3438, 2023. doi: 10.1109/GLOBECOM54140.2023.10436928.

25 S. H. A. Kazmi, R. Hassan, F. Qamar, K. Nisar, and A. A. A. Ibrahim. Security concepts in emerging 6g communication: Threats, countermeasures, authentication techniques and research directions. *Symmetry*, 15(6), 2023.

ISSN 2073-8994. doi: 10.3390/sym15061147. URL https://www.mdpi.com/2073-8994/15/6/1147.

26 C. Kuhn, M. Beck, S. Schiffner, E. Jorswieck, and T. Strufe. On privacy notions in anonymous communication. *Computer Science*, 2019(2):105–125, 2019. doi: 10.2478/popets-2019-0022.

27 M. Letafati and S. Otoum. Global differential privacy for distributed metaverse healthcare systems. In *2023 International Conference on Intelligent Metaverse Technologies & Applications (iMETA)*, pages 01–08, 2023. doi: 10.1109/iMETA59369.2023.10294469.

28 M. Letafati and S. Otoum. On the privacy and security for e-health services in the metaverse: An overview. *Ad Hoc Networks*, 150:103262, 2023. ISSN 1570-8705. doi: 10.1016/j.adhoc.2023.103262. URL https://www.sciencedirect.com/science/article/pii/S1570870523001828.

29 M. Letafati and S. Otoum. Digital healthcare in the metaverse: Insights into privacy and security. *IEEE Consumer Electronics Magazine*, 13(3):80–89, 2023. doi: 10.1109/MCE.2023.3334494.

30 M. Letafati, A. Kuhestani, H. Behroozi, and D. W. K. Ng. Jamming-resilient frequency hopping-aided secure communication for Internet-of-Things in the presence of an untrusted relay. *IEEE Transactions on Wireless Communications*, 19(10):6771–6785, 2020. doi: 10.1109/TWC.2020.3006012.

31 M. Letafati, H. Behroozi, B. H. Khalaj, and E. A. Jorswieck. Deep learning for hardware-impaired wireless secret key generation with man-in-the-middle attacks. In *2021 IEEE Global Communications Conference (GLOBECOM)*, pages 1–6, 2021. doi: 10.1109/GLOBECOM46510.2021.9685537.

32 M. Letafati, H. Behroozi, B. H. Khalaj, and E. A. Jorswieck. Content-based medical image transmission against randomly-distributed passive eavesdroppers. In *2021 IEEE International Conference on Communications Workshops (ICC Workshops)*, pages 1–7, 2021. doi: 10.1109/ICCWorkshops50388.2021.9473492.

33 M. Letafati, A. Kuhestani, K.-K. Wong, and M. J. Piran. A lightweight secure and resilient transmission scheme for the Internet of Things in the presence of a hostile jammer. *IEEE Internet of Things Journal*, 8(6):4373–4388, 2021. doi: 10.1109/JIOT.2020.3026475.

34 M. Letafati, H. Behroozi, B. H. Khalaj, and E. A. Jorswieck. Hardware-impaired PHY secret key generation with man-in-the-middle adversaries. *IEEE Wireless Communications Letters*, 11(4):856–860, 2022. doi: 10.1109/LWC.2022.3147952.

35 M. Letafati, H. Behroozi, B. H. Khalaj, and E. A. Jorswieck. On learning-assisted content-based secure image transmission for delay-aware systems with randomly-distributed eavesdroppers. *IEEE Transactions on Communications*, 70(2):1125–1139, 2022. doi: 10.1109/TCOMM.2021.3128423.

36 M. Letafati, H. Behroozi, B. H. Khalaj, and E. A. Jorswieck. Wireless-powered cooperative key generation for e-health: A reservoir learning approach. In *2022 IEEE 95th Vehicular Technology Conference: (VTC2022-Spring)*, pages 1–7, 2022. doi: 10.1109/VTC2022-Spring54318.2022.9860947.

37 M. Letafati, H. Behroozi, B. H. Khalaj, and E. A. Jorswieck. Learning-based secret key generation in relay channels under adversarial attacks. *IEEE Open Journal of Vehicular Technology*, 4:749–764, 2023. doi: 10.1109/OJVT.2023.3315216.

38 G. Li, C. Sun, J. Zhang, E. Jorswieck, B. Xiao, and A. Hu. Physical layer key generation in 5G and beyond wireless communications: Challenges and opportunities. *Entropy*, 21(5):497, May 2019. ISSN 1099-4300. doi: 10.3390/e21050497.

39 G. Li, C. Sun, E. A. Jorswieck, J. Zhang, A. Hu, and Y. Chen. Sum secret key rate maximization for TDD multi-user massive MIMO wireless networks. *IEEE Transactions on Information Forensics and Security*, 16:968–982, 2021. doi: 10.1109/TIFS.2020.3026466.

40 G. Li, Y. Xu, W. Xu, E. Jorswieck, and A. Hu. Robust key generation with hardware mismatch for secure MIMO communications. *IEEE Transactions on Information Forensics and Security*, 16:5264–5278, 2021. doi: 10.1109/TIFS.2021.3127021.

41 Y. Liang, H. V. Poor, and S. Shamai (Shitz). Information theoretic security. *Foundations and Trends® in Communications and Information Theory*, 5(4–5):355–580, 2009. ISSN 1567-2190. doi: 10.1561/0100000036.

42 P.-H. Lin, C. R. Janda, E. A. Jorswieck, and R. F. Schaefer. Stealthy secret key generation. *Entropy*, 22(6):679, Jun 2020. ISSN 1099-4300. doi: 10.3390/e22060679.

43 P.-H. Lin, E. A. Jorswieck, R. F. Schaefer, M. Mittelbach, and C. R. Janda. New capacity results for fading Gaussian multiuser channels with statistical CSIT. *IEEE Transactions on Communications*, 68(11):6761–6774, 2020. doi: 10.1109/TCOMM.2020.3013614.

44 P. Linning, G. Li, J. Zhang, R. Woods, M. Liu, and A. Hu. An investigation of using loop-back mechanism for channel reciprocity enhancement in secret key generation. *IEEE Transactions on Mobile Computing*, 18(3):507–519, 2018.

45 U. Maurer. Secret key agreement by public discussion from common information. *IEEE Transactions on Information Theory*, 39(3):733–742, May 1993.

46 L. Mucchi, S. Jayousi, S. Caputo, E. Paoletti, P. Zoppi, S. Geli, and P. Dioniso. How 6G technology can change the future wireless healthcare. In *2020 2nd 6G Wireless Summit (6G SUMMIT)*, pages 1–6, 2020. doi: 10.1109/6GSUMMIT49458.2020.9083916.

47 L. Mucchi et al. Physical-layer security in 6G networks. *IEEE Open Journal of the Communications Society*, 2:1901–1914, 2021. doi: 10.1109/OJCOMS.2021.3103735.

48 M. Naito, S. Watanabe, R. Matsumoto, and T. Uyematsu. Secret key agreement by reliability information of signals in Gaussian Maurer's model. In F. R. Kschischang and E.-H. Yang, editors, *2008 IEEE International Symposium on Information Theory*, ISIT 2008, Toronto, ON, Canada, July 6–11, 2008, pages 727–731. IEEE, 2008. doi: 10.1109/ISIT.2008.4595082.

49 V.-L. Nguyen, P.-C. Lin, B.-C. Cheng, R.-H. Hwang, and Y.-D. Lin. Security and privacy for 6G: A survey on prospective technologies and challenges. *IEEE Communications Surveys & Tutorials*, 23(4):2384–2428, 2021. doi: 10.1109/COMST.2021.3108618.

50 D. C. Nguyen, Q.-V. Pham, P. N. Pathirana, M. Ding, A. Seneviratne, Z. Lin, O. Dobre, and W.-J. Hwang. Federated learning for smart healthcare: A survey. *ACM Computing Surveys*, 55(3), Feb 2022. ISSN 0360-0300. doi: 10.1145/3501296.

51 J. Nötzel, M. Wiese, and H. Boche. The arbitrarily varying wiretap channel—secret randomness, stability, and super-activation. *IEEE Transactions on Information Theory*, 62(6):3504–3531, 2016. doi: 10.1109/TIT.2016.2550587.

52 P. Porambage, G. Gür, D. P. M. Osorio, M. Liyanage, A. Gurtov, and M. Ylianttila. The roadmap to 6G security and privacy. *IEEE Open Journal of the Communications Society*, 2:1094–1122, 2021. doi: 10.1109/OJCOMS.2021.3078081.

53 A. M. Sayeed. Deconstructing multiantenna fading channels. *IEEE Transactions on Signal Processing*, 50(10):2563–2579, 2002. doi: 10.1109/TSP.2002.803324.

54 L. Sun, P. Ren, Q. Du, and Y. Wang. Fountain-coding aided strategy for secure cooperative transmission in industrial wireless sensor networks. *IEEE Transactions on Industrial Informatics*, 12(1):291–300, 2016. doi: 10.1109/TII.2015.2509442.

55 M. Tahmasbi and M. R. Bloch. Covert secret key generation with an active warden. *IEEE Transactions on Information Forensics and Security*, 15:1026–1039, 2020. doi: 10.1109/TIFS.2019.2932906.

56 D. Tse and P. Viswanath. *Fundamentals of Wireless Communication*. Cambridge University Press, 2005.

57 Y. Wang, W. Xu, H. Zhang, and X. You. Wideband mmWave channel estimation for hybrid massive MIMO with low-precision ADCs. *IEEE Wireless Communications Letters*, 8(1):285–288, 2019. doi: 10.1109/LWC.2018.2870360.

58 T. Wang, Y. Shen, L. Gao, Y. Jiang, T. Ma, and X. Zhu. Energy consumption minimization with throughput heterogeneity in wireless-powered body area networks. *IEEE Internet of Things Journal*, 8(5):3369–3383, 2021. doi: 10.1109/JIOT.2020.3022325.

59 S. Warnat-Herresthal et al. Swarm learning for decentralized and confidential clinical machine learning. *Nature*, 594: 265–270, 2021. doi: https://doi.org/10.1038/s41586-021-03583-3.

60 M. Wiese, J. Nötzel, and H. Boche. A channel under simultaneous jamming and eavesdropping attack—correlated random coding capacities under strong secrecy criteria. *IEEE Transactions on Information Theory*, 62(7):3844–3862, 2016. doi: 10.1109/TIT.2016.2565482.

61 Q. Wu, S. Zhang, B. Zheng, C. You, and R. Zhang. Intelligent reflecting surface-aided wireless communications: A tutorial. *IEEE Transactions on Communications*, 69(5):3313–3351, 2021.

62 A. D. Wyner. The wiretap channel. *The Bell System Technical Journal*, 54:1355–1387, 1975.

63 Q. Xie, Z. Wang, and Z. Yang. Polar decomposition of mutual information over complex-valued channels. *IEEE Transactions on Information Theory*, 60(6):3163–3171, 2014. doi: 10.1109/TIT.2014.2313081.

64 L. You, X. Gao, G. Y. Li, X.-G. Xia, and N. Ma. BDMA for millimeter-wave/terahertz massive MIMO transmission with per-beam synchronization. *IEEE Journal on Selected Areas in Communications*, 35(7):1550–1563, 2017. doi: 10.1109/JSAC.2017.2699100.

65 M. Zoli, A. N. Barreto, S. Köpsell, P. Sen, and G. Fettweis. Physical-Layer-Security Box: a concept for time-frequency channel-reciprocity key generation. *EURASIP Journal on Wireless Communications and Networking*, 2020(1):114, Jun 2020. ISSN 1687-1499. doi: 10.1186/s13638-020-01712-6.

13

The Role of Non-terrestrial Networks: Features and Physical-Layer Security Concerns

Marco Giordani, Francesco Ardizzon, Laura Crosara, Nicola Laurenti, and Michele Zorzi

Department of Information Engineering, University of Padova, Padova, Italy

Despite the growing penetration of fifth generation (5G) commercial deployments, researchers are working on sixth generation (6G) networks, together with their use cases, requirements, and technologies. In the 6G framework, non-terrestrial networks (NTNs) play a key role as an approach to provide cost-effective, high-capacity, ubiquitous wireless connectivity via aerial/space nodes. However, there are still many issues to be addressed for proper network design, in view of the unique latency, coverage, and security constraints of these systems. In this chapter, we introduce the concept of NTN for 6G (Section 13.1), in particular enabling applications, requirements, technologies, and open research questions. Then, we focus on the issues related to physical-layer security in NTN (Section 13.2), especially secrecy, authentication, and position integrity.

13.1 Non-terrestrial Networks for 6G

The Internet has been growing at an astounding rate, and today, there are almost five billion active Internet users worldwide. Even so, current networks are mainly deployed in urban areas and fall short to provide adequate coverage to rural regions [18]. Furthermore, cellular infrastructures are vulnerable to external attacks, which may create connectivity outages during, e.g., natural or human-made disasters. Additionally, even in the most technologically advanced countries, existing networks may lack the level of reliability and responsiveness requested by future 6G applications [29].

Physical-Layer Security for 6G, First Edition. Edited by Parthajit Mohapatra, Nikolaos Pappas, Arsenia Chorti, and Stefano Tomasin.

One solution to enhance network coverage would be to increase the density of cellular infrastructures worldwide, which, however, especially in rural regions, involves exorbitant deployment costs for network operators, e.g., in areas with adverse weather conditions and/or rough terrain, as well as in the absence of transport connectivity and access to power sources. In light of this, the research community identifies NTN as a crucial paradigm to provide cost-effective and high-capacity connectivity in 6G networks [28], as depicted in Figure 13.1. While previous network generations have been deployed to provide connectivity for a quasi-two-dimensional space, 6G envisions a three-dimensional network where terrestrial users are served by non-terrestrial platforms like unmanned aerial vehicles (UAVs), high altitude platforms (HAPs), and satellites.

In particular, UAVs operate at a height of a few hundred meters and are ideal candidates to provide on-demand, wide-scale, flexible connectivity on hot-spot areas, e.g., during temporary events or in case of emergency. Moreover, given their physical proximity to the ground, UAVs promote low-latency and high-resolution sensing. Still, UAVs suffer from severe energy consumption for propulsion and hovering, which raise severe power management constraints.

HAPs, instead, operate in the stratosphere at a height of around 20 km and stand out as quick-to-deploy solutions to provide ultra-flexible, cost-effective, wide geographical coverage of a few tens of km of radius. However, HAPs may require frequent refueling and energy for stabilization in the air, even though energy harvesting from solar panels can partially alleviate those constraints.

As far as satellites are concerned, Geostationary Earth Orbit (GEO) satellites surely offer a number of advantages to ground users, including extremely large coverage regions of thousands of km of radius, favorable line of sight (LOS) connectivity, and continuous visibility from the Earth. In recent years, however,

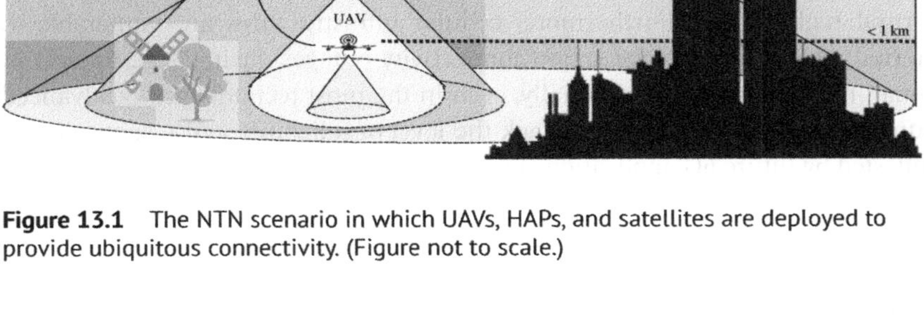

Figure 13.1 The NTN scenario in which UAVs, HAPs, and satellites are deployed to provide ubiquitous connectivity. (Figure not to scale.)

Low Earth Orbit (LEO) satellites, operating between 200 and 2000 km, have gained special attention to provide connectivity at much lower propagation delay compared to GEO systems. However, LEO satellites must operate in constellations to maintain service continuity, which poses critical challenges in terms of mobility, routing, and architecture management.

In the past, NTNs, especially satellites, have been studied to support basic services like meteorology, video surveillance, positioning, television broadcasting, weather forecasting, and navigation. Lately, the aerial/space industry is promoting more advanced use cases toward 6G, as we will describe in Section 13.1.1. Based on that, in Section 13.1.2, we will present enabling technologies for NTN in the 6G landscape, and, in Section 13.1.3, we will shed light on the challenges in the field that are still open for future research, in particular those related to the usage of the spectrum, medium access control, coverage and mobility management, and network security.

13.1.1 Use Cases

NTN promotes a positive impact to our society and encourages the support of several use cases in line with 6G requirements, as summarized below.

13.1.1.1 Continuous and Ubiquitous Network Coverage

NTNs can provide on-demand connectivity in unserved areas (e.g., oceans and deserts), or where network deployments like installation of base stations and/or fiber optic, connection to power plants, and network management are difficult and costly (e.g., in underdeveloped or rural regions). In these areas, connectivity via NTN will offer new opportunities for local communities who have not yet access to digital technologies, providing better education and social engagement, additional job opportunities, and more efficient health systems. At the same time, NTN can be seen as an alternative or complementary connectivity solution when already deployed terrestrial towers are out of service (e.g., after natural disasters), or overloaded (e.g., during public events in hot-spot areas), respectively.

13.1.1.2 Support for the Internet of Things

The Internet of Things (IoT) era will provide advantages in many fields but may be a source of congestion for resource-constrained terrestrial networks that may fail to fulfill service requirements efficiently [61]. In this context, NTNs can act as aerial/space gateways operating from above and provide very large continuous and effective coverage on the ground. Even in the absence of pre-existing terrestrial infrastructures, NTNs can support global connectivity for IoT applications, thus, supporting the transition of remote areas toward smart cities [57].

13.1.1.3 Integration Between Communication and Computation

Besides providing connectivity, NTNs can also host computing servers for data processing, storage, and/or caching, which is particularly useful for power-constrained devices with limited computational capability [8, 53]. While computation in terrestrial infrastructures may be costly and inefficient, if not entirely precluded in uncovered areas, the favorable propagation conditions, the massive payload capability, and the large coverage footprint make computation in NTN an ideal solution for faster, more reliable, and privacy-preserving processing of data [54].

13.1.1.4 Energy-Efficient Service

The deployment of a large number of 5G small cells, the need to provide multi-Gbps data rates to battery-constrained mobile users, and the need for sensors to be completely standalone and able to run for years without battery replacement pose new challenges for cellular networks in terms of power consumption. UAVs also incur significant energy consumption for propulsion and hovering and fly for a limited time (i.e., a few tens of minutes) on a full charge. In turn, HAPs and satellites can count on renewable and clean energy from solar panels in addition to traditional power sources like batteries, thereby promoting energy efficiency.

13.1.2 Enabling Technologies

The evolution of NTN will be encouraged by several innovations in the aerial/space industry, as summarized in [28] and described below.

13.1.2.1 Novel Network Solutions

The space industry is developing new solutions to reduce the operational costs of satellite launch, deployment, and maintenance. Toward this goal, miniaturization of satellites, e.g., via Gallium Nitride (GaN) technologies, allows smaller hardware components compared to their silicon counterparts, which may save fuel and area on the payload of satellites, thereby improving operational efficiency. The recent design of solid-state lithium batteries, compared to standard lithium-ion batteries, further promotes energy efficiency and longer lifetime, specially for battery-powered devices like UAVs. Architecture optimization is also favored by software defined networks (SDNs) which, in combination with recent paradigms such as network slicing, facilitates the transition from physical to virtual platforms, which promotes improved flexibility, automation, and agility in NTN service operations. Finally, while terrestrial networks tend to operate as standalone systems, NTNs can natively be structured as multilayered hierarchical networks, where intermediate aerial/space platforms act as relays of an upstream device, which provides superior coverage and resilience [57].

13.1.2.2 Novel Antenna Solutions

In the antenna domain, new antenna technologies for satellites, e.g., based on reconfigurable phased arrays, metasurfaces, and intelligent structures, enable lower power consumption and reduced size and weight compared to mechanical antennas. Inflatable and fractal antennas with unique geometrical designs can further improve directivity, thus, promoting spectrum efficiency and spatial diversity.

13.1.2.3 Novel Spectrum Solutions

NTNs generally operate in the legacy sub-6 GHz frequency bands which, however, may not address the stringent requirements of 6G. Indeed, capacity issues can be addressed by transitioning to high-frequency communications, e.g., in the millimeter wave (mmWave) bands, where the huge bandwidths available may offer the opportunity of ultrafast connections. Additionally, the vast amount of unlicensed spectrum at mmWaves can promote better return of investment for service delivery, especially for small providers, and improve network elasticity. The use of this technology has been successful in 5G for both cellular and vehicular networks. In the NTN domain, instead, the mmWave technology has been, so far, mainly reserved for satellite services like weather forecasting, and it is still unclear whether it can be used to support broadband communication [27]. One of the major barriers is the severe path loss experienced at high frequencies, combined with the larger propagation delay and the stronger effect of atmospheric perturbations than in terrestrial networks.

NTNs can also operate in the optical spectrum for the feeder link, i.e., the link between the gateway (thus the core network) and the space/airborne platform, for the backhaul, or for inter-connecting satellites, which allows an aggregate capacity in the order of terabits-per-second at lower power and mass. At these frequencies, however, the signal is attenuated by perturbations and interference from sunlight, even though wavefront correctors and deformable mirrors, as well as error control coding, can partially alleviate signal distortion.

13.1.3 Open Research Questions

Despite current efforts and research trends toward NTN, there are many challenges to be solved before commercial deployment, as described below.

13.1.3.1 Physical-Layer Procedures

The large propagation delays in NTN, as well as the severe Doppler effect for high-speed aerial/space stations, may complicate physical-layer control procedures. For example, time division duplexing (TDD), the most popular choice for terrestrial networks, may be infeasible in NTN since the length of the guard period, which ensures that consecutive uplink and downlink transmissions do not interfere, must be proportional to the propagation delay and the cell size, which are both particularly large especially for satellite communication. Similarly, Hybrid Automatic Repeat reQuest (HARQ) comes with delayed acknowledgments due to the long round trip time (RTT), which may affect retransmissions in the NTN scenario. Notice that, in the case of hierarchical networks (see Section 13.1.2), different network nodes at different layers may process signals at different times and rates, which also increases the overall communication delay.

13.1.3.2 Synchronization

NTNs move fast (except for GEO satellites) with respect to the Earth and shape larger cells compared to terrestrial networks. At low elevation, the propagation delay for users at the center of the cell may be different with respect to users at the cell edge (up to around 10 ms for GEO [1]), which may raise synchronization issues.

13.1.3.3 Channel Estimation and Random Access

Channel estimation requires terrestrial terminals to measure the quality of link connections with non-terrestrial stations by detecting synchronization signals. This is particularly challenging in NTN, where the channel response may change quickly over time (mainly because of the Doppler shift introduced by non-geostationary moving platforms), as initial estimates may rapidly become obsolete. As such, channel estimation has to be periodically and continuously repeated, which results in a large overhead. Also, in multilayered networks, intermediate nodes generally select the gateway to associate with based on unilateral benefits, while underestimating the potential disadvantages on the whole end-to-end network.

As far as random access is concerned, differential delay could be experienced by two or more terrestrial devices within the same cell. As a result, the preambles sent by different users in the same random access channel (RACH) opportunity (RO) may reach the network at different times, making it difficult for the network to link the preambles to the corresponding RO. Moreover, after transmitting the random access preamble (RAP), the device monitors the physical downlink control channel (PDCCH) for the random access response (RAR) message. If no valid response is received within a predefined response window, a new RAP is sent. While in terrestrial communication, this procedure takes a few milliseconds, in NTN, the propagation delay is much larger, which requires random access to be modified accordingly.

13.1.3.4 Mobility Management

Satellites may operate through high-directivity pencil beams to achieve sufficient link budget, especially at mmWaves (see Section 13.1.2). In this case, beam alignment may affect several control operations, including user tracking and handover. These issues are particularly critical for NTN, as the nodes moving at high speed could lose beam alignment before data exchange is completed. Additionally, air/spaceborne stations are fast moving (e.g., the angular velocity of a LEO satellite is around 7.8 km/s, with a visibility period of a few minutes) and may create coverage holes in certain regions. In this case, a constellation is needed to maintain service continuity, which in turn increases the cost and complexity of networks.

Moreover, considering the large cell size of NTNs, many devices may be served within a single cell. Depending on constellation assumptions (e.g., propagation delay or satellite altitude and speed) and the density of terrestrial devices, a potentially very large number of devices may need to perform handover at a given time, leading to large signaling overhead and service continuity challenges. For example, the 3rd Generation Partnership Project (3GPP) estimates that, for a cell diameter if 1,000 km and assuming around 65,000 connected devices (i.e., the average device density is 0.08 devices/km^2), the average time required for handover may be as large as 132 seconds [1].

Finally, while coordination in terrestrial networks, as well as the backhaul, is realized through fiber connections or via a central entity, in the NTN scenario, coordination needs to be implemented wirelessly, which makes constellation management even more difficult.

13.1.3.5 Resource Saturation

NTNs create larger coverage areas than terrestrial cells and, as such, serve a larger number of terminals on the ground. For example, for a single LEO satellite at an altitude of 550 km and with an elevation of 40°, the approximate cell radius is 580 km (versus 5 km for a 4G-LTE eNB), i.e., around 1 million km^2 (versus 80 km^2 for a 4G-LTE eNB) [11]: with an average density of 100 devices/km^2 [1, Table B.2-1], the total number of nodes under the coverage of the LEO satellite can be as large as 100 million (versus around 8000 for a 4G-LTE eNB), which may lead to saturation of the available resources, with strong implications in terms of latency and throughput. To partially solve this issue, satellites tend to operate through spot beams, so as to cover only a smaller geographic area on Earth and dedicate communication resources to a limited set of devices.

13.1.3.6 Higher-Layer Protocol (Re)design

Routing information may quickly become obsolete (especially in case of topology changes due to mobility, e.g., for LEO satellites) and requires frequent and periodic updates, thus, increasing the communication overhead. Furthermore, the long RTTs in NTN may increase the duration of the slow start phase of transmission control protocol (TCP) and require large transmission buffers at the mobile terminal's side. Similarly, the length of TCP timers, especially the retransmission timeout (RTO), may be comparable to the RTT in NTN, which implies that TCP will often be in slow start, with negative implications in terms of delay. Finally, sudden drops in the link quality, which may be common in NTN, make the congestion window at the transport layer to inflate and deflate continuously, which prevents the system from operating at full capacity.

13.1.3.7 The Role of the Uplink

In downlink, NTNs (especially based on satellites) have little energy and space constraints and can mount massive circular aperture antennas offering sufficient gain to compensate for the very long transmission links. In uplink, instead, terrestrial handsets mount very small aperture terminal (VSAT) antennas or MIMO-like phased arrays, where the limited antenna gain and transmit power may prevent high-capacity direct handset-to-satellite communication. At the same time, uplink scheduling is currently based on a combination of scheduling requests (from the end users to the gNB) and scheduling grants (from the gNB to the end users), which may take up to two RTTs. While this delay is tolerable for terrestrial networks, it may become prohibitively large in NTN.

13.1.3.8 Security and Privacy

NTN communication from the ground to satellites and vice versa may involve several security risks, including accidental or illegitimate destruction, loss, alteration, or unauthorized access to sensitive data, which requires the supervision by a trusted central authority to maintain network security. More details on these issues will be provided in Section 13.2.

13.2 Physical-Layer Security in Non-terrestrial Networks

Due to the broadcasting nature of wireless channels, NTNs are subject to such threats as side-channel attacks, spoofing, network eavesdropping, message modification/injection, and many others [4, 7, 16, 55, 59]. Therefore, guaranteeing privacy and security in NTNs is a challenging issue, which has attracted a lot of attention in the scientific community. This can be addressed by using physical-layer security, thus, exploiting the unique physical properties of the satellite-to-ground and inter-satellite channels to enhance the security level.

Conventional cryptography-based algorithms could cause excessive transmission overhead, communication delay, and power consumption, which is infeasible for devices such as IoT sensors, UAVs, and even some satellites with small form factor. On the other hand, physical-layer security requires fewer resources and computation capabilities and can offer an additional level of security if combined with cryptographic features.

In the rest of this section, we focus on physical-layer security in NTNs, specifically secrecy (Section 13.2.1), authentication (Section 13.2.2), and position integrity (Section 13.2.3) methods.

13.2.1 Physical-Layer Secrecy in NTNs

Physical-layer secrecy solutions were proposed as early as the 1970s in a general abstract transmission scenario, starting from the seminal works by Wyner [60], Csiszár and Körner [20], and Leung-Yan-Cheong and Hellman [45]. For the wiretap channel model with one input and two diverse outputs (to the legitimate receiver and to the eavesdropper, respectively), they introduced the notion of secrecy capacity as the supremum of all communication rates for which physical-layer secrecy is possible. However, since a necessary condition to make secrecy possible is that the legitimate channel be better (e.g., higher signal-to-noise ratio (SNR) or lower bit error rate [BER]) than the eavesdropper channel, this theory did not find practical applications for decades.

Since the late 2000s, physical-layer secrecy (PLS) found a renewed interest as the inherent randomness and diversity of wideband wireless channels to distinctly located receivers allow the creation of virtual subchannels in the time (slow fading) frequency (dispersive channels and selective fading) or spatial (multiple input multiple output [MIMO] diverse channels) domains. Schemes can then be designed that exploit those time intervals, subchannel bands, and spatial directions where the legitimate channel is better. And even if the eavesdropper channel state is generally not known, one can still leverage the knowledge of the legitimate channel state and provide an outage-based design, relying on the lack of correlation between the two channels.

On the other hand, the randomness and diversity of dispersive terrestrial radio channels are unlikely to be found in a satellite-to-ground scenario. In fact, as the satellite channel is more often Gaussian with flat fading, in the corresponding wiretap channel with input x, legitimate output y and eavesdropper output z we have

$$y = g_B x + w_B, \tag{13.1}$$

$$z = g_E x + w_E. \tag{13.2}$$

The secrecy capacity C_s is null whenever the eavesdropper has a better SNR than the legitimate receiver, e.g., when it is equipped with a larger-gain antenna (when both receivers are on the ground), or is physically located closer to the transmitter (e.g., in the uplink). In order to provide appropriate PLS solutions for NTNs, it is therefore necessary to establish an advantage for the legitimate receiver over the eavesdropper by exploiting alternative means, e.g., a reverse feedback channel, one or more collaborative relays, or by assuming geographical constraints on the eavesdropper position.

13.2.1.1 Two-Way Protocols

The use of a two-way protocol for a satellite channel was proposed in [34], similarly to what is done with privacy amplification in information theoretic secret key agreement protocols. In fact, the secret communication protocol is split into two phases: first, the sharing of correlated random variables via the reverse channel, with a proper post-selection in order to distill a correlation advantage for the legitimate parties; then, the random binning encoding of the confidential information in the uplink with the help of the shared random variables. This scheme is able to achieve reliability and secrecy at the same time, with a positive secrecy rate, even when the legitimate main channel is stochastically degraded with respect to the eavesdropper channel. This is akin to the well-known physical-layer key agreement scenario where the presence of a feedback channel and the possibility of reverse reconciliation guarantee positive secrecy rates in Markovian channels [3]. However, in this case, the information reconciliation step is not required, as it is sufficient that the random variables shared between the legitimate parties be correlated, not necessarily identical. Yet, as expected, adequate secrecy performance can only be obtained if the eavesdropper is not too close to the source of random variable, i.e., the receiver for the confidential message (ground station (GS) for downlink communication, satellite for the uplink), and its SNR advantage over the legitimate transmitter is not too large.

13.2.1.2 Geographical Constraints

In land mobile satellite networks, where geographic coverage is wide, the advantage of legitimate receivers over an eavesdropper placed at a different location can be created by exploiting multiple transmission beams and corresponding beamforming. If both the legitimate and the eavesdropper channel state information (CSI) are known or at least estimated with a good degree of precision, as in the initial proposals [43], reasonable secrecy capacity can be obtained. More often, only partial or statistical CSI should be assumed for both the legitimate and the eavesdropper channels, based on geographical information such as angle of departure and receiver distance, and performance metrics that take into account imperfect CSI knowledge, such as outage secrecy rates and secrecy outage probability, have been evaluated and discussed in more recent works [30, 46] with promising results.

13.2.1.3 Use of Relays and Friendly Jamming Helpers

In a hybrid satellite–terrestrial relay network scenario, one or more GS antennas can act as a helper to the confidential communication terminals. Besides acting as a relay (e.g., by amplify-and-forward or decode-and-forward techniques) to improve the chance of reliable decoding even in the presence of possible physical obstacles or a strongly attenuated channel, the GS can also improve the secrecy by confusing the eavesdropper with the transmission of artificial noise and/or intended interference, that is designed so as not to impact the legitimate channel (e.g., by lying in its null space, or by keeping it orthogonal). This model has been extended to the case when the relay is another non-terrestrial source, such as a lower altitude satellite (LEO versus GEO), a HAP, or a UAV [35]. In this case, the maximization of the secrecy rate is known to be a complex task, that can only be solved numerically, due to the non-concavity of the objective function. However, several results on the achievable secrecy rates for the relay channel can also be found in the PLS literature, e.g., [6, 13, 23, 37], and applied to the hybrid satellite–terrestrial context.

On the other hand, when UAVs are used [33], the lower altitude and higher angular velocity, viewed from a ground receiver, provide more diversity and fast fading, which can then be leveraged for PLS, obtaining significant ergodic secrecy capacity and achievable secrecy rates, which are the proper metrics to employ in a fast-fading environment.

Recently, the PLS paradigm has been extended to the envisioned scenario of cognitive satellite–terrestrial integrated networks [49], where the satellite and terrestrial links share the spectrum as primary and secondary communication, respectively. The design target becomes, for the terrestrial link, to maximize the achievable secrecy rate for the secondary user, under the constraint of a limited interference on the primary satellite users. The same design can be applied to fully non-terrestrial cognitive networks, where the secondary links are provided by several UAVs [44] with the target of maximizing the secrecy sum rate over all the secondary transmissions.

13.2.2 Physical-Layer Authentication for NTNs

Physical-layer authentication (PLA) is a technique that allows a receiver, namely, Bob, to distinguish between the legitimate signals, transmitted by Alice, and the ones transmitted by an attacker, Eve, by exploiting the statistical properties of the received signals. To figure out if the received signal, $\boldsymbol{r}$, has been transmitted by Alice or by Eve, Bob takes advantage of a prior knowledge, e.g., the statistical distribution of a certain signal feature. While the use of PLA for NTNs is rather new, we can find several proposals for PLA in different contexts, such as in IoT [31], cellular [39], and Wi-Fi [22] networks.

The authentication problem can be framed in binary hypothesis testing, considering as hypotheses $\mathcal{H}_0$ and $\mathcal{H}_1$, where:

$$\mathcal{H}_0 : \boldsymbol{r} = \boldsymbol{x} + \boldsymbol{\omega}, \text{ the signal is transmitted by Alice;} \tag{13.3}$$

$$\mathcal{H}_1 : \boldsymbol{r} = \boldsymbol{y} + \boldsymbol{\omega}, \text{ the signal is transmitted by Eve,} \tag{13.4}$$

where $\boldsymbol{w}$ is the channel noise while $\boldsymbol{x}$ and $\boldsymbol{y}$ are Alice's and Eve's signals after the respective channel distortions.

Figure 13.2 depicts a general PLA scheme, including signal generation at Alice's and Eve's, according to arbitrary distributions p_u and $p_{\tilde{u}}$, respectively, and the channel introducing both nonlinear effects (e.g., Doppler spread) and Additive White Gaussian Noise (AWGN) noise. Next, Bob assesses the authenticity of the signal using a verification and a decision functions, leading to decision $\hat{\mathcal{H}}_0$ or $\hat{\mathcal{H}}_1$, if the signal is considered to be legitimate or spoofed, respectively.

In this context, the optimal approach to minimize missed detection for a fixed false alarm probability is to use the likelihood ratio test (LRT) [48], i.e.,

$$\text{LRT: } \frac{p(\boldsymbol{r}|\mathcal{H}_0)}{p(\boldsymbol{r}|\mathcal{H}_1)} \underset{\mathcal{H}_0}{\overset{\mathcal{H}_1}{\gtrless}} \lambda \, . \tag{13.5}$$

Threshold λ is set by the user to match a predefined false alarm probability. Notice that this operation implicitly includes both the verification, represented by the computation of the LRT, and the decision, implemented by thresholding the LRT result.

We remark that, often, instead of working on the whole signal $\boldsymbol{r}$, it may be convenient to work with features of the signal, eventually extracted with proper feature extraction functions, e.g., discrete wavelet transform (DWT). This approach achieves a trade-off between solution complexity and classification accuracy, i.e., false alarm and missed detection.

However, there are two main issues when using LRT in PLA. First, knowledge of the attacker's signal distribution $p(\boldsymbol{r}|\mathcal{H}_1)$ is required. Still, the distribution of $\boldsymbol{r}$

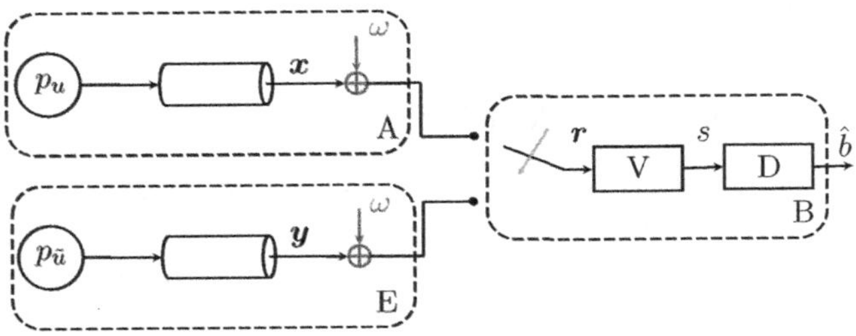

Figure 13.2 General model for a PLA scheme. Alice and Eve generate signals with distributions p_u and $p_{\tilde{u}}$. The signals go through a channel introducing nonlinearities and AWGN noise. The receiver, Bob, uses verification and a decision function to distinguish between legitimate and false signals.

depends on several factors, such as the attacker's positions or the actual attacker's strategy. Indeed, in many realistic scenarios, this knowledge is not a given. Thus, we have to consider the authentication as a one-class classification (OCC). A possible solution could be to use a generalized likelihood ratio test (GLRT) [17]. However, in general, there is no proof of the optimality of the GLRT-based approach for OCC.

A second issue is that, often, not even the distribution of the legitimate signal is known. This is due to multiple factors, e.g., environmental changes or Alice's and/or Bob's movement. A popular solution is to resort to machine learning (ML) methods, such as deep neural networks (DNNs) and support vector machines (SVMs). These approaches are data driven, thus, the authentication function is trained directly on an (authenticated) dataset that may eventually be periodically updated via continual learning. ML can also be used when only the training dataset for one of the hypotheses is available, using, for instance, autoencoders (AEs) or one-class SVM (OC-SVM). It has been proven that if the dataset is sufficiently large and the DNNs are complex enough, in a two-class classification setting, the performance of an LRT converges to one of the optimal LRTs [9]. Conversely, in an OCC scenario, OC-SVM and a DNN trained with a suitable artificial dataset both converge to the GLRT [5].

In general, PLA techniques can be distinguished into two categories, namely, device-based PLA and channel-based PLA. In the following paragraphs, we will describe both methods for NTN scenarios and present research challenges.

13.2.2.1 Device-Based PLA

The intuition behind device-based PLA is that each transmitter is different and is bound to introduce some artifacts to the channel. These imperfections are due to the hardware mounted on the transmitting device, which are difficult to replicate. Hence, in a device-based PLA scheme, the receiver identifies the transmitter by checking if the artifacts are present on the channel. For this reason, this method is also called fingerprinting. Examples of these hardware impairments are phase noises, introduced by the transmitter's clock instabilities, and nonlinearities introduced by the power amplifier or the digital-to-analog converter [12, 21, 26, 51].

In [50], the authors propose a device-based PLA technique to authenticate LEO satellites, specifically Iridium. They considered two different scenarios, which they call intra-constellation satellite authentication and satellite authentication in the wild. In the former, they assume that the receiver has a training set containing I/Q samples from all the satellites. This is treated as a multi-class labeling problem and is tackled using a convolutional neural network (CNN). In the latter, they assume that only the dataset for the satellite to be authenticated is used for training. Thus, this is an OCC problem, and they tackle it using sparse AEs.

In [56], a fingerprinting procedure for authenticating global navigation satellite system (GNSS) signals is proposed. In particular, the authors focus on pre-correlation data, i.e., before the signal acquisition, where each satellite signal is identified and decoupled from the others using the associated spreading code sequence.[1] More in detail, they use kurtosis as a feature extraction method, the Teager–Kaiser energy operator, and the spectrogram. As a classification tool, they use SVM. The work was further extended in [58], where the authors considered both pre- and post-correlation data. The performance was evaluated considering experimental data, collected using a Universal Software Radio Peripheral (USRP). Results confirmed that this approach was able to achieve an average correct classification probability (i.e., the probability of identifying whether any of the received signals was spoofed) up to 99.99% using the pre-correlation data, and 87.72% for post-correlation data. Notice that using pre-correlation data allows only to check if the spoofer tampered with any of the signals. On the other hand, working with post-correlation data makes it possible to distinguish between legitimate and spoofing signals. Thus, even if the latter may allow the defender to compute a legitimate position by removing the tampered signals, working in the post-correction domain is inherently more challenging, since it requires the identification of the specific spoofed signals.

13.2.2.2 Channel-Based PLA

Channel-based PLA schemes exploit the unique characteristics of the channel to check whether the transmission was authentic or not. Indeed, non-terrestrial channels are challenging due, for instance, to the multipath fading and the strong Doppler effect. Thus, it is hard for a transmitter in a different position to replicate the same channel.

In [25], the authors propose a channel-based PLA technique using the Doppler frequency as a feature for authentication. By using the ephemeris, which can be broadcast by an authenticated side channel, it is possible to compute the positions and velocities of the transmitter and the receiver at a given time. Thus, it is possible to have a prediction of the Doppler frequency as

$$f_{\mathrm{d}}(t) = \frac{f_{\mathrm{c}}}{c} \cdot v_{\mathrm{r}}(t) \,, \tag{13.6}$$

where f_{c} is the carrier frequency, c is the speed of light, and $v_{\mathrm{r}}(t)$ is the radial component of the velocity with respect to the receiver at time t. More in detail the radial velocity component is

$$v_{\mathrm{r}}(t) = \left(\boldsymbol{v}_{\mathrm{tx}}(t) - \boldsymbol{v}_{\mathrm{rx}}(t)\right)^{\mathrm{T}} \boldsymbol{e}(t) \,, \tag{13.7}$$

1 Details about this step can be found in [38, Ch. 5].

where $\boldsymbol{v}_{\text{tx}}(t)$ and $\boldsymbol{v}_{\text{rx}}(t)$ are the 3D velocity of the transmitter and the receiver at time t, respectively, and $\boldsymbol{e}(t)$ is the unit vector pointing from the transmitter to the receiver satellite. As an authentication test, the authors compare the absolute difference between the measured and the predicted Doppler frequencies with a threshold, that is set to match a predefined false alarm probability.

Similarly, the authors in [52] propose a channel-based PLA technique for inter-satellite (LEO) communication links. As in [25], they use the ephemeris to have a prediction of the Doppler frequency of each transmitter–receiver channel. Hence, multiple receivers check the authenticity of the received signal, trying to match the Doppler frequency of the received samples with the predicted ones. The final decision is computed by fusing all the local decisions using several methods. Notably, the majority rule was proved to achieve the best results in terms of false alarm and missed detection probability.

In [2], an ML-based PLA solution is proposed to authenticate the signals transmitted by LEO satellites. The authors consider the Doppler frequency and the received power as features for authentication, where the latter is given by

$$P_{\text{rx}} = \frac{P_{\text{tx}}\, g_{\text{tx}}\, g_{\text{rx}}}{(4\pi\, d/\lambda)^2}, \tag{13.8}$$

where P_{tx} is the power at the transmitter, $(4\pi\, d/\lambda)^2$ is the free space pathloss at wavelength λ, while g_{tx} and g_{rx} are the gains at the transmitter and receiver antennas, respectively. Notice that gains are a function of the direction of the antenna. Thus, assuming that the receiver points the antenna toward the transmitter, any signal coming from a different position will experience a lower gain, and thus will be received with lower power. Hence, the received power can be used to distinguish between Alice and Eve. Next, the authors consider the problem as an OCC, and solve it using OC-SVM as classifier, trained only with Alice's data.

13.2.2.3 Challenges and Future Works for PLA

In this paragraph, we highlight some of the challenges and future improvements for PLA in the NTN context. Both device- and channel-based PLA methods consider a naive attacker model, assuming that it is hard for the attacker to replicate features to be authenticated.

The first issue mainly concerns channel-based PLA, observing that all the proposed schemes limit the analysis to only one or few features. Indeed, it may be interesting to consider more complex systems where a higher number of features is considered. Moreover, when considering more complex network architectures, where multiple receivers cooperate to evaluate the authenticity of the received signals, it may be interesting to investigate methods to (i) merge the local information, e.g., by giving properly chosen weights to each local decision, and (ii) compress the local information, allowing the central node to leverage more inputs than the local decision.

Another issue is that a sophisticated Eve may exploit either the knowledge about the feature vector or even the test function used by Bob for authentication to design a signal fooling the defender. Indeed, this requires prior knowledge (partial, at least) of the channel between Eve and Bob. The spoofing signal could be designed by exploiting multitask learning [15], considering two objective functions: (i) to check if the signal could actually fool the authentication test and (ii) to evaluate how well the signal is causing the intended spoofing. Indeed, transmitting a signal which is able to satisfy (i) but not (ii), e.g., by transmitting a signal which fools the attacker but has the same information content as the legitimate one, is meaningless for the attacker. Thus, the attacker may exploit multitask learning to find the most suitable trade-off between the two losses. A possible way to implement this architecture is to exploit and adversarial-like neural network (NN), such as an adversarial neural network (AAE), using the authentication test as the discriminator, e.g., as proposed in [63].

13.2.3 Position Integrity for NTNs

Position integrity for NTNs refers to the ability to accurately and reliably determine the position of objects using signals transmitted by anchor nodes (ANs), whose position is known, when the transmitters or the receiver are not located on the Earth's surface [4]. Positioning schemes typically rely on timing-based techniques that take advantage of the phase of the received radio frequency (RF) signals to calculate the distance between a transmitter and a receiver, e.g., by estimating the time of arrival (TOA) or the time difference of arrival (TDOA) of signals. Then, using this information, a trilateration estimation algorithm can be used to estimate the position of the user [38, Ch. 7], [41].

However, while computing the position, a user makes two assumptions that may lead to vulnerability: (i) the data broadcast by the ANs and later used in the position computation are authentic; (ii) the propagation time of the signal is strictly related to the geometric range between the user and each AN. If either of these assumptions is violated, the associated position is invalid. An adversary may manipulate the position computation by exploiting the message content and/or the signal travel time. Integrity protection of navigation data ensures the authenticity of the message content and is achieved by inserting unpredictable data (i.e., a signature) inside the message. Therefore, to achieve reliable positioning, the receiver should have confidence that the measured time-of-flight of the signal transmitted by the ANs is strictly related to the geometric range between each AN and its receiver. So, the aim of the legitimate receiver is to authenticate the AN–receiver channels by exploiting the characteristics of the received signal.

In this case, the attacker is anyone who attempts to forge a signal directed to the receiver, pretending to be a satellite transmitter, or someone who intercepts and re-transmits signals, thereby modifying their arrival times at the receiver.

In this section, we analyze the problem of position integrity from a general perspective, considering a scenario where a receiver has to estimate its position through the signals sent by ANs, and an attacker tries to induce a wrong position computation at the receiver, transmitting appropriately tampered signals.

13.2.3.1 System Model

We examine the system illustrated in Figure 13.3, where a constellation of m ANs (Alice, denoted as block A in Figure 13.3) transmit signals to a receiver (Bob, denoted as block B in Figure 13.3). Bob determines its position by estimating the relative delays among the signals received from the m ANs and using the positions of the ANs that can be transmitted in the signal payload. For instance, A can represent a LEO, Medium Earth Orbit (MEO) or GEO satellite constellation, while A may be a receiver on Earth or a UAV [4, 32].

An attacker (Eve, denoted as block E in Figure 13.3) receives the sum of the signals from the ANs, manipulates it, and sends a counterfeit signal to Bob. Bob aims to distinguish whether the received signal originates from the ANs constellation or from Eve. In turn, Eve's goal is to impersonate Alice, thereby transmitting a signal that appears authentic to Bob but induces him to calculate an incorrect TDOA among the AN signals, leading to the computation of a wrong position.

A reliable asynchronous side channel, that cannot be used for position estimation, allows the transmission of authenticated data from Alice to Bob, which will be used for signal verification. We assume that the information exchanged on this side channel is unknown to Eve, preventing her from utilizing it in the attack.

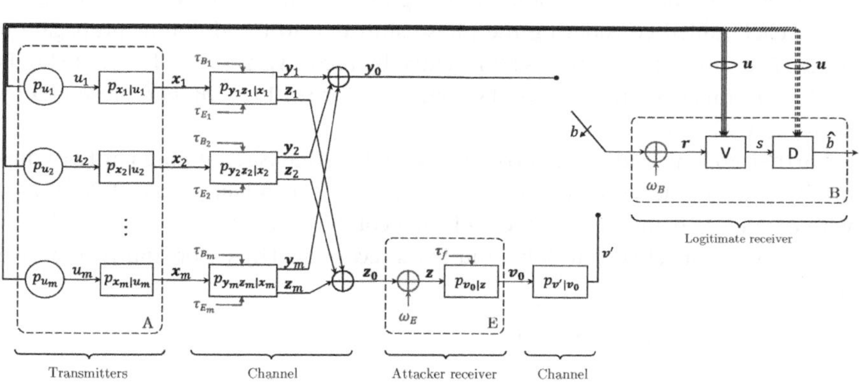

Figure 13.3 General system model for the position integrity problem.

One viable method to implement this channel is through delayed authentication techniques, a common solution in the context of GNSS-based positioning.

Let $\boldsymbol{x}_i \in \mathbb{C}^n$ represent the complex baseband discrete-time equivalent form of the radio signal transmitted by the ith AN, $i = 1, 2, \dots, m$, where n denotes the signal length in terms of the number of samples. Additionally, we assume that samples corresponding to distinct messages are statistically independent. The transmitted word $\boldsymbol{x}$ is formed by concatenating the signals transmitted by all ANs, $\boldsymbol{x} = [\boldsymbol{x}_1, \boldsymbol{x}_2, \dots, \boldsymbol{x}_m] \in \mathbb{C}^{mn}$. Since each $\boldsymbol{x}_i$ is n-dimensional, $\boldsymbol{x}$ is an (mn)-dimensional vector. In our model, each discrete time signal $\boldsymbol{x}_i$ is generated from $\boldsymbol{u}_i$, which is a binary *information word* of length n' that represents an equivalent compressed version of $\boldsymbol{x}_i$. The (possibly probabilistic) encoding of $\boldsymbol{u}_i$ into $\boldsymbol{x}_i$ is represented by the conditional probability density function (pdf) $p_{\boldsymbol{x}_i|\boldsymbol{u}_i}$. Word $\boldsymbol{u} = [\boldsymbol{u}_1, \boldsymbol{u}_2, \dots, \boldsymbol{u}_m]$ is also sent to Bob via the side channel so that he can use it during the verification step of the authentication procedure.

Additionally, $\tau_{\mathrm{B},i}$ and $\tau_{\mathrm{E},i}$ denote the propagation times of $\boldsymbol{x}_i$ from the ith AN, respectively, to Bob and to Eve. For simplicity, we assume that $\min_i\{\tau_{\mathrm{B},i}\} = \min_i\{\tau_{\mathrm{E},i}\} = 0$, without loss of generality. Thus, vectors $\boldsymbol{\tau}_\mathrm{B} = [\tau_{\mathrm{B},1}, \tau_{\mathrm{B},2}, \dots, \tau_{\mathrm{B},m}]$ and $\boldsymbol{\tau}_\mathrm{E} = [\tau_{\mathrm{E},1}, \tau_{\mathrm{E},2}, \dots, \tau_{\mathrm{E},m}]$ gather the relative delays among the signals transmitted by the m ANs, i.e., the TDOA, which is used to estimate the m distances (pseudoranges) between the ANs and the user position. We also define $\delta_\mathrm{B} = \max_i\{\tau_{\mathrm{B},i}\}$ and $\delta_\mathrm{E} = \max_i\{\tau_{\mathrm{E},i}\}$.

Each signal $\boldsymbol{x}_i$, $i \in \{1, 2, \dots, m\}$, is transmitted through an independent channel having $\boldsymbol{x}_i$, τ_{E_i}, and τ_{B_i} as inputs, and two outputs, namely, $\boldsymbol{y}_i \in \mathbb{C}^{n+\tau_{B_i}}$ and $\boldsymbol{z}_i \in \mathbb{C}^{n+\tau_{E_i}}$. Specifically, $\boldsymbol{y}_i$ and $\boldsymbol{z}_i$, $i \in \{1, 2, \dots, m\}$ are the signals directed to Bob and Eve, respectively. The channel is modeled by the joint conditional pdf $p_{\boldsymbol{y}_i\boldsymbol{z}_i|\boldsymbol{x}_i}$ determined by the relative signal delays, channel gains, signal distortion, the fading environment, and fluctuations in the atmospheric parameters. Under nominal conditions, Bob receives the sum of the signals coming from the m AN (or satellites), each added considering the phase shift due to the different propagation times. The sum of the signals is represented by vector $\boldsymbol{y}_0 \in \mathbb{C}^{n+\delta_B}$. The same holds for Eve; in this case, the sum of the signals $\boldsymbol{z}_i$, for $i = 1, \dots, m$, is represented by vector $\boldsymbol{z}_0 \in \mathbb{C}^{n+\delta_E}$.

Furthermore, $\boldsymbol{y}_0$ and $\boldsymbol{z}_0$ are corrupted by AWGN, that is represented by two complex Gaussian random vectors $\boldsymbol{\omega}_\mathrm{B} \sim \mathcal{CN}(\boldsymbol{0}, 2\sigma_\mathrm{B}^2\boldsymbol{I}_{n+\delta_\mathrm{B}})$ and $\boldsymbol{\omega}_\mathrm{E} \sim \mathcal{CN}(\boldsymbol{0}, 2\sigma_\mathrm{E}^2\boldsymbol{I}_{n+\delta_\mathrm{E}})$, with σ_B^2 and σ_E^2 being the variances of each component, either real or imaginary, of the noise respectively at Bob's and Eve's receivers. Thus, Bob and Eve receive the signals

$$\boldsymbol{y} = \boldsymbol{y}_0 + \boldsymbol{\omega}_\mathrm{B}, \quad \boldsymbol{z} = \boldsymbol{z}_0 + \boldsymbol{\omega}_\mathrm{E}, \tag{13.9}$$

respectively. So, $\boldsymbol{y} \in \mathbb{C}^{n+\delta_B}$ and $\boldsymbol{z} \in \mathbb{C}^{n+\delta_E}$ are a reduced size and noisy version of $\boldsymbol{x}$. We denote with M_x the transmitting power, so $\Lambda_{AB} = \frac{M_x}{\sigma_B^2}$ and $\Lambda_{AE} = \frac{M_x}{\sigma_E^2}$ denote the SNR in the channels from Alice to Bob and Eve, respectively.

13.2.3.2 Attack Model

Eve's objective is to generate and transmit a spoofing signal to deceive Bob and make him compute a *target position* different from his actual location. The determination of Bob's position relies on the estimated signal propagation times, so Eve tampers with the propagation times of the original signals, introducing the relative fake delays corresponding to a specified target position, denoted as $\boldsymbol{\tau}\text{f} = [\tau\text{f}, 1, \tau_{\text{f},2}, \dots, \tau_{\text{f},m}]$. We define $\delta_\text{f} = \max_i \tau_{\text{f},i}$.

We make the following assumptions: (i) $\boldsymbol{x}$ is secret to Eve, who only observes $\boldsymbol{z}$; (ii) the joint distributions of $\boldsymbol{x}$, $\boldsymbol{y}$, and $\boldsymbol{z}$ are known by Eve; (iii) Eve is aware of the statistical characterization of channels A → B, A → E, and E → B, and the associated noise statistics. In practice, Eve derives these values by estimating them based on her knowledge of both Bob's and her positions. Note that Eve is unable to process each signal individually, as this would require her to know every word $\boldsymbol{x}_i$, for $i = 1, \dots, m$.

Eve's attack strategy leverages the information contained in her observations $\boldsymbol{z}$. To enhance generality, we consider Eve to employ a probabilistic strategy, defined by the conditional pdf $p_{\boldsymbol{v}_0|\boldsymbol{z}}$. The signal $\boldsymbol{v}_0 \in \mathbb{C}^{n+\delta_f}$ is transmitted through the channel whose statistical description is given by $p_{\boldsymbol{v}'|\boldsymbol{v}_0}$. Therefore, under attack conditions, the counterfeit signal that Bob receives is

$$\boldsymbol{v} = \boldsymbol{v}' + \boldsymbol{\omega}_\text{B}. \tag{13.10}$$

Additionally, Eve knows the statistics of the AWGN at Bob's receiver, so her strategy is characterized by the pdf $p_{\boldsymbol{v}|\boldsymbol{z}}$. As $\boldsymbol{z}$ encompasses all the information exploitable by Eve for deceiving Bob, the attack signal $\boldsymbol{v}$ is conditionally independent of the secret word $\boldsymbol{x}$, given the observation $\boldsymbol{z}$.

Let $\boldsymbol{r}$ represent the signal received by Bob:

$$\boldsymbol{r} = \begin{cases} \boldsymbol{y} & \text{if Bob is locked on the legitimate signal} \, (b = 0); \\ \boldsymbol{v} & \text{if Bob is locked on the spoofing signal} \, (b = 1), \end{cases} \tag{13.11}$$

where the b is a binary variable indicating the legitimate ($b = 0$) or attack ($b = 1$) state. In the under-attack scenario, Bob acquires and tracks only the tampered signal $\boldsymbol{v}$ [36]. Hence, the goal of Eve is to prevent Bob from distinguishing between the attack $\boldsymbol{v}$ and a nominal signal $\boldsymbol{y}$, received when the propagation delays are those related to the target position of the attack, thus $\boldsymbol{\tau}_\text{B} = \boldsymbol{\tau}_\text{f}$.

13.2.3.3 Authentication Procedure

The legitimate receiver, Bob, aims to determine whether the received signal $\boldsymbol{r}$ comes from Alice or Eve. To make his decision, Bob leverages the knowledge of the transmitted word $\boldsymbol{x}$, derived from its compressed version $\boldsymbol{u}$ disclosed by Alice via the side channel. We analyze this problem within the binary hypothesis testing framework, as described in Section 13.2.2: Bob performs an authentication test to detect the spoofing attack, where, given $\boldsymbol{u}$ (or, equivalently, $\boldsymbol{x}$) and $\boldsymbol{r}$, Bob chooses between two hypotheses:

$$\mathcal{H}_0 : \boldsymbol{r} = \boldsymbol{y}, \text{ the signal is from Alice,} \tag{13.12}$$

$$\mathcal{H}_1 : \boldsymbol{r} = \boldsymbol{v}, \text{ the signal is forged.} \tag{13.13}$$

In Figure 13.3, hypothesis $\mathcal{H}_0$ or $\mathcal{H}_1$ is modeled by $\hat{b} = 0$ or 1, respectively. Correct verification is accomplished when $\hat{b} = b$. The legitimate receiver consists of two blocks: the first block P corresponds to the processing phase, while the second block D corresponds to the decision phase. Block P takes as input the received signal $\boldsymbol{r}$ and the vector $\boldsymbol{u}$, and outputs the sufficient statistic s. The task of Block D is to analyze s and decide between hypotheses $\mathcal{H}_0$ or $\mathcal{H}_1$, and then return $\hat{b} = 0$ or $\hat{b} = 1$: $s = P(\boldsymbol{u}, \boldsymbol{r})$, $\hat{b} = D(\boldsymbol{u}, s)$.

From binary hypothesis testing theory, we have that when both the statistics of the legitimate and attack signals are known (i.e., if Bob knows the particular attack strategy adopted by Eve) then the optimal test is given by the Neyman–Pearson criterion [40, 47], i.e., considering the natural logarithm of the LRT in Eq. (13.5), i.e.,

$$L(\boldsymbol{r}) = \ln \frac{p_{\boldsymbol{uv}}(\boldsymbol{u}, \boldsymbol{r})}{p_{\boldsymbol{uy}}(\boldsymbol{u}, \boldsymbol{r})} = \ln \frac{p_{\boldsymbol{v}|\boldsymbol{u}}(\boldsymbol{r}|\boldsymbol{x})}{p_{\boldsymbol{y}|\boldsymbol{u}}(\boldsymbol{r}|\boldsymbol{x})} \underset{H_0}{\overset{H_1}{\gtrless}} \theta, \tag{13.14}$$

where θ is a particular threshold value. However, in the more realistic case in which the legitimate receiver is solely aware of the statistics of the authentic signal, and cannot make any assumption on the attack strategy, and it has no information on the channels A → E and E → B, the LRT does no longer apply. Therefore, we consider the GLRT method [40, 62], i.e.,

$$G(\boldsymbol{r}) = \ln \frac{\max_{v} p_{\boldsymbol{v}|\boldsymbol{u}}(\boldsymbol{v}|\boldsymbol{u})}{p_{\boldsymbol{y}|\boldsymbol{u}}(\boldsymbol{r}|\boldsymbol{u})} \underset{H_0}{\overset{H_1}{\gtrless}} \theta, \tag{13.15}$$

where the numerator of $G(\boldsymbol{r})$ is taken considering all potential attack statistics. Moreover, during the authentication test, the legitimate receiver considers only a portion of the received signal with duration T_{I}, representing the integration time at the receiver.

13.2.3.4 Performance Metrics

The task of authentication for position integrity can be viewed as a special case of binary hypothesis testing, as described in Eqs. (13.12) and (13.13). The performance of a binary hypothesis testing scheme is evaluated based on: (i) false alarm error probability α, i.e., the probability that Bob discards a legitimate message as a forged one; (ii) missed detection error probability β, i.e., the probability that Bob accepts a message transmitted by Eve as legitimate. A properly designed authentication scheme aims to make both error probabilities α and β as small as possible. The optimal achievable performance can in principle be determined analytically, given the statistics of the observation variables under the nominal and the attack hypotheses. However, an analytical evaluation is typically hardly feasible, especially when the decision vector contains many correlated variables, and, for practical schemes, it is most often obtained through numerical simulations. So, theoretical bounds on the achievable error probability region are useful to bound the performance of practical schemes.

We obtain a first bound on the achievable error region, described as the set of reachable points in the (α, β) plane for a given attack strategy, from the application of the data processing inequality (DPI) on the Kullback–Leibler (K–L) divergence $\mathbb{D}(\cdot|\cdot)$ [42], computed for the binary hypothesis decision scheme. From [10] and [47], we have

$$\mathbb{D}(p_{\hat{b}|\mathcal{H}_1} \| p_{\hat{b}|\mathcal{H}_0}) \leq \mathbb{D}(p_{\boldsymbol{ru}|\mathcal{H}_1} \| p_{\boldsymbol{ru}|\mathcal{H}_0}) = \mathbb{D}(p_{\boldsymbol{uv}} \| p_{\boldsymbol{uy}}) = \mathbb{D}(p_{\boldsymbol{xv}} \| p_{\boldsymbol{xy}}). \tag{13.16}$$

In Eq. (13.16), we have considered the joint pdf $p_{\boldsymbol{rx}}$ since while performing the authentication test, the legitimate receiver knows $\boldsymbol{u}$, and the decision $\hat{b}$ is made considering both inputs. Moreover, the first inequality follows from the Markov relation $\boldsymbol{x} \rightarrow \boldsymbol{z} \rightarrow \boldsymbol{v}$, while the last equality in (13.16) holds since $\boldsymbol{u}$ is a lossless compressed version of $\boldsymbol{x}$. Let the function $h(\beta, \alpha)$ be defined as

$$h(\beta, \alpha) \triangleq \beta \log \frac{\beta}{1-\alpha} + (1-\beta) \log \frac{1-\beta}{\alpha}, \tag{13.17}$$

with $\alpha, \beta \in [0, 1]$. Observing that $p_{\hat{b}|\mathcal{H}_0}(1) = \alpha$ and $p_{\hat{b}|\mathcal{H}_1}(0) = \beta$, Eq. (13.16) can be rewritten as

$$h(\beta, \alpha) \leq \mathbb{D}(p_{\boldsymbol{xv}} \| p_{\boldsymbol{xy}}). \tag{13.18}$$

Thus, $\mathbb{D}(p_{\boldsymbol{xv}} \| p_{\boldsymbol{xy}})$ gives an outer bound on the region of achievable values of α and β, that does not depend on the particular choice of the detection mechanism.[2]

2 The bound in Eq. (13.18) is tight when Bob is aware of Eve's attack strategy and Eve remains unaware of Bob's defense strategy. However, Eq. (13.18) is valid in the case of interest where Bob does not know Eve's attack strategy but Eve knows the defense strategy.

The K–L divergence $\mathbb{D}(p_{xv}||p_{xy})$ is formulated in terms of the attack strategy and the pdf of the transmitted word p_x as

$$\begin{aligned}\mathbb{D}(p_{xv}||p_{xy}) &= \iint p_x(a)p_{v|x}(b|a)\log\frac{p_{v|x}(b|a)}{p_{y|x}(b|a)}da\,db\\ &= \iint p_x(a)\int p_{v|z}(b|c)p_{z|x}(c|a)dc\log\\ &\quad\times\frac{\int p_{v|z}(b|c)p_{z|x}(c|a)dc}{p_{y|x}(b|a)}da\,db.\end{aligned}\tag{13.19}$$

To highlight the contribution of the attack strategy, we define

$$f(p_x, p_{v|z}) \triangleq \mathbb{D}(p_{xv}||p_{xy}),\tag{13.20}$$

fixing channels $p_{y|x}$ and $p_{z|x}$.

Therefore, from Eqs. (13.18) and (13.20), the value of $f(p_x, p_{v|z})$ gives an outer bound to the detection error trade-off (DET) curve, determined by α and β, i.e., a common metric to evaluate the attack detection capability at the receiver. The attacker wants to narrow the achievable region, reducing the value of $f(p_x, p_{v|z})$ as much as possible. On the other hand, Alice wants to enlarge the achievable region by making $f(p_x, p_{v|z})$ as large as possible. So, we address the tasks of attack and defense as the maximin problem

$$\max_{p_x}\min_{p_{v|z}} f(p_x, p_{v|z}).\tag{13.21}$$

In NTN, the communication between the ANs and the receivers is usually in LOS. In this case, we can consider, with good approximation, the channels $A \to B$ and $A \to E$ to be linear, and we have that $\boldsymbol{y}_i = \boldsymbol{A}_i\boldsymbol{x}_i$ and $\boldsymbol{z}_i = \boldsymbol{F}_i\boldsymbol{x}_i$. Matrices $\boldsymbol{A}_i$ and $\boldsymbol{F}_i$ are determined by the joint conditional pdf $p_{\boldsymbol{y}_i\boldsymbol{z}_i|\boldsymbol{x}_i}$.

Moreover, $\boldsymbol{A}_i$ and $\boldsymbol{F}_i$ include proper padding to ensure that each channel output vector has the same length. The concatenation of the matrices $\boldsymbol{A}_i$ and $\boldsymbol{F}_i$, for $i = 1, \dots, m$, is denoted by $\boldsymbol{A} = [\boldsymbol{A}_1, \boldsymbol{A}_2, \dots, \boldsymbol{A}_m] \in \mathbb{C}^{(n+\delta_B)\times(mn)}$ and $\boldsymbol{F} = [\boldsymbol{F}_1, \boldsymbol{F}_2, \dots, \boldsymbol{F}_m] \in \mathbb{C}^{(n+\delta_E)\times(mn)}$. In legitimate conditions, Bob and Eve receive, respectively, $\boldsymbol{y}_0 = \boldsymbol{A}\boldsymbol{x}$ and $\boldsymbol{z}_0 = \boldsymbol{F}\boldsymbol{x}$. When channels are linear, vectors $\boldsymbol{x}$, $\boldsymbol{y}$, and $\boldsymbol{z}$ are jointly Gaussian distributed. In this scenario, the optimal attack and defense strategies that solve the problem in Eq. (13.21), $p^\star_{v|z}$ and $p^\star_x$, respectively, are derived in [19, 24]. The optimal attack strategy is found to be a proper linear transformation of the signal received at the attacker's position, summed with an appropriately tuned independent AWGN vector.

Figure 13.4 shows the missed detection probability β, when the false alarm probability is fixed to $\alpha = 0.01$, for the proposed optimal attack strategy for different values of Λ_{AB}, with LRT detection mechanism, compared with the bound given

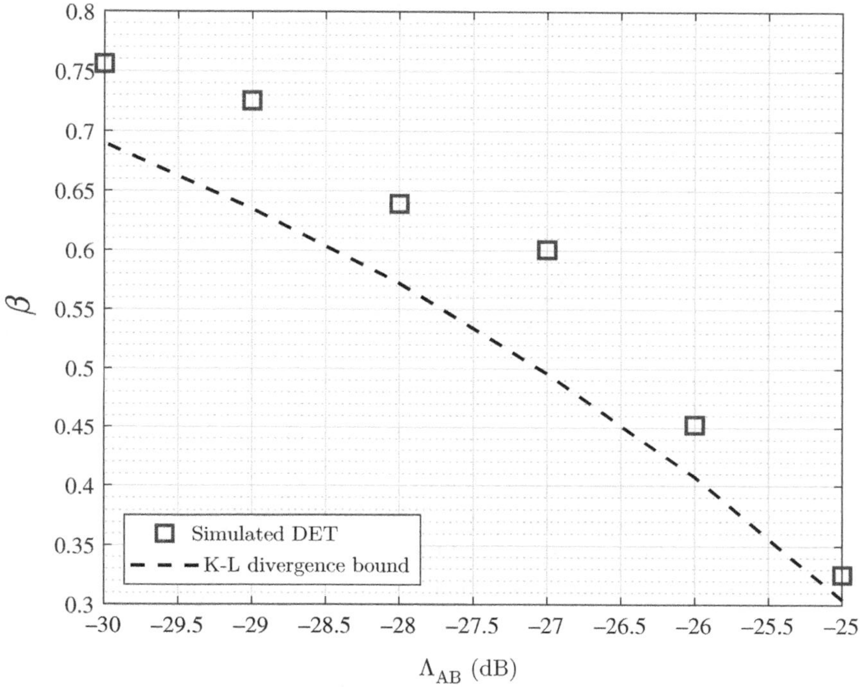

Figure 13.4 Missed detection probability β versus the SNR on the legitimate channel Λ_{AB}, when $\alpha = 0.01$, for the proposed optimal attack strategy, with LRT detection mechanism, compared with the bound given by the K–L divergence, with $m = 4$, $n = 1000$, $M_x = 1$, $\Lambda_{AE} = -20$ dB, and p_x is a Binary distribution.

by the K–L divergence. The figure shows that the bound on the achievable error region given by the K–L divergence holds and is tight, thus, it is representative of the actual performance.

As an example, we compare the performance of the proposed optimal attack to a reference scheme analogous to the one in [14], where Eve knows the channel matrices, estimates the binary word transmitted by each AN from the received signal through hard decision and iterative cancelation, ordered by decreasing received power, then transmits a signal forged according to the legitimate channel as seen from the target spoofed position. From Figure 13.5, we see that when the detection test is the LRT, the K–L divergence bound in Eq. (13.18) is tight and therefore is representative of the actual performance given by the DET curve. Moreover, when the GLRT is applied at Bob's receiver, the optimal attack derived in [19] performs better than the reference attack, since the DET curve related to it

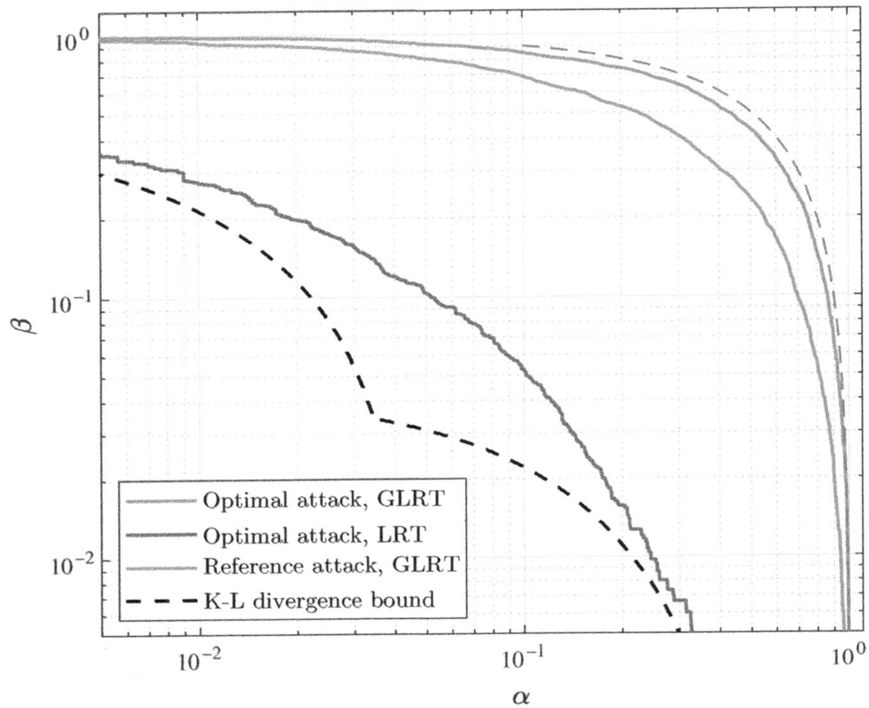

Figure 13.5 Performance of the optimal attack strategy, with LRT and GLRT detection mechanism, and of the reference attack strategy, with GLRT detection mechanism, when $m = 4$, $n = 1000$, $M_x = 1$, $\Lambda_{AB} = -25$ dB, $\Lambda_{AE} = -20$ dB, and p_x is a binary distribution. The gray dashed line describes the trivial limit case wherein the decision is made by tossing a biased coin.

is closer to the gray dashed line, which describes the trivial limit case, wherein the decision is made without considering the received signal, but instead by tossing a biased coin.

13.3 Conclusions

NTNs have been studied as a key innovation to provide global and continuous connectivity (e.g., in rural/unserved areas or to complement 5G coverage), as well as to support 6G applications. In the first part of this chapter, we described recent technologies in the aerial/space industry to enable NTN and summarized current open challenges for the deployment of NTN, especially those associated to latency, coverage, and security constraints. In particular, NTNs can be exposed to security threats, such as channel attacks, spoofing, network eavesdropping, or message modification/injection. Therefore, in the second part of this chapter, we shed light on the issues related to physical-layer security, specifically physical-layer secrecy, authentication, and integrity. Then, we discussed how to guarantee privacy and security in NTNs.

References

1 3GPP. Solutions for NR to support Non-Terrestrial Networks (NTN). TR 38.821 (Release 16), 2020.

2 M. Abdrabou and T. A. Gulliver. Physical layer authentication for satellite communication systems using machine learning. *IEEE Open Journal of the Communications Society*, 3:2380–2389, Dec 2022.

3 R. Ahlswede and I. Csiszár. Common randomness in information theory and cryptography. I. Secret sharing. *IEEE Transactions on Information Theory*, 39:1121–1132, Jul 1993.

4 I. Ahmad, J. Suomalainen, P. Porambage, A. Gurtov, J. Huusko, and M. Höyhtyä. Security of satellite-terrestrial communications: Challenges and potential solutions. *IEEE Access*, 10:96038–96052, Sept 2022.

5 F. Ardizzon and S. Tomasin. Generalized likelihood ratio test with one-class classifiers. *arXiv preprint arXiv:2210.12494*, 2022.

6 Z. H. Awan, A. Zaidi, and L. Vandendorpe. Secure communication over parallel relay channel. *IEEE Transactions on Information Forensics and Security*, 7(2):359–371, Apr 2012.

7 V. Bankey and P. K. Upadhyay. Physical layer security of multiuser multirelay hybrid satellite-terrestrial relay networks. *IEEE Transactions on Vehicular Technology*, 68(3):2488–2501, Jan 2019.

8 M. Bordin, M. Giordani, M. Polese, T. Melodia, and M. Zorzi. Autonomous driving from the sky: Design and end-to-end performance evaluation. In *IEEE Globecom Workshops (GC Wkshps)*, 2022.

9 A. Brighente, F. Formaggio, G. M. Di Nunzio, and S. Tomasin. Machine learning for in-region location verification in wireless networks. *IEEE Journal on Selected Areas in Communications*, 37(11):2490–2502, Aug 2019.

10 C. Cachin. An information-theoretic model for steganography. In *Proceedings of the 2nd International Workshop Information Hiding (IH)*, volume 1525, pages 306–318, 1998.

11 S. Cakaj. The parameters comparison of the "Starlink" LEO satellites constellation for different orbital shells. *Frontiers in Communications and Networks*, 2:643095, May 2021.

12 D. Calero and E. Fernandez. Characterization of chip-scale atomic clock for GNSS navigation solutions. In *Proceedings of the International Association of Institutes of Navigation World Congress (IAIN)*, 2015.

13 K. Cao, Y. Cai, Y. Wu, and W. Yang. Secure communication for amplify-and-forward relay networks with finite alphabet input. *IEEE Transactions on Information Forensics and Security*, 13(9):2269–2278, Sept 2018.

14 G. Caparra, S. Ceccato, N. Laurenti, and J. Cramer. Feasibility and limitations of self-spoofing attacks on GNSS signals with message authentication. In *30th International Technical Meeting of the Satellite Division of the Institute of Navigation (ION GNSS+)*, 2017.

15 R. Caruana. *Multitask learning*. PhD thesis, School of Computer Science, Carnegie Mellon University, Pittsburgh, PA, 1997.

16 M. Ceccato, F. Formaggio, and S. Tomasin. Spatial GNSS spoofing against drone swarms with multiple antennas and wiener filter. *IEEE Transactions on Signal Processing*, 68:5782–5794, Oct 2020.

17 M. Ceccato, F. Formaggio, N. Laurenti, and S. Tomasin. Generalized likelihood ratio test for GNSS spoofing detection in devices with IMU. *IEEE Transactions on Information Forensics and Security*, 16:3496–3509, May 2021.

18 A. Chaoub et al. 6G for bridging the digital divide: Wireless connectivity to remote areas. *IEEE Wireless Communications*, 29(1):160–168, Jul 2021.

19 L. Crosara, F. Ardizzon, S. Tomasin, and N. Laurenti. Worst-case spoofing attack and robust countermeasure in satellite navigation systems. *IEEE Transactions on Information Forensics and Security*, 19:2039–2050, Dec 2023.

20 I. Csiszár and J. Korner. Broadcast channels with confidential messages. *IEEE Transactions on Information Theory*, 24(3):339–348, May 1978.

21 M. D'Apuzzo, M. D'Arco, A. Liccardo, and M. Vadursi. Modeling DAC output waveforms. *IEEE Transactions on Instrumentation and Measurement*, 59(11):2854–2862, Apr 2010.

22 H. Dai, W. Shi, Z. Zhou, and J. Jiang. Authentication method for WiFi connection of devices based on channel state information. In *Proceedings of the 6th International Conference on Computer and Communications (ICCC)*, 2020.

23 E. Ekrem and S. Ulukus. Secrecy in cooperative relay broadcast channels. *IEEE Transactions on Information Theory*, 57(1):137–155, Jan 2011.

24 A. Ferrante, N. Laurenti, C. Masiero, M. Pavon, and S. Tomasin. On the error region for channel estimation-based physical layer authentication over Rayleigh fading. *IEEE Transactions on Information Forensics and Security*, 10(5):941–952, Jan 2015.

25 Q.-Y. Fu, Y.-H. Feng, H.-M. Wang, and P. Liu. Initial satellite access authentication based on Doppler frequency shift. *IEEE Wireless Communications Letters*, 10(3):498–502, Nov 2021.

26 R. Giofré, P. Colantonio, L. González, F. De Arriba, L. Cabría, D. L. Molina, E. C. Garrido, and F. Vitobello. Design realization and tests of a space-borne GaN solid state power amplifier for second generation Galileo navigation system. *IEEE Transactions on Aerospace and Electronic Systems*, 54(5):2383–2396, Mar. 2018.

27 M. Giordani and M. Zorzi. Satellite communication at millimeter waves: A key enabler of the 6G era. *IEEE International Conference on Computing, Networking and Communications (ICNC)*, 2020.

28 M. Giordani and M. Zorzi. Non-terrestrial networks in the 6G era: Challenges and opportunities. *IEEE Network*, 35(2):244–251, Dec 2021.

29 M. Giordani, M. Polese, M. Mezzavilla, S. Rangan, and M. Zorzi. Toward 6G networks: Use cases and technologies. *IEEE Communications Magazine*, 58(3):55–61, Mar 2020.

30 K. Guo, K. An, Y. Huang, and B. Zhang. Physical layer security of multiuser satellite communication systems with channel estimation error and multiple eavesdroppers. *IEEE Access*, 7:96253–96262, Jul 2019.

31 T. Guo, K. Yu, G. Srivastava, W. Wei, L. Guo, and N. N. Xiong. Latent discriminative low-rank projection for visual dimension reduction in green Internet of Things. *IEEE Transactions on Green Communications and Networking*, 5(2):737–749, Jun 2021.

32 H. Guo, J. Li, J. Liu, N. Tian, and N. Kato. A survey on space-air-ground-sea integrated network security in 6G. *IEEE Communications Surveys & Tutorials*, 24(1):53–87, Firstquarter 2022.

33 C. Han, L. Bai, T. Bai, and J. Choi. Joint UAV deployment and power allocation for secure space-air-ground communications. *IEEE Transactions on Communications*, 70(10):6804–6818, Oct 2022.

34 M. Hayashi and Á. Vázquez-Castro. Two-way physical layer security protocol for Gaussian channels. *IEEE Transactions on Communications*, 68(5):3068–3078, May 2020.

35 D. He, Z. Sha, H. Liu, T. Mao, and Z. Wang. UAV-assisted satellite-terrestrial secure communication using large-scale antenna array with one-bit ADCs/DACs. *IEEE Transactions on Communications*, 71(1):580–594, Jan 2022.

36 T. E. Humphreys, B. M. Ledvina, M. L. Psiaki, B. W. O'Hanlon, and P. M. Kintner Jr. Assessing the spoofing threat. *GPS World*, 20(1):28–38, Jan 2009.

37 F. Jameel, S. Wyne, G. Kaddoum, and T. Q. Duong. A comprehensive survey on cooperative relaying and jamming strategies for physical layer security. *IEEE Communications Surveys & Tutorials*, 21(3):2734–2771, Thirdquarter 2019.

38 E. D Kaplan and C. Hegarty. *Understanding GPS/GNSS: Principles and Applications*. Artech House, 2017.

39 L. Karaçay, Z. Bilgin, A. B. Gündüz, P. Çomak, E. Tomur, E. U. Soykan, U. Gülen, and F. Karakoç. A network-based positioning method to locate false base stations. *IEEE Access*, 9:111368–111382, Aug 2021.

40 S. M. Kay. *Fundamentals of Statistical Signal Processing - Volume II: Detection Theory*. Prentice-Hall, Inc., USA, 1993.

41 R. Keating, M. Säily, J. Hulkkonen, and J. Karjalainen. Overview of positioning in 5G new radio. In *16th International Symposium on Wireless Communication Systems (ISWCS)*, 2019.

42 S. Kullback. *Information Theory and Statistics*. Dover Publications Inc., New York, NY, 1967.

43 J. Lei, Z. Han, M. Á. Vazquez-Castro, and A. Hjørungnes. Secure satellite communication systems design with individual secrecy rate constraints. *IEEE Transactions on Information Forensics and Security*, 6(3):661–671, Sept 2011.

44 C. Lei, W. Feng, Y. Chen, and N. Ge. Secrecy rate maximization for 6G cognitive satellite-UAV networks. *China Communications*, 20(1):246–260, Jan 2023.

45 S. K. Leung-Yan-Cheong and M. E. Hellman. The Gaussian wire-tap channel. *IEEE Transactions on Information Theory*, 24(4):451–456, Jul 1978.

46 Y. Li, K. An, T. Liang, and X. Yan. Secrecy performance of land mobile satellite systems with imperfect channel estimation and multiple eavesdroppers. *IEEE Access*, 7:31751–31761, Mar 2019.

47 U. M. Maurer. Authentication theory and hypothesis testing. *IEEE Transactions on Information Theory*, 46(4):1350–1356, Jul 2000.

48 J. Neyman and E. S. Pearson. On the problem of the most efficient tests of statistical hypotheses. *Philosophical Transactions of the Royal Society of London, Series A*, 231:289–337, Jan 1933.

49 H. Niu, Z. Lin, K. An, X. Liang, Y. Hu, D. Li, and G. Zheng. Active RIS-assisted secure transmission for cognitive satellite terrestrial networks. *IEEE Transactions on Vehicular Technology*, 72(2):2609–2614, Feb 2022.

50 G. Oligeri, S. Sciancalepore, S. Raponi, and R. Di Pietro. PAST-AI: Physical-layer authentication of satellite transmitters via deep learning. *IEEE Transactions on Information Forensics and Security*, 18:274–289, Nov 2023.

51 S. Thombre, J. Raasakka, H. Hurskainen, J. Nurmi, M. Valkama, and S. Lohan. Local oscillator phase noise effects on phase angle component of GNSS code correlation. In *Proceedings of the International Conference on Localization and GNSS (ICL-GNSS)*, 2011.

52 O. A. Topal and G. K. Kurt. Physical layer authentication for LEO satellite constellations. In *Wireless Communications and Networking Conference (WCNC)*, 2022.

53 A. Traspadini, M. Giordani, and M. Zorzi. UAV/HAP-assisted vehicular edge computing in 6G: Where and what to offload? In *European Conference on Networks and Communications 6G Summit*, 2022.

54 A. Traspadini, M. Giordani, G. Giambene, and M. Zorzi. Real-time hap-assisted vehicular edge computing for rural areas. *IEEE Wireless Communications Letters*, 12(4):674–678, Jan 2023.

55 Z. Tu, H. Zhou, K. Li, M. Li, and A. Tian. An energy-efficient topology design and DDoS attacks mitigation for green software-defined satellite network. *IEEE Access*, 8:211434–211450, Nov 2020.

56 W. Wang, I. A. Sanchez, G. Caparra, A. McKeown, T. Whitworth, and E. S. Lohan. A survey of spoofer detection techniques via radio frequency fingerprinting with focus on the GNSS pre-correlation sampled data. *Sensors*, 21(9):3012, Apr 2021.

57 D. Wang, A. Traspadini, M. Giordani, M.-S. Alouini, and M. Zorzi. On the performance of non-terrestrial networks to support the Internet of Things. In *Asilomar Conference on Signals, Systems, and Computers*, 2022.

58 W. Wang, E. S. Lohan, I. A. Sanchez, and G. Caparra. Pre-correlation and post-correlation RF fingerprinting methods for GNSS spoofer identification with real-field measurement data. In *10th Workshop on Satellite Navigation Technology (NAVITEC)*, 2022.

59 Z. Wu, Y. Zhang, Y. Yang, C. Liang, and R. Liu. Spoofing and anti-spoofing technologies of global navigation satellite system a survey. *IEEE Access*, 8:165444–165496, Sept 2020.

60 A. D. Wyner. The wire-tap channel. *The Bell System Technical Journal*, 54(8):1355–1387, Oct 1975.

61 A. Zanella, N. Bui, A. Castellani, L. Vangelista, and M. Zorzi. Internet of Things for smart cities. *IEEE Internet of Things Journal*, 1(1):22–32, Feb 2014.

62 O. Zeitouni, J. Ziv, and N. Merhav. When is the generalized likelihood ratio test optimal? *IEEE Transactions on Information Theory*, 38(5):1597–1602, Sept 1992.

63 J. Zhou and X. Zeng. Physical layer secret key generation for spatially correlated channels based on multi-task autoencoder. In *International Conference on Intelligent Computing and Signal Processing (ICSP)*, 2022.

14

Quantum Hardware-Aware Security for 6G Networks

Matthias Frey[1], *Igor Bjelakovic*[2,3], *Janis Nötzel*[4], *Juliane Krämer*[5], *and Sławomir Stańczak*[2,3]

[1] *Department of Electrical and Electronic Engineering, The University of Melbourne, Parkville, Victoria, Australia*
[2] *Technische Universität Berlin, Berlin, Germany*
[3] *Fraunhofer Heinrich-Hertz-Institut, Berlin, Germany*
[4] *Emmy-Noether group Theoretical Quantum Systems Design, Technische Universität München, München, Bayern, Germany*
[5] *Universität Regensburg, Regensburg, Bayern, Germany*

14.1 Introduction

Advances in the real-world implementation of quantum computers in recent years have made the need for security methods that are resilient to attacks with quantum hardware increasingly apparent, both in communication and in data storage. Wireless communication networks such as fifth generation of mobile networks (5G) have always been particularly susceptible to adversarial attacks since the wireless communication medium is by its very nature much more difficult to protect against unauthorized access than other communication media such as wire and fiber. Current estimates predict that by the time the next generation of mobile networks (6G) is deployed, there will be quantum computers and hardware with sufficient capabilities to perform various quantum attacks. In particular, it is possible that such quantum attacks on wireless communications will have evolved significantly in terms of their practicability. With this in mind, it is important that 6G security be built from the ground up on methods that are resistant to quantum attacks.

6G is expected to feature more open interfaces between the various network components due to rapidly progressing modularization and the goal that components of many different vendors should be able to interoperate as part of the network infrastructure. This increases the attack surface significantly

Physical-Layer Security for 6G, First Edition. Edited by Parthajit Mohapatra,
Nikolaos Pappas, Arsenia Chorti, and Stefano Tomasin.

compared to 5G and therefore makes the network more susceptible to any kind of attack, including attacks that utilize quantum measurement or computation equipment. This is an additional risk factor besides the technological advancement until the rollout of 6G which attackers can exploit. While these risks apply in principle to any kind of distributed communication network, we expect that 6G, due to its ubiquitousness and openness, will present a particularly large attack surface, and due to its widespread use, will be a particularly attractive target for attackers. Therefore, we believe that security against quantum attacks, while an important aspect in the design of any future communication network, is particularly relevant in the design of 6G networks.

To date, there are three ways to combat this threat that have received significant attention in research:

- *Quantum cryptography (QC)* is mainly concerned with exploiting quantum effects to make processing and communication between the legitimate communication parties more efficient, but QC schemes take security against quantum attacks into account.
- *Post-quantum cryptography (PQC)* proposes and analyzes schemes in which the legitimate transmitter and receiver can only perform classical computations to protect their communication. There is an assumption that the adversary is limited by a computational bound, but it may conduct any attacks that can be efficiently performed using classical computation as well as computations on a quantum computer. However, there is an inherent assumption that any signal that an eavesdropper intercepts is classically represented, typically as a sequence of bits.
- *Physical-layer security (PLS)* is concerned with schemes which are secure against eavesdropping under assumptions on the signal strength of the attacker. Such schemes are information-theoretically secure in the sense that they do not rely on a bound on the attacker's computational capabilities for their security guarantees. This means in particular that they are secure against attacks conducted with the help of quantum computers. However, similarly to PQC, it is assumed that the attacker intercepts a classically represented signal which is typically a series of complex numbers representing phase and amplitude information about electromagnetic waves.

Practical implementations of cryptography and communication systems are often vulnerable to side-channel attacks (SCAs). An attacker could for instance measure the time or energy it takes to perform certain cryptographic operations, or it could analyze emissions of heat, or electromagnetic waves. In the context of secure communications where the legitimate parties process their signals on purely classical devices, most research on SCAs has focused on attackers that intercept the outputs of such side channels in a classically represented way. In this case, the system models commonly used in PLS are suitable to model SCAs that exploit emissions of the analog parts of transceiver chains.

It is, however, important to note that attackers will not always be bound by such constraints. Only very few works [16] so far have investigated quantum SCAs on cryptography and communication systems in which the legitimate parties process their signals exclusively on classical devices. However, in the field of quantum cryptography, there are extensive experimental studies that investigate side channels with quantum outputs. For example, Huang et al. [25] list a number of attacks on quantum cryptographic protocols that rely on violations of the theoretical assumptions of these protocols due to imperfections in their hardware implementation. Besides various attacks that require an adversary to transmit its own signals, there are also side channels that can be exploited by passive eavesdroppers, both in the preparation [43] and in the detection [29] of quantum states. Many of the mitigation techniques discussed in these works will attenuate the side channel but not completely eliminate it. Moreover, there is no reason to assume that such side channel attacks are limited to communication systems that process quantum information. In fact, we expect that quantum side channels which can potentially be exploited by attackers are also relevant in classical communication systems.

Several works [16, 28] have shown that photon emission from integrated circuits during the execution of encryption can be exploited to reveal the secret keys used. This can be seen as a quantum side channel since the properties of the received photons exhibit a strong dependence on the data being processed and the operation that the device performs. The same phenomenon is expected to occur in the semiconductor hardware that is used both in computation and in the analog components of transceiver chains needed for signal processing in communications.

Therefore, for PLS to be used on a large scale, a comprehensive approach to security is necessary. The expected increased usage of PLS in the near future will give rise to attacks not only against the communication channel, but also against the hardware used to implement PLS encoder and decoder. For example, a rewarding strategy from the point of view of the attacker could be to perform an SCA to determine the messages or randomization parameters used in the process of PLS encoding.

Hence, it is necessary to include protection against all conceivable and realistic attacks against these devices from the beginning. Especially the susceptibility of signal processing techniques toward side-channel and fault attacks has to be studied early on to develop efficient protection measures.

We believe that *quantum physical-layer security* (*QPLS*) can provide an important building block in this endeavor. QPLS is concerned with a quantum extension of Wyner's wiretap channel [42] where, instead of a classical signal, the eavesdropper has access to quantum systems at the channel output. The legitimate communication parties either use purely classical means of communication or also take advantage of the underlying quantum nature of the communication medium.

In this chapter, we survey works in the area of QPLS which we believe may be relevant to quantum hardware aware security for 6G networks. The security guarantees in this context can be of two types: in *secret communication*, the eavesdropper should not be able to draw useful conclusions about the transmitted messages. This is a sufficient security guarantee for many types of communication, but under certain circumstances, it can be necessary to prevent the eavesdropper from detecting that communication is taking place at all; one example is a system provided by law enforcement that can transmit a distress signal without being detected by an adversary. For such applications, *covert communication* provides guarantees that a message transmission cannot be distinguished from background noise.

The remainder of this chapter has the following structure: we introduce necessary terminology and notations for the wiretap channel models we consider as well as for the information measures that are used to express capacity bounds in Section 14.2. Then, we survey QPLS results which we believe could lead the way and aid in developing quantum-secure communication protocols for future 6G networks, both for the case of secret communication requirements in Section 14.3 and for covert communication requirements in Section 14.4. In Section 14.5, the chapter is concluded with the discussion of some research questions we believe should be addressed to pave the way for adoption of quantum-secure PLS in practical systems.

14.2 Preliminaries

In this section, we introduce the basic models for the transmission of classical messages over quantum channels. In addition, we augment these models to include the presence of an eavesdropper with access to a version of the input which is affected by quantum noise. We are only able to give a very brief summary of the mathematical description of quantum systems and channels. A more detailed introduction can be found, e.g., in [10, 22, 23].

14.2.1 Quantum States and Observables

The starting point in the Hilbert space formulation of quantum theory is a separable (possibly infinite dimensional) Hilbert space $\mathcal{H}$ over $\mathbb{C}$. The *state* of a quantum system is described in mathematical terms by a *density operator*. The set of all density operators is

$$\mathcal{S}(\mathcal{H}) := \{\rho \in \mathcal{T}(\mathcal{H}) : \rho \geq 0, \quad \mathrm{tr}(\rho) = 1\}, \tag{14.1}$$

where $\mathcal{T}(\mathcal{H})$ denotes the set of trace class operators (cf. [[10], Section 3.4]) defined by

$$\mathcal{T}(\mathcal{H}) := \{a \in \mathcal{B}(\mathcal{H}) : \mathrm{tr}\sqrt{a^*a} < \infty\}, \tag{14.2}$$

with a^* denoting the adjoint of a, and $\mathcal{B}(\mathcal{H})$ being the set of all bounded linear operators on $\mathcal{H}$. A density operator can be understood as the quantum analog of a classical probability measure.

An *observable* can be understood as the quantum analog of a random variable and is mathematically described with a *positive operator-valued measure (POVM)* (cf. [[10], Sections 4.6 and 9.3]). We will be dealing with observables with finitely many possible outcomes associated with the set of messages to be sent over the channel. Therefore, a POVM in our case is a collection of operators $(F_t)_{t\in\mathcal{F}}$, where $\mathcal{F}$ is a finite set and

$$\forall t \in \mathcal{F}\ F_t \in \mathcal{B}(\mathcal{H}), \quad \forall t \in \mathcal{F}\ F_t \geq 0, \quad \sum_{t\in\mathcal{F}} F_t = \mathbf{1}, \tag{14.3}$$

where $\mathbf{1}$ denotes the identity operator on $\mathcal{H}$.

If the observation represented by the POVM $(F_t)_{t\in\mathcal{F}}$ is performed on a system prepared in state $\rho \in \mathcal{S}(\mathcal{H})$, the probability of the outcome $t \in \mathcal{F}$ is given by Born's rule as follows:

$$p(t) := \mathrm{tr}(\rho F_t).$$

Since $\rho \geq 0$ and $F_t \geq 0$, we have $p(t) \geq 0$. Combining (14.3) and (14.1) with linearity of trace ensures that

$$\sum_{t\in\mathcal{F}} p(t) = 1,$$

so p is indeed a probability mass function on $\mathcal{F}$.

14.2.2 Quantum Channels

The case in which the legitimate communication parties use a classical channel $W_\mathrm{B} : \mathcal{X} \to \mathcal{Y}$ with continuous input and output alphabets $\mathcal{X}$ and $\mathcal{Y}$ (endowed with a suitable measurable structure, i.e., σ-algebras Σ and Σ' on $\mathcal{X}$ and $\mathcal{Y}$ respectively)[1] is depicted in Figure 14.1a. There the eavesdropper observes the output of a

1 In mathematical terms, the channel is defined as the stochastic kernel $W_\mathrm{B} : \mathcal{X} \times \Sigma' \to [0,1]$ such that (i) for every $x \in \mathcal{X}$, $W_\mathrm{B}(x,\cdot)$ is a probability measure on $(\mathcal{Y}, \Sigma')$ and (ii) for every $B \in \Sigma'$, the map $x \mapsto W_\mathrm{B}(x, B)$ is measurable.

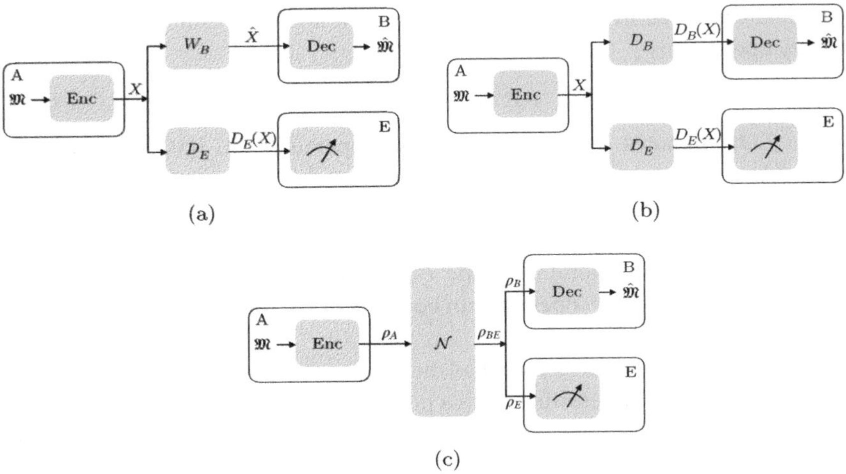

Figure 14.1 Quantum wiretap channel models. (a) classical/classical-quantum (c/cq) wiretap channel, (b) classical/quantum (c/q) wiretap channel, and (c) quantum/quantum (q/q) wiretap channel.

classical-quantum (cq) channel $D_E : \mathcal{X} \to \mathcal{S}(\mathcal{H}_E)$, where $\mathcal{H}_E$ is a separable Hilbert space over $\mathbb{C}$. Endowed with the *trace norm*

$$\|a\|_{\text{tr}} := \text{tr}\sqrt{a^*a},$$

the trace class $\mathcal{T}(\mathcal{H}_E)$ turns out to be a Banach space. This Banach space structure induces the σ-algebra of Borel sets and we tacitly assume in the following that the cq channel $D_E : \mathcal{X} \to \mathcal{S}(\mathcal{H}_E)$ and any other cq channel under consideration is a measurable map with respect to the measurable structures on the input alphabet and the set of quantum states at the output. In Figure 14.1b, the classical channel between the legitimate parties is replaced with a cq channel $D_B : \mathcal{X} \to \mathcal{S}(\mathcal{H}_B)$ mapping into the set of states at the receiver. A general notion of a quantum channel with terminals for the legitimate communication parties and an eavesdropper is depicted in Figure 14.1c. It transforms input states into output states and is mathematically described by a completely positive trace-preserving map $\mathcal{N} : \mathcal{T}(\mathcal{H}_A) \to \mathcal{T}(\mathcal{H}_B \otimes \mathcal{H}_E)$ (cf. [[10], Sections 7.1 and 7.6]). For any input state $\rho_A \in \mathcal{S}(\mathcal{H}_A)$, the output state $\rho_{BE} := \mathcal{N}(\rho_A) \in \mathcal{S}(\mathcal{H}_B \otimes \mathcal{H}_E)$ on the composite system induces the marginal versions $\rho_B := \text{tr}_E(\rho_{BE}) \in \mathcal{S}(\mathcal{H}_B)$ and $\rho_E := \text{tr}_B(\rho_{BE}) \in \mathcal{S}(\mathcal{H}_E)$ which correspond to the states of systems the legitimate receiver and the eavesdropper have access to respectively. The partial trace $\text{tr}_E : \mathcal{S}(\mathcal{H}_B \otimes \mathcal{H}_E) \to \mathcal{S}(\mathcal{H}_B)$ is defined by

$$\text{tr}(\text{tr}_E(\rho_{BE})a) = \text{tr}(\rho_{BE}(a \otimes \mathbf{1}_E)), \tag{14.4}$$

for all $a \in \mathcal{B}(\mathcal{H}_B)$, where $\mathbf{1}_E$ denotes the identity operator acting on $\mathcal{H}_E$, and with an analogous definition for tr_B.

The marginal or complementary channels $\mathcal{N}_B : \mathcal{T}(\mathcal{H}_A) \to \mathcal{T}(\mathcal{H}_B)$ and $\mathcal{N}_E : \mathcal{T}(\mathcal{H}_A) \to \mathcal{T}(\mathcal{H}_E)$ are defined by $\mathcal{N}_B := \mathrm{tr}_E \circ \mathcal{N}$ and $\mathcal{N}_E := \mathrm{tr}_B \circ \mathcal{N}$.

14.2.3 Bosonic Systems

Based on the chosen parameter regime, bosonic channels model a variety of communication systems in the radio and optical domain. These systems are described on the Hilbert space $\mathcal{H}$ which has a distinguished orthonormal basis $\{|n\rangle\}_{n\in\mathbb{N}}$ of the so-called *number states*. A bosonic system in state $|n\rangle$ consists of exactly n photons. Following [[23], Sections 12.1.3 and 12.1.4], the basic bosonic cq channel is given by the mapping

$$\mathbb{C} \ni \mu \mapsto S_N(\mu) := \frac{1}{\pi N} \int |\xi\rangle\langle\xi| \exp\left\{-\frac{|\mu-\xi|^2}{N}\right\} d\xi, \tag{14.5}$$

modeling transmission of the classical signal $\mu \in \mathbb{C}$ subject to additive quantum Gaussian noise of power N. The states $|\xi\rangle\langle\xi|$ are the so-called coherent states (cf. [[23], Section 12.1.3]), where the normalized vector $|\xi\rangle \in \mathcal{H}$ is labeled by $\xi \in \mathbb{C}$ and defined by

$$|\xi\rangle := \exp\{-|\xi|^2/2\} \sum_{n=0}^{\infty} \frac{\xi^n}{\sqrt{n!}} |n\rangle. \tag{14.6}$$

In physics, a coherent state is the standard model for a narrowband laser pulse [2]. The states $S_N(\mu)$ defined in (14.5) are called the displaced thermal states.

The most elementary wiretap channel model in this domain is a beam splitter of transmittivity $\eta \in [0, 1)$, which splits the state $S_N(\mu)$ into the part $D_B(\mu) = S_{\eta N}(\sqrt{\eta}\mu)$ and $D_E(\mu) = S_{(1-\eta)N}(\sqrt{1-\eta}\mu)$. The resulting branches of the cq channel to the legitimate receiver and eavesdropper, respectively, are given as the maps $\mathbb{C} \ni \mu \mapsto D_B(\mu) \in \mathcal{S}(\mathcal{H}_B)$ and $\mathbb{C} \ni \mu \mapsto D_E(\mu) \in \mathcal{S}(\mathcal{H}_E)$. In this communication setting, the transmitted energy is taken to be $|\mu|^2$ [23].

In general, a two-mode beam splitter is described by a unitary operation on $\mathcal{H}^{\otimes 2}$. For cases where states other than displaced thermals states are to be transmitted, a more complete description of the beam splitter is needed (cf. [[23], Section 12.6.1, p. 323, and p. 328] where it is called attenuation channel). In such cases, the description is via the relations

$$\hat{b} = \sqrt{\eta}\hat{a} + \sqrt{1-\eta}\hat{f}, \tag{14.7}$$

$$\hat{e} = \sqrt{1-\eta}\,\hat{a} + \sqrt{\eta}\hat{f}, \tag{14.8}$$

where $\hat{a}$, $\hat{b}$, and $\hat{e}$ are the annihilation operators of the transmitter, receiver, and eavesdropper, respectively. $\hat{f}$ is the annihilation operator of the environment which is in its vacuum state, and $\eta \in [0, 1)$.

More information on bosonic states and channels is contained in the monograph [[23], Chapter 12].

14.2.4 Information Measures

In order to be able to express capacity bounds for wiretap channels, we introduce a few information quantities. Given any quantum state $\rho \in \mathcal{S}(\mathcal{H})$, we define the *von Neumann entropy* as

$$S(\rho) := -\text{tr}(\rho \log \rho).$$

Note that in case of infinite-dimensional $\mathcal{H}$, it is possible that $\rho \log \rho \notin \mathcal{T}(\mathcal{H})$. In these cases, we adopt the convention $S(\rho) = \infty$.

Given quantum states ρ, σ, we define the *quantum relative entropy* as

$$H(\rho \| \sigma) := \text{tr}(\rho \log \rho) - \text{tr}(\rho \log \sigma),$$

where we again use the convention that the relative entropy is ∞ if one of these operators is not trace class.

For a cq channel D and a probability distribution P on the input alphabet $\mathcal{X}$ of D, we define the *Holevo information*

$$\begin{aligned}\chi(P; D) :&= \int_{\mathcal{X}} \text{tr}(D(x) \log D(x))P(dx) - \text{tr}\left(\int_{\mathcal{X}} D(x)P(dx) \log \int_{\mathcal{X}} D(x)P(dx)\right) \\ &= S(\mathbb{E}_X D(X)) - \mathbb{E}_X S(D(X)).\end{aligned}$$

We note that in case of infinite $\mathcal{X}$, the existence of these integrals needs to be checked in order for the Holevo information to be well defined. For the infinite-dimensional case, we adopt the convention that $\chi(P; D) = \infty$ whenever $S(\mathbb{E}_X D(X)) = \infty$ or $S(D(X)) = \infty$ on a non-null set.

For channels $\mathcal{N}_\text{B}$ between quantum systems, we define the *coherent information* as

$$I_c(\rho; \mathcal{N}_\text{B}) := S(\mathcal{N}_\text{B}(\rho)) - S(\mathcal{N}_\text{E}(\rho)),$$

where $\mathcal{N}_\text{E}$ and $\mathcal{N}_\text{B}$ are complementary channels.

In order to maintain notational consistence, we denote the (classical) mutual information between the input and the output of a classical channel W with $I(P; W)$.

14.3 Secret Communication

In this section, we discuss the various measures of secrecy that are commonly used and survey results for secret communication through wiretap channels with a quantum output.

14.3.1 Semantic Security and Its Operational Significance

For both the c/cq and c/q channels in Figure 14.1a and 14.1b, the encoding for n channel uses is a stochastic map $\mathrm{Enc} : \mathcal{M} \to \mathcal{X}^n$, where $\mathcal{M} := \{1, \dots, M\}$ denotes the set of messages.[2] The decoding operation for the c/cq channel (Figure 14.1a) is given by a (measurable) deterministic map $\mathrm{Dec} : \mathcal{Y}^n \to \mathcal{M}$, where $\mathcal{Y}$ denotes the (continuous) output alphabet of the channel W_B in Figure 14.1a.

The decoding process for the c/q channel is described by a POVM $\mathrm{Dec} := (F_m)_{m\in\mathcal{M}}$ on $\mathcal{H}_B^{\otimes n}$. The corresponding decoding success probabilities for c/cq and c/q channels for block length n are given by

$$P_{\mathrm{dec}}^{(n)}(W_B, \mathrm{Enc}, \mathrm{Dec}) := \frac{1}{M}\sum_{m=1}^{M} \int W_B^{\otimes n}(x^n, \mathrm{Dec}^{-1}(m))\mathrm{Enc}(m, dx^n), \tag{14.9}$$

and

$$P_{\mathrm{dec}}^{(n)}(D_B, \mathrm{Enc}, \mathrm{Dec}) := \frac{1}{M}\sum_{m=1}^{M} \int \mathrm{tr}(F_m D_B^n(x^n))\mathrm{Enc}(m, dx^n), \tag{14.10}$$

respectively, where $D_B^n(x^n) := D_B(x_1) \otimes \cdots \otimes D_B(x_n)$ with $x^n := (x_1, \dots, x_n) \in \mathcal{X}^n$.

For the q/q channel in Figure 14.1c, the encoding for n uses of the channel $\mathcal{N}$ is a map $\mathrm{Enc} : \mathcal{M} \to \mathcal{S}(\mathcal{H}_A^{\otimes n})$ and the decoding operation is, again, given by a POVM $\mathrm{Dec} := (F_m)_{m\in\mathcal{M}}$ on $\mathcal{H}_B^{\otimes n}$. The corresponding decoding success probability is given by

$$P_{\mathrm{dec}}^{(n)}(\mathcal{N}_B, \mathrm{Enc}, \mathrm{Dec}) := \frac{1}{M}\sum_{m=1}^{M} \mathrm{tr}(F_m \mathcal{N}_B^{\otimes n}(\mathrm{Enc}(m))), \tag{14.11}$$

where $\mathcal{N}_B^{\otimes n} := \mathcal{N}_B \otimes \cdots \otimes \mathcal{N}_B$.

Semantic security is an operational security measure which quantifies the advantage of an adversary's observation-based inference ability about the sent message over pure guessing, i.e., without access to the channel output. Given an arbitrary partition Π of the message set $\mathcal{M} = \{1. \dots, M\}$, the goal of the eavesdropper is to decide to which partition element $\pi \in \Pi$ the actually transmitted message belongs to. For example, the partition corresponding to the goal of reconstruction of the entire message is given by $\Pi = \{\{1\}, \dots, \{M\}\}$, while the objective of reconstruction of the first bit of the message sent is given by the

2 Explicitly, Enc is a stochastic kernel $\mathrm{Enc} : \mathcal{M} \times \Sigma^{\otimes n} \to [0, 1]$ such that for each $m \in \mathcal{M}$, $\mathrm{Enc}(m, \cdot)$ is a probability measure on $(\mathcal{X}^n, \Sigma^{\otimes n})$ and the requirement of measurable dependence on $m \in \mathcal{M}$ is superfluous as $\mathcal{M}$ is a finite set.

partition $\Pi = \{\pi_0, \pi_1\}$ with

$$\pi_b := \{m \in \mathcal{M} : m \text{ has a binary representation starting with } b\}, \quad (14.12)$$

for $b \in \{0, 1\}$.

Let $\rho_E^n = \rho_E^n(\mathfrak{M})$ denote the state of the eavesdropper's quantum system in each of the channel models depicted in Figure 14.1a–14.1c after performing the encoding operation Enc and after using the corresponding channel n times. We assume that the eavesdropper can perform any measurement represented by any Π-valued POVM $(F_\pi)_{\pi\in\Pi}$ on the system in state $\rho_E^n(\mathfrak{M})$. The state $\rho_E^n(\mathfrak{M})$ and the POVM $(F_\pi)_{\pi\in\Pi}$ induce a conditional probability distribution (conditioned on $\mathfrak{M} = m$) on Π given by

$$p(\pi|m) := \mathrm{tr}(F_\pi \rho_E^n(m)), \quad (14.13)$$

for $\pi \in \Pi$, which is the probability that the message sent is contained in the partition element π when measuring the observable $(F_\pi)_{\pi\in\Pi}$ on a system in state $\rho_E^n(m)$.

The maximum success probability of the eavesdropper when a measurement of a Π-valued quantum observable is performed on a system in the quantum state $\rho_E^n(\mathfrak{M})$ is defined as

$$\mathrm{Succ}(\Pi, P_{\mathfrak{M}}, \rho_E^n(\mathfrak{M})) := \sup \left\{ \mathbb{E}_{\mathfrak{M}} \sum_{\pi\in\Pi:\,\mathfrak{M}\in\pi} \mathrm{tr}(F_\pi \rho_E^n(\mathfrak{M})) : (F_\pi)_{\pi\in\Pi} \text{ is a POVM} \right\}.$$

The maximum achievable success probability for partition Π and probability distribution $P_{\mathfrak{M}}$ of the message *without access* to the channel output state $\rho_E^n(\mathfrak{M})$ is given by

$$\mathrm{Guess}(\Pi, P_{\mathfrak{M}}) := \sup \{\mathbb{P}(\mathfrak{M} \in \pi) : \pi \in \Pi\}.$$

The *eavesdropper's advantage* associated with the eavesdropper's channel $C_E \in \{D_E, \mathcal{N}_E\}$ and the encoding operation Enc resulting in the output state $\rho_E^n(\mathfrak{M})$ after n memoryless channel uses is defined as

$$\mathrm{Adv}^{(n)}(C_E, \mathrm{Enc}) := \sup_{\Pi, P_{\mathfrak{M}}} \left\{\mathrm{Succ}(\Pi, P_{\mathfrak{M}}, \rho_E^n(\mathfrak{M})) - \mathrm{Guess}(\Pi, P_{\mathfrak{M}})\right\}, \quad (14.14)$$

and is a natural extension of the notion of *semantic security* (cf. [5]) to quantum wiretap channels. The code is said to have *semantic security level* δ if

$$\mathrm{Adv}^{(n)}(C_E, \mathrm{Enc}) \leq \delta, \quad (14.15)$$

for $C_E \in \{D_E, \mathcal{N}_E\}$ and $\delta \geq 0$.

A rate $R \in [0, \infty)$ is called *achievable* with semantic security if for every block length $n \in \mathbb{N}$, there is an encoding–decoding pair $(\mathrm{Enc}_n, \mathrm{Dec}_n)$ with $M_n = |\mathcal{M}|$ and

$$\liminf_{n\to\infty} \frac{1}{n} \log M_n \geq R, \tag{14.16}$$

such that

$$\lim_{n\to\infty} \mathrm{Adv}^{(n)}(C_\mathrm{E}, \mathrm{Enc}_n) = 0 \quad \text{and} \quad \lim_{n\to\infty} P^{(n)}_{\mathrm{dec}}(C_\mathrm{B}, \mathrm{Enc}_n, \mathrm{Dec}_n) = 1, \tag{14.17}$$

where $C_\mathrm{E} \in \{D_\mathrm{E}, \mathcal{N}_\mathrm{E}\}$ and $C_\mathrm{B} \in \{W_\mathrm{B}, D_\mathrm{B}, \mathcal{N}_\mathrm{B}\}$.

The *secrecy capacity* C_s for $s \in \{c/cq, c/q, q/q\}$ is defined as the least upper bound of semantically secure achievable rates.

14.3.2 Other Security Measures Used in the Analysis of Secret Communication

While the semantic security defined via the advantage over guessing has a clear and natural operational interpretation, it is apparently challenging to derive bounds on semantic security levels based directly on the definition in (14.14). Therefore, we introduce some alternative security notions here and briefly discuss implications between them.

For a given encoding operation Enc, $m \in \mathcal{M}$, and the corresponding output state $\rho^n_\mathrm{E}(m)$ after n uses of either D_E or $\mathcal{N}_\mathrm{E}$, we define the *message distinguishability* as

$$\mathrm{Dis}^{(n)}(C_\mathrm{E}, \mathrm{Enc}) := \max_{m_1, m_2 \in \mathcal{M}} \left\| \rho^n_\mathrm{E}(m_1) - \rho^n_\mathrm{E}(m_2) \right\|_{\mathrm{tr}}, \tag{14.18}$$

where $C_\mathrm{E} \in \{D_\mathrm{E}, \mathcal{N}_\mathrm{E}\}$. The code is said to have a *distinguishing security level* of δ if

$$\mathrm{Dis}^{(n)}(C_\mathrm{E}, \mathrm{Enc}) \leq \delta, \tag{14.19}$$

for $\delta \geq 0$ and $C_\mathrm{E} \in \{D_\mathrm{E}, \mathcal{N}_\mathrm{E}\}$.

Alternatively, we can look at the information that the eavesdropper's channel output has with the transmitted message under a message distribution $P_\mathfrak{M}$ and define

$$\mathrm{MIS}^{(n)}(P_\mathfrak{M}, C_\mathrm{E}, \mathrm{Enc}) := \chi(P_\mathfrak{M}; \rho^n_\mathrm{E}).$$

Based on this, the code is said to have *mutual information security level* δ if

$$\mathrm{MIS}^{(n)}_{\max}(C_\mathrm{E}, \mathrm{Enc}) := \max_{P_\mathfrak{M}} \mathrm{MIS}(P_\mathfrak{M}, C_\mathrm{E}, \mathrm{Enc}) \leq \delta,$$

where $P_\mathfrak{M}$ ranges over all probability distributions on $\{1, \ldots, M\}$. We say that the code has *strong secrecy level* δ if

$$\mathrm{MIS}^{(n)}_{\mathrm{unif}}(C_\mathrm{E}, \mathrm{Enc}) := \mathrm{MIS}(P_\mathfrak{M}, C_\mathrm{E}, \mathrm{Enc}) \leq \delta,$$

where $P_{\mathfrak{M}}$ is the uniform distribution on $\{1, \ldots, M\}$. Finally, we say that the code has *weak secrecy level* δ if

$$\frac{1}{n}\,\mathrm{MIS}^{(n)}_{\mathrm{unif}}(C_{\mathrm{E}}, \mathrm{Enc}) \leq \delta.$$

We say that a sequence of codes (for growing block length n) satisfies weak secrecy (strong secrecy, semantic security, distinguishing security, mutual information security) if the corresponding secrecy or security level vanishes as n tends to infinity.

Then, the following implications hold: .

Lemma 14.1 For $C_{\mathrm{E}} \in \{D_{\mathrm{E}}, \mathcal{N}_{\mathrm{E}}\}$ and any sequence of codes for block length n tending to infinity, we have the following implications:

1. If the sequence satisfies strong secrecy, it also satisfies weak secrecy.
2. If the sequence satisfies mutual information security, it also satisfies strong secrecy.
3. If the sequence satisfies mutual information security, it also satisfies distinguishing security.
4. The sequence satisfies distinguishing security iff it satisfies semantic security.
5. If the eavesdropper's output is finite dimensional and $n\delta$ vanishes where δ is the distinguishing security level, then the sequence of codes satisfies mutual information security.

Items 1 and 2 are trivially clear, and the remaining items hold in parallel to the classical case [5, 6]. Detailed proofs of the statements in Lemma 14.1 can be found in [[18], Section III-B]. For item 5, it should also be noted that [[18], Lemma 7] gives sufficient conditions besides finite dimensionality and shows in particular that the implication holds for certain infinite-dimensional bosonic channels as long as the average input energy is bounded.

Semantic security is not always defined the same way in the literature. For instance, the authors of [17, 30, 39–41] show distinguishing security while the authors of [7, 21] show mutual information security. Lemma 14.1 tells us that both these notions imply semantic security in the sense defined in this chapter, and at least in a subset of cases, we have equivalence of mutual information security, distinguishing security, and semantic security.

14.3.3 Survey of Results

To the best of our knowledge, the q/q wiretap channel model was first proposed in [31] for the special case $n = 1$. Later works have considered a regime where n grows to arbitrarily large values and analyzed how many different values M the message $\mathfrak{M}$ can take depending on n while ensuring reliability in decoding the message at the receiver and security against the eavesdropper. This leads to the standard notions of rate in (14.16) and capacity, and we survey results for this regime in the remainder of this section.

14.3.3.1 Finite-Dimensional Case

The authors of [11, 14] independently investigated block coding for quantum wiretap channels, deriving very similar results. In [[14], Theorem 1], the capacity $C_{c/q}$ of the c/q wiretap channel is shown to be

$$C_{c/q} = \lim_{n\to\infty} \frac{1}{n} \max_{\tilde{P},V} \left(\chi(\tilde{P}; D_{B}^{\otimes n} \circ V) - \chi(\tilde{P}; D_{E}^{\otimes n} \circ V)\right), \tag{14.20}$$

where the maximum ranges over all auxiliary distributions on finite alphabets and auxiliary channels V from the auxiliary alphabet to $\mathcal{X}^n$.

Cai et al. [[11], Theorem 4] state a single-letter version of the direct part of this result, namely, that

$$C_{c/q} \geq \max_{P} \left(\chi(P; D_{B}) - \chi(P; D_{E})\right), \tag{14.21}$$

where the maximum ranges over all possible distributions on the input alphabet of the channel.

It should be noted that by restricting the maximization of $\tilde{P}$ to i.i.d. distributions on $\mathcal{X}^n$ and fixing the auxiliary channel V at identity, it is straightforward to argue that (14.20) implies (14.21), and similar arguments apply to other multi-letter capacities based on Holevo information below. However, it is known that there exist q/q, c/q, and c/cq wiretap channels for which this single-letter bound is strictly less than the capacity [35].

The result for the c/q wiretap channel is extended to the case of the q/q wiretap channel where the legitimate receiver's and the eavesdropper's channels are complementary in [[14], eq. (30)], giving the capacity as

$$C_{q/q} = \lim_{n\to\infty} \frac{1}{n} \max_{\rho_A \in \mathcal{H}_A^{\otimes n}} \Bigg(I_c(\rho_A; \mathcal{N}_B^{\otimes n}) - \min \Bigg\{ \sum_{u\in\mathcal{U}} \tilde{P}(u) I_c(\rho_u; \mathcal{N}_B^{\otimes n}) : \\ \mathcal{U} \text{ finite, } \tilde{P} \text{ p.m.f. on } \mathcal{U},\ \forall u \in \mathcal{U}\ \rho_u \in \mathcal{S}(\mathcal{H}_A^{\otimes n}),\ \sum_{u\in\mathcal{U}} \tilde{P}(u)\rho_u = \rho_A \Bigg\} \Bigg).$$

It is noted that the subtracted min term is always nonpositive, since any decomposition of ρ_A into pure states realizes 0.

Cai et al. [[11], Theorem 3] extend the c/q wiretap channel result to the q/q wiretap channel where the receiver's and eavesdropper's channels do not need to be complementary. The capacity is in this case determined to be

$$C_{q/q} = \lim_{n\to\infty} \frac{1}{n} \sup_{\tilde{P},V} \left(\chi(\tilde{P}; \mathcal{N}_{B}^{\otimes n} \circ V) - \chi(\tilde{P}; \mathcal{N}_{E}^{\otimes n} \circ V)\right), \tag{14.22}$$

where the supremum ranges over all probability distributions $\tilde{P}$ on finite alphabets and all cq channels mapping to $\mathcal{S}(\mathcal{H}_A^{\otimes n})$, the set of possible quantum channel inputs over n uses of $\mathcal{N}$.

All preceding results are stated for security guarantees in the form of weak or strong secrecy, but the authors of [11, 14] prove (as an intermediate step) that they hold even under the semantic security criterion.

Later, the results for the c/q wiretap channel were extended in [21] to cases in which only randomness of lower quality is available at the transmitter, and equivocation rates were analyzed for the case that the rate of communication is larger than the secrecy capacity. It is also shown that secret communication is possible using only linear codes. Furthermore, this work showed that any rate that is achievable for the c/q wiretap channel under strong secrecy is also achievable under semantic security.

The relationship between semantic security and strong secrecy was further explored in [7], providing explicit constructions which transform any code which achieves strong secrecy for a c/q or q/q wiretap channel to a code that achieves semantic security and has asymptotically the same rate.

14.3.3.2 Infinite-Dimensional Case

In the context of 6G, it is of particular interest to explore the quantum properties of channels that use electromagnetic waves as a medium. Finite-dimensional models can only capture particular aspects of these physical systems, such as polarization of photons. Therefore, we will consider models that are based on infinite-dimensional Hilbert spaces. An example of properties of physical systems that require a description by infinite-dimensional Hilbert spaces is the photon number or energy of the quantized electromagnetic field.

In this section, we survey results for general infinite-dimensional channels as well as results that are specific to bosonic channels, which also include electromagnetic waves as a special case.

In [20] , the results from [14] are extended to the single-mode noiseless bosonic channel of transmittivity η. The authors argue that this channel is degradable and use [[34], Theorem 2] to prove that under the condition of strong secrecy and a mean input photon-number constraint $\bar{N}_a$, the single-letter capacity expression

$$C_{\text{nlb}} = \begin{cases} g(\eta\bar{N}_a) - g((1-\eta)\bar{N}_a), & \eta > 1/2 \\ 0, & \eta \leq 1/2, \end{cases} \tag{14.23}$$

holds, where

$$g : [0,\infty) \to [0,\infty), t \mapsto \begin{cases} 0, & t = 0 \\ (t+1)\log(t+1) - t\log t, & t > 0, \end{cases}$$

denotes the Gordon function. However, the converse part of this result holds only under an unproven conjecture, the entropy photon-number inequality (EPNI).

In [[41], Theorem 7], this result is extended to a scenario where a trade-off of private classical communication, public classical communication, and secret key generation is of interest. The result is under condition of semantic security and recovers (14.23) for the special case of only private classical communication, and the proof of the converse part still uses the EPnI.

The same trade-off is explored for general q/q wiretap channels in [[39], Theorem 1]. The specialization of this result to private communication only recovers (14.22) for the case of not necessarily finite-dimensional channels.

A direct result for the thermal-noise bosonic channel, which also follows from (14.7), (14.8), but where the environment is in a thermal state with mean photon number $\bar{N}_f$, is given in [[41], Theorem 10]. The result is also for a trade-off between secret key generation, public classical communication, and private classical communication. Specialized for private communication only, it determines a capacity bound

$$C_{\text{thb}} \geq g(\eta\bar{N}_a + (1-\eta)\bar{N}_f) - g\left(\frac{1}{2}(D + (1-\eta)(\bar{N}_a - \bar{N}_f) - 1)\right) - g\left(\frac{1}{2}(D - (1-\eta)(\bar{N}_a - \bar{N}_f) - 1)\right),$$

where

$$D^2 = ((1+\eta)\bar{N}_a + (1-\eta)\bar{N}_f + 1)^2 - 4\eta\bar{N}_a(\bar{N}_a + 1).$$

The authors remark that they expect that equality does not hold here in general for this capacity bound.

In [[30], Section V], the authors extend these results to the quantum amplifier channel, described by

$$\hat{b} = \sqrt{\kappa}\hat{a} + \sqrt{\kappa - 1}\hat{f}^*$$
$$\hat{e}^* = \sqrt{\kappa - 1}\hat{a} + \sqrt{\kappa}\hat{f}^*,$$

where the environment is in its vacuum state and $\kappa \in [1, \infty)$ is called the *gain* of the channel. The derived capacity region then specializes to

$$C_{\text{amp}} = g(\kappa\bar{N}_a + \kappa - 1) - g((\kappa - 1)(\bar{N}_a + 1)). \tag{14.24}$$

In [40], these results are extended to general quantum channels under a uniform energy constraint. The uniform energy constraint is given by a Gibbs observable G and a positive real number P and it demands that for every channel input ρ_{A} that corresponds to a transmitted message

$$\text{tr}(\bar{G}_n\rho_{\text{A}}) \leq P,$$

is satisfied, where

$$\bar{G}_n := \frac{1}{n}(G \otimes \mathbf{1} \otimes \cdots \otimes \mathbf{1} + \cdots + \mathbf{1} \otimes \cdots \otimes \mathbf{1} \otimes G),$$

where there are n factors in each tensor product. Under a finite output entropy condition, the capacity for a degradable q/q wiretap channel under semantic security is given in [[40], Theorem 3] as

$$C_{\mathrm{q/q,d},G,P} = \sup_{\rho:\mathrm{tr}G\rho\leq P} I_{\mathrm{c}}(\rho;\mathcal{N}_{\mathrm{B}}),$$

where the legitimate receiver's channel is $\mathcal{N}_{\mathrm{B}}$ and the eavesdropper observes the output of a complementary channel. For the case of a not necessarily degradable channel, Wilde and Qi [[40], Theorem 4] give a multi-letter converse that can be understood as an infinite-dimensional version of the converse part of (14.22) under an energy constraint. Wilde and Qi [[40], Theorem 7] also give the capacity for degradable Gaussian phase-insensitive channels.

The authors note in particular [[40], Section X-A] that these results recover (14.23) and (14.24) as special cases, and in contrast to the former works that first established them, the proofs in [40] do not rely on unproven conjectures.

Ding and Guha [15] investigate a scenario under weak secrecy where in addition to a thermal bosonic q/q wiretap channel, there is a noiseless feedback channel from the receiver to the transmitter and the eavesdropper only gets a noisy version of the feedback channel's output.

The argument in [21] that any rate that is achievable for a c/q wiretap channel under strong secrecy can also be achieved under semantic security is equally valid in the infinite-dimensional case, as was noted in [7].

The work [17] looks at the c/cq channel and develops a new proof method based on symmetrization arguments from statistics and the proof idea in [14] to prove that

$$C_{\mathrm{c/cq}} \geq I(P;W) - \chi(P;D).$$

What is special here compared to many of the other works discussed here is that the derived error bounds are explicit meaning that they can be numerically evaluated at finite block lengths.

Finally, in [36], information-theoretic secure data transmission in the sense of [14] is investigated in the context of a satellite downlink. The connection to free space propagation models leads the authors to define a so-called "exclusion zone". As long as the legitimate communication parties can guarantee that no eavesdropper can execute a measurement on the signals traveling in this exclusion zone, the system is information-theoretically secure.

14.4 Covert Communication

The key idea of covert communication is to communicate without getting detected. This idea goes back to the work [33]. Essentially, successful covert communication

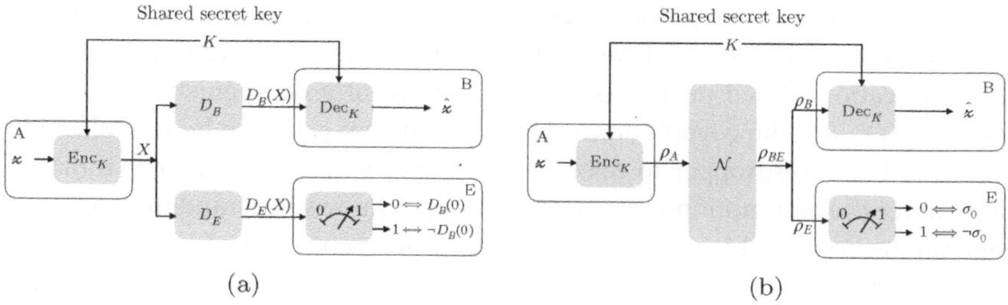

Figure 14.2 The basic idea of covert communication is to prevent the attacker from reliably deciding whether communication is taking place or not. (a) Covert communication over c/q channel and (b) covert communication over q/q channel.

implies secure communication. The topic was initially studied with an emphasis on steganography, the art of hiding messages in digital objects such as images [1, 13, 26, 27], leading to the observation that the number of bits in the digital object that can be changed without getting detected scales as $\mathcal{O}(\sqrt{n})$, where n is the size of the digital object. In [3], a purely information-theoretic study of the subject was initiated, leading to a corresponding square-root law being established for covert communication over zero-mean Gaussian noise channels. Here, we focus on the description of the task as initiated in [4], which describes covert communication in the context of elementary communication systems such as bosonic systems. For such systems, the covertness constraint induces a limitation on the signal energy at the transmitter side. This constraint makes the topic of covert communication particularly suited for the application of quantum communication strategies, which are known to be superior to their conventional analogs for low received signal energy [2] and low signal energy at the transmitter side [24].

14.4.1 System Model

The model underlying the information-theoretic approach to covert communication is a communication system consisting of three parties A, B, and E which represent the sender, the legitimate receiver, and the eavesdropper, respectively. A is connected to B and E either via a c/q channel $D_{\mathrm{BE}} : \mathcal{X} \to \mathcal{S}(\mathcal{H}_{\mathrm{B}} \otimes \mathcal{H}_{\mathrm{E}})$ (Figure 14.2a) or via a q/q channel $\mathcal{N} : \mathcal{S}(\mathcal{H}_{\mathrm{A}}) \to \mathcal{S}(\mathcal{H}_{\mathrm{B}} \otimes \mathcal{H}_{\mathrm{E}})$ (Figure 14.2b).

In the c/q case, on input $x \in \mathcal{X}$, the state $D_{\mathrm{B}}(x) := \mathrm{tr}_{\mathrm{E}}(D_{\mathrm{BE}}(x))$ is accessible by B and $D_{\mathrm{E}}(x) := \mathrm{tr}_{\mathrm{B}}(D_{\mathrm{BE}}(x))$ is accessible by E. The goal for the eavesdropper is to detect communication between the legitimate parties. To formalize this act, a particular state, indicated by $x = 0 \in \mathcal{X}$, has to be chosen which corresponds to the hypothesis that A and B are not communicating. For practical systems, where each signal contains a certain amount of energy, the state $D_{\mathrm{BE}}(0)$ will typically be the one with the minimum energy among all states $D_{\mathrm{BE}}(x)$, $x \in \mathcal{X}$. There are several ways of formalizing this problem further. Our focus will be on a version which quickly conveys the essential thoughts leading to the square-root law of covert communication. Based on the assumed existence of a distinguished "vacuum" state $\rho_0 = D_{\mathrm{BE}}(0)$, the covert communication problem can then be formalized by requiring the eavesdropper to reject the hypothesis that the sender is transmitting, under the assumption that the eavesdropper knows the (random) codebook generation process which will eventually turn out to be defined by some probability distribution P_n of the codewords, but not the codebook itself. Assuming n transmissions over the channel, the resulting state for E is

$$\rho_{P_n} := \int D_{\mathrm{E}}(x_1) \otimes \cdots \otimes D_{\mathrm{E}}(x_n) P_n(dx_1, \ldots, dx_n). \tag{14.25}$$

In the q/q case, on input $\sigma \in \mathcal{S}(\mathcal{H}_{\mathrm{A}})$, the state $\mathrm{tr}_{\mathrm{B}}(\mathcal{N}(\sigma))$ will be accessible by B and $\mathrm{tr}_{\mathrm{E}}(\mathcal{N}(\sigma))$ will be accessible by E. In parallel to the c/q case, there is some state (usually a state of minimum energy) $\sigma_0 \in \mathcal{S}(\mathcal{H}_{\mathrm{A}})$ which corresponds to the hypothesis that A and B are not communicating, and as in the c/q case, we denote the corresponding eavesdropper output state with $\rho_0 := \mathrm{tr}_{\mathrm{B}}(\mathcal{N}(\sigma_0))$. The encoder depends on a secret key K drawn from an auxiliary alphabet $\mathcal{K}_n$ according to a probability distribution P_n. Given a message $\mathfrak{M}$ and key K, the encoder output is denoted $\mathrm{Enc}_K(\mathfrak{M}) \in \mathcal{S}(\mathcal{H}_{\mathrm{A}})$. The channel state for E resulting from a transmission is therefore

$$\rho_{P_n} := \int \mathrm{tr}_{\mathrm{B}}(\mathcal{N}(\mathrm{Enc}_K(\mathfrak{M}))) P_n(dK), \tag{14.26}$$

where Enc_K is such that ρ_{P_n} is independent of $\mathfrak{M}$.

In the following, we formalize this further for the c/q and q/q cases jointly, where the states ρ_0 and ρ_{P_n} are as defined above for the respective case. Operationally, the eavesdropper will use a binary hypothesis test defined by a POVM $F, \mathbf{1} - F$, with the goal being to maximize its probability of detecting transmissions. Following [32], this can be formalized as finding F solving the optimization problem

$$\max_{0 \le T \le 1} \mathrm{tr}(T(\rho_{P_n} - \rho_0^{\otimes n})). \tag{14.27}$$

Due to the variational characterization of trace distance, the above difference in probabilities equals $\frac{1}{2}\|\rho_{P_n} - \rho_0^{\otimes n}\|_{\mathrm{tr}}$. By Pinsker's inequality, we have $\|\rho_{P_n} - \rho_0^{\otimes n}\|_{\mathrm{tr}}^2 \le H(\rho_{P_n} \| \rho_0^{\otimes n})$. This observation motivates the use of the relative entropy also for the general case: requiring, e.g., that $H(\rho_{P_n} \| \rho_0^{\otimes n}) < \delta$ lets the detection probabilities $\mathrm{tr}(F\rho_{P_n})$ and $\mathrm{tr}(F\rho_0^{\otimes n})$ differ by less than $\sqrt{\delta}/2$.

Conversely, since the legitimate parties would like to communicate without getting detected, their task is to maximize the number M of messages they can transmit with at most some given error λ. We say that a decoding POVM $(\tilde{F}(m))_{m \in \mathcal{M}}$ on

$\mathcal{H}_{\mathrm{B}}^{\otimes n}$ (which may depend on C resp. K) satisfies the error λ if, in the c/q case,

$$P_{\mathrm{dec}}^{(n)}(P_n) := \mathbb{E}_C\left(\frac{1}{M}\sum_{m=1}^{M}\mathrm{tr}\left(\tilde{F}(m)\,D_{\mathrm{B}}(C(m))\right)\right) \geq 1-\lambda, \tag{14.28}$$

or, in the q/q case,

$$P_{\mathrm{dec}}^{(n)}(P_n) := \mathbb{E}_K\left(\frac{1}{M}\sum_{m=1}^{M}\mathrm{tr}\left(\tilde{F}(m)\,\mathrm{tr}_{\mathrm{E}}\left(\mathcal{N}\left(\mathrm{Enc}_K(m)\right)\right)\right)\right) \geq 1-\lambda. \tag{14.29}$$

Taking into account the fact that the goal of the information-theoretic analysis is to extract the correct scaling of information transmission tasks under the given constraints, a rate $R \geq 0$ is then defined as achievable under scaling $f : \mathbb{N} \to \mathbb{R}$ with covertness parameter $\delta > 0$, where f is required to be a monotone increasing function, if a sequence $(P_n)_{n\in\mathbb{N}}$ of probability distributions over codes exists such that

$$P_{\mathrm{dec}}^{(n)}(P_n) \to 1,\ \limsup_{n\to\infty} H(\rho_{P_n}\|\rho_0^{\otimes n}) \leq \delta,\ \liminf_{n\to\infty}\frac{\log M_n}{f(n)} \geq R. \tag{14.30}$$

The covert capacity of the channel under a given scaling f is then defined as the supremum over all R which are achievable under scaling f. As it turns out in the analysis, three types of scaling have been found to be of importance so far: the case $f(n) = n$, the case $f(n) = \sqrt{n}\log n$, and the case $f(n) = \sqrt{n}$.

14.4.2 Survey of Results

The content of this subsection is drawn from the works [4, 9, 19, 32, 37, 38].

In [4, 9, 19, 38], covert communication over bosonic q/q channels is treated. For this channel model, Theorem 4 of [4] proves that covert communication at a scaling of $\mathcal{O}(\sqrt{n})$ is possible. In addition, Bash et al. [[4], Theorem 5] show that any scaling $f(n) \in \omega(\sqrt{n})$ implies zero capacity for all systems with photon number variance scaling as $\mathcal{O}(n)$. Here, the definition $\omega(g(n)) := \{f(n) :$ for any positive constant $c > 0$, there exists a constant $n_0 > 0$ such that $0 \leq c \cdot g(n) < f(n)\,\forall n \geq n_0\}$ from [12] is applied. Gagatsos et al. [[19], Theorem 1] show that the capacity expression of a beam splitter channel with thermal noise is given by

$$\delta\frac{\sqrt{2\eta N(1+\eta N)}\eta\log\left(1+\frac{1}{(1-\eta)N}\right)}{1-\eta}, \tag{14.31}$$

where η is the transmittivity of the beam splitter, N is the mean photon number of the noise, and the scaling is $f(n) = \sqrt{n}$. To this end, the authors prove both a direct capacity bound and a matching converse bound with the help of converse results from [9]. The authors of [38] consider trace distance as a metric for covertness in contrast to quantum relative entropy which is used in the other works. This means that direct results are weaker and converse results are stronger than in the rest of the literature. As a direct bound, [[38], Prop. III.2] recovers the expression (14.31), while a converse bound is given in [[38], Prop. III.3] as

$$Q^{-1}\left(\frac{1-\delta}{2}\right)\frac{2\sqrt{\eta N(1+\eta N)}\eta\log\left(1+\frac{1}{(1-\eta)N}\right)}{1-\eta},$$

where Q^{-1} is the inverse of the survival function of the standard normal distribution

$$Q : t \mapsto \int_t^{\infty}\frac{1}{\sqrt{2\pi}}\exp\left(\frac{-\hat{t}^2}{2}\right)d\hat{t}.$$

In [32] and in [37], the problem of covert communication over generic finite-dimensional c/q systems is analyzed. While Sheikholeslami et al. [32] treat the case a of binary input alphabet ($|\mathcal{X}| = 2$), Wang [37] proves results for general (finite) input alphabet sizes. The result of [37] can be summarized as follows. Let $\mathcal{X}$ be any finite input alphabet, and let $D_{\mathrm{BE}} : \mathcal{X} \to \mathcal{S}(\mathcal{H}_{\mathrm{B}} \otimes \mathcal{H}_{\mathrm{E}})$ a c/q channel for a finite-dimensional Hilbert spaces $\mathcal{H}_{\mathrm{B}}$ and $\mathcal{H}_{\mathrm{E}}$. Then, the covert capacity shows a dichotomic behavior as follows: if there is a probability distribution p on $\mathcal{X}$ for which

$$D_{\mathrm{E}}(0) = \sum_x p(x)D_{\mathrm{E}}(x) \quad \text{and} \quad \max_x p(x)\|D_{\mathrm{B}}(x) - D_{\mathrm{B}}(0)\|_{\mathrm{tr}} > 0, \tag{14.32}$$

holds, then the channel capacity on the linear scale ($f(n) = n$) is given by

$$\max\left\{\chi(p;\mathcal{N}) : p \in \mathcal{P}(\mathcal{X}) \text{ and } \sum_x p(x)D_{\mathrm{E}}(x) = D_{\mathrm{E}}(0)\right\}. \tag{14.33}$$

Here and in what follows, we denote by $\mathcal{P}(\mathcal{X})$ the set of probability distributions on the (finite) alphabet $\mathcal{X}$.

If, however, (14.32) does not hold, then the capacity is finite on the $\sqrt{n}$ scale and in particular if one uses the scaling $f(n) = \sqrt{\delta \cdot n}$, the capacity is given by

$$\max\left\{\frac{\sqrt{2}\sum_x p(x)H(D_{\mathrm{E}}(x)\|D_{\mathrm{E}}(0))}{\sqrt{\eta(D_{\mathrm{E}}^1(p)\|D_{\mathrm{E}}(0))}} : p \in \mathcal{P}(\mathcal{X}) \text{ and } p(0) = 0\right\}. \tag{14.34}$$

Here, $\eta(A\|B)$ is defined as $-1 + \int_0^{\infty} trA(B+s)^{-1}A(B+s)^{-1}ds$ [37].

In [8], covert communication is analyzed for entanglement-assisted bosonic systems. It is found that $f(n) = \sqrt{n}\log(n)$ provides the correct scaling. Thus, the presence of entanglement between the legitimate communication parties changes the scaling.

14.5 Conclusion

In this chapter, we have summarized existing research which we believe is relevant for securing classical wireless communication systems such as 6G against eavesdropping with QPLS methods. Although there is already an extensive body of literature on this topic, there are also many open questions, some of which we summarize in the following:

1. We are not aware of any practically feasible coding schemes for wiretap channels with a quantum output for the eavesdropper. If QPLS is to be implemented in any kind of practical system, suitable coding schemes will be crucial. Ideally, such practical schemes would retain strong security guarantees such as semantic security or even covertness of communications, but even coding with much weaker guarantees such as a minimum bit error rate guarantee for the eavesdropper could be an important first step.
2. The QPLS results we are aware of focus on memoryless channels. The channel being memoryless is an assumption that may not always be warranted especially in wireless systems. Research on QPLS for channels with memory is a challenge both for theoretical results and for practically feasible coding results.
3. So far, there are only very few works about quantum side channels of classical communication systems. However, the existence of a large number of potential quantum side channels in quantum cryptography and communication systems is already experimentally confirmed. For the application of QPLS to classical communication systems, it would be important to have a good characterization of relevant quantum side channels of such systems. Together with mitigation techniques that can be used to attenuate these side channels, this could be a way to guarantee the channel strength requirements on which QPLS results rely. An important aspect with regard to these side channels is the asymmetry between the hardware on the encoder and the decoder side. Since the decoder processes signals that have already passed through the communication channel, it is easier to protect against SCAs with means of PLS than the encoder. Cryptographic protections, in contrast, are equally applicable on the encoder and decoder side and may therefore be a suitable means to protect the encoder side against SCAs.
4. So far QPLS is focused on a specific attack vector, namely, passive eavesdropping by a computationally unbounded adversary which has access to a low-quality quantum signal from a side channel of the communication. However, there is a multitude of other relevant types of attacks in case the adversary can transmit its own signals. Moreover, it may not be possible to guarantee the channel strength requirements that QPLS needs under all

circumstances. Besides possible extensions of QPLS methods, we believe that it is also essential to combine the use of QPLS and cryptographic methods such as PQC and find good ways to integrate them into communication systems.

In conclusion, the need to secure future communication systems against attackers with quantum capabilities has resulted in a multitude of interesting and important open problems, and we believe that research in these directions should be an integral part of the greater 6G research efforts.

Acknowledgments

IB, JK, and SS acknowledge the financial support by the Federal Ministry of Education and Research of Germany in the program of "Souverän. Digital. Vernetzt." Joint project 6G-RIC, project identification numbers: 16KISK020K, 16KISK030, and 16KISK033.

MF and SS acknowledge support by the German Research Foundation (DFG) under Grant STA 864/15-1.

JN acknowledges funding via the DFG Emmy-Noether program under Grant number NO 1129/2-1 and the Federal Ministry of Education and Research of Germany in the program of "Souverän. Digital. Vernetzt." Joint project 6G-life, project identification number: 16KISK002, and under grant number 16KISQ039. Further, the support of the Munich Center for Quantum Science and Technology (MCQST) is acknowledged.

References

1 R. Anderson. *Stretching the Limits of Steganography*. volume 1174. Springer-Verlag, Berlin Heidelberg, 1996.

2 K. Banaszek, L. Kunz, M. Jachura, and M. Jarzyna. Quantum limits in optical communications. *Journal of Lightwave Technology*, 38(10):2741–2754, 2020.

3 B. A. Bash, D. Goeckel, and D. Towsley. Limits of reliable communication with low probability of detection on AWGN channels. *IEEE Journal on Selected Areas in Communications*, 31(9):1921–1930, 2013.

4 B. A. Bash, A. H. Gheorghe, M. Patel, J. L. Habif, D. Goeckel, D. Towsley, and S. Guha. Quantum-secure covert communication on bosonic channels. *Nature Communications*, 6:8626, 2015.

5 M. Bellare, S. Tessaro, and A. Vardy. Semantic security for the wiretap channel. In *Advances in Cryptology–CRYPTO 2012*, pages 294–311. Springer, 2012.

6 M. Bellare, S. Tessaro, and A. Vardy. A cryptographic treatment of the wiretap channel. *arXiv preprint arXiv:1201.2205*, 2012.

7 H. Boche, M. Cai, C. Deppe, R. Ferrara, and M. Wiese. Semantic security for quantum wiretap channels. *Journal of Mathematical Physics*, 63(9):092204, 2022.

8 M. S. Bullock, C. N. Gagatsos, S. Guha, and B. A. Bash. Entanglement-assisted quantum-secure covert communication. In *OSA Quantum 2.0 Conference*, page QM6B.5. Optica Publishing Group, 2020.

9 M. S. Bullock, C. N. Gagatsos, S. Guha, and B. A. Bash. Fundamental limits of quantum-secure covert communication over bosonic channels. *IEEE Journal on Selected Areas in Communications*, 38(3):471–482, 2020.

10 P. Busch, P. Lahti, J.-P. Pellonpää, and K. Ylinen. *Quantum Measurement.* Theoretical and Mathematical Physics, 1st edn. Springer, 2016.

11 N. Cai, A. Winter, and R. W. Yeung. Quantum privacy and quantum wiretap channels. *Problems of Information Transmission*, 40(4):318–336, 2004.

12 T. H. Cormen, C. E. Leiserson, R. L. Rivest, and C. Stein. *Introduction to Algorithms*, 2nd edn. MIT Press, Cambridge, London, 2001.

13 S. Craver. On public-key steganography in the presence of an active warden. In D. Aucsmith, editor, *Information Hiding*, volume 1525, pages 355–368. Springer-Verlag, Berlin Heidelberg, 1998.

14 I. Devetak. The private classical capacity and quantum capacity of a quantum channel. *Transactions on Information Theory*, 51(1):44–55, 2005.

15 D. Ding and S. Guha. Noisy feedback and loss unlimited private communication. In *2018 IEEE International Symposium on Information Theory (ISIT)*, pages 586–590. IEEE, 2018.

16 J. Ferrigno and M. Hlaváč. When AES blinks: Introducing optical side channel. *IET Information Security*, 2:94–98(4), Sept 2008.

17 M. Frey, I. Bjelaković, J. Nötzel, and S. Stańczak. Semantic security with infinite dimensional quantum eavesdropping channel. In *2022 IEEE Information Theory Workshop (ITW)*, pages 720–725. IEEE, 2022.

18 M. Frey, I. Bjelaković, J. Nötzel, and S. Stańczak. Semantic security with infinite dimensional quantum eavesdropping channel. *Submitted to IEEE Transactions on Information Theory*, 2023. Preprint available at arXiv:2205.07663v2.

19 C. N. Gagatsos, M. S. Bullock, and B. A. Bash. Covert capacity of bosonic channels. *IEEE Journal on Selected Areas in Information Theory*, 1(2):555–567, 2020.

20 S. Guha, J. H. Shapiro, and B. I. Erkmen. Capacity of the bosonic wiretap channel and the entropy photon-number inequality. In *2008 IEEE International Symposium on Information Theory*, pages 91–95. IEEE, 2008.

21 M. Hayashi. Quantum wiretap channel with non-uniform random number and its exponent and equivocation rate of leaked information. *IEEE Transactions on Information Theory*, 61(10):5595–5622, 2015.

22 T. Heinosaari and M. Ziman. *The Mathematical Language of Quantum Theory: From Uncertainty to Entanglement*. Cambridge University Press, Cambridge, 2011.

23 A. S. Holevo. *Quantum Systems, Channels, Information*, 2nd edn. De Gruyter, 2019.

24 A. S. Holevo and R. F. Werner. Evaluating capacities of bosonic Gaussian channels. *Physical Review A*, 63:032312, Feb 2001.

25 A. Huang, S. Barz, E. Andersson, and V. Makarov. Implementation vulnerabilities in general quantum cryptography. *New Journal of Physics*, 20(10):103016, 2018.

26 A. D. Ker. Batch steganography and pooled steganalysis. In J. L. Camenisch, C. S. Collberg, N. F. Johnson, and P. Sallee, editors, *Information Hiding*. Springer-Verlag, Berlin Heidelberg, 2007.

27 A. D. Ker. The square root law in stegosystems with imperfect information. In R. Böhme, P. W. L. Fong, and R. Safavi-Naini, editors, *Information Hiding*, pages 145–160. Springer-Verlag, Berlin, Heidelberg, 2010. ISBN 978-3-642-16435-4.

28 J. Krämer, M. Kasper, and J.-P. Seifert. The role of photons in cryptanalysis. In *2014 19th Asia and South Pacific Design Automation Conference (ASP-DAC)*, pages 780–787, 2014.

29 P. V. P. Pinheiro, P. Chaiwongkhot, S. Sajeed, R. T. Horn, J.-P. Bourgoin, T. Jennewein, N. Lütkenhaus, and V. Makarov. Eavesdropping and countermeasures for backflash side channel in quantum cryptography. *Optics Express*, 26(16):21020–21032, 2018.

30 H. Qi and M. M. Wilde. Capacities of quantum amplifier channels. *Physical Review A*, 95:012339, Jan 2017.

31 B. Schumacher and M. D. Westmoreland. Quantum privacy and quantum coherence. *Physical Review Letters*, 80(25):5695, 1998.

32 A. Sheikholeslami, B. A. Bash, D. Towsley, D. Goeckel, and S. Guha. Covert communication over classical-quantum channels. In *2016 IEEE International Symposium on Information Theory (ISIT)*, pages 2064–2068. IEEE, 2016.

33 G. J. Simmons. The prisoners' problem and the subliminal channel. In D. Chaum, editor, *Advances in Cryptology: Proceedings of Crypto 83*, pages 51–67. Springer, Boston, MA, 1984.

34 G. Smith. Private classical capacity with a symmetric side channel and its application to quantum cryptography. *Physical Review A*, 78(2):022306, 2008.

35 A. Tikku, M. Berta, and J. M. Renes. Non-additivity in classical-quantum wiretap channels. *IEEE Journal on Selected Areas in Information Theory*, 1(2):526–535, 2020.

36 A. Vázquez-Castro, D. Rusca, and H. Zbinden. Quantum keyless private communication versus quantum key distribution for space links. *Physical Review Applied*, 16:014006, 2021.

37 L. Wang. Optimal throughput for covert communication over a classical-quantum channel. In *2016 IEEE Information Theory Workshop (ITW)*, pages 364–368. IEEE, 2016.

38 S.-Y. Wang, T. Erdoğan, and M. Bloch. Towards a characterization of the covert capacity of bosonic channels under trace distance. In *2022 IEEE International Symposium on Information Theory (ISIT)*, pages 318–323. IEEE, 2022.

39 M. M. Wilde and M.-H. Hsieh. Public and private resource trade-offs for a quantum channel. *Quantum Information Processing*, 11(6):1465–1501, 2012.

40 M. M. Wilde and H. Qi. Energy-constrained private and quantum capacities of quantum channels. *IEEE Transactions on Information Theory*, 64(12):7802–7827, 2018.

41 M. M. Wilde, P. Hayden, and S. Guha. Quantum trade-off coding for bosonic communication. *Physical Review A*, 86:062306, Dec 2012.

42 A. D. Wyner. The wire-tap channel. *The Bell System Technical Journal*, 54(8):1355–1387, 1975.

43 H.-L. Yin et al. Experimental quantum digital signature over 102 km. *Physical Review A*, 95(3):032334, 2017.

15

Leveraging the Physical Layer to Achieve Practically Feasible Confidentiality and Authentication

Marco Baldi and Linda Senigagliesi

Department of Information Engineering, Università Politecnica delle Marche, Ancona, Italy

15.1 Introduction

In an ever-changing landscape of information security and data protection, the quest for robust, efficient, and practically feasible methods of ensuring confidentiality and authentication remains a paramount challenge.

Opposed to paradigms based on computational security, i.e., the difficulty for an attacker to compromise the security of communications due to limitations of the attacker's computing power, the physical-layer secrecy (PLS) paradigm does not require computational assumptions about attackers and exploits techniques inherent in the transmissions themselves, which makes it suitable for use even in resource-constrained contexts. This makes PLS particularly attractive for use in a variety of application contexts, ranging from IoT to 6G. However, the adoption of PLS solutions in practical communication systems is often limited for two main reasons:

- *Ideality of assumptions*: Many studies adopt PLS notions and metrics that make use of ideal assumptions, such as the use of random coding and continuous modulations. Real systems, on the other hand, use standard transmission schemes, such as based on linear block codes and discrete modulations. These practical constraints must be taken into account if the achievable PLS is to be studied in real-world contexts.
- *Weakness of security metrics*: Many classic PLS metrics do not provide a significant advantage to legitimate users over attackers nor do they prevent part of a confidential message from being leaked to attackers. It is clear that a real system cannot afford these shortcomings, so it is necessary to use security metrics and

Physical-Layer Security for 6G, First Edition. Edited by Parthajit Mohapatra, Nikolaos Pappas, Arsenia Chorti, and Stefano Tomasin.

notions that provide more robust guarantees, e.g., inspired by those used in cryptographic systems.

To achieve confidentiality at the physical layer, some already established tools are the use of channel coding [7] and the exploitation of channel quality fluctuations – or fading – and Multiple-Input Multiple-Output (MIMO) transmissions [20]. Another tool that has proven useful in coping with fluctuations in the channel quality of legitimate users is also the adoption of on–off transmission (OOT) protocols [18], which rely on using the wireless channel only during intervals when it has acceptable quality, with the possible transmission of fake packets during the other intervals [10].

Differently from physical layer confidentiality, physical-layer authentication (PLA) aims at distinguishing legitimate from malicious users by exploiting the uniqueness and reciprocity of wireless channels as a sort of user fingerprint [28]. In the case of PLA, channel variations are a problem, as they make new estimates of a transmission channel more difficult to trace back to the initial ones. Similarly, the possible correlation of communication channels of spatially proximate users may cause one user to be confused for another.

In this chapter, we first provide a system model and some security metrics and notions suitable for application in real systems, both with regard to confidentiality and authentication at the physical layer. Then, we describe some techniques that enable a sufficiently robust level of confidentiality and authentication at the physical layer in practical scenarios involving wireless communications, taking into account the constraints imposed by the use of real systems and devices.

The chapter is organized as follows: in Section 15.2, we introduce the system model and recall some background notions; in Section 15.3, we describe techniques for achieving confidentiality at the physical layer in practical settings; in Section 15.4, we describe techniques for authentication at the physical layer in practical settings; in Section 15.5, we provide some examples of application of the techniques presented; and Section 15.6 concludes the chapter.

15.2 System Model

For both confidentiality and authentication at the physical layer, we can refer to the classic simplified system model involving three actors: (i) Alice, the legitimate transmitter; (ii) Bob, the legitimate receiver or authenticator; and (iii) Eve, the opponent.

An example of such a model suitable for describing both confidentiality and authentication applications at the physical layer is shown in Figure 15.1. In such a model, Eve can behave either as a passive attacker (receiving Alice's communications) or an active one (making transmissions to Bob). The former case is of

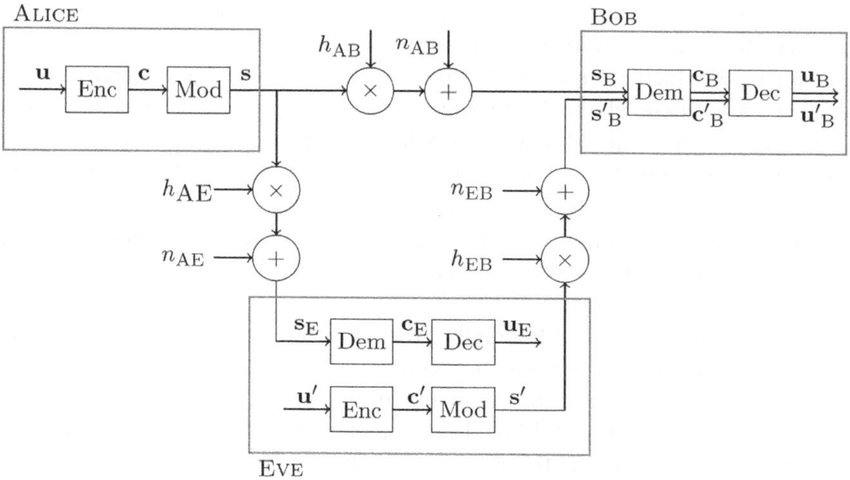

Figure 15.1 General system model for confidentiality and authentication at the physical layer.

interest for confidentiality applications, where Eve behaves as an eavesdropper, while the latter is of interest for authentication applications, where Eve behaves as an impersonator.

For the case of confidentiality, such a system model boils down to the so-called *wiretap channel*, introduced by Wyner in 1975 [27]. According to the wiretap channel model, Alice aims at transmitting some secret message vector $\mathbf{u}$ to Bob, where $\mathbf{u}$ is formed by k entries that take values in $\mathbb{F}_q$, the finite field with q elements. The message $\mathbf{u}$ is encoded by Alice into a codeword $\mathbf{c}$, which is a vector of n entries taking values in $\mathbb{F}_q$, through the use of some channel coding technique known to everyone (including Eve). The codeword $\mathbf{c}$ is then modulated into a signal vector $\mathbf{s}$, whose elements take complex values that depend on the modulation format chosen. Usually, real systems use discrete modulation formats, such as quadrature amplitude modulation (QAM). Finally, the modulated signal is transmitted by Alice over the channel. We achieve confidentiality when Bob is able to decode the secret message correctly, while Eve is unable to intercept it completely or even partially.

As we see in Figure 15.1, the way channel impairments are usually modeled is to consider, for each communication channel, some multiplicative channel gain (expressed through the random variables h) and some additive noise (expressed through the random variables n). Since such impairments make each communication channel unique as well as reciprocal, upon transmission of $\mathbf{c}$, Bob and Eve receive two different noisy versions of $\mathbf{c}$, $\mathbf{c}_B$ and $\mathbf{c}_E$, from which they decode the two messages $\mathbf{u}_B$ and $\mathbf{u}_E$ that are generally different. Such a setting can easily be extended to the case of transmission over N parallel channels, as it occurs, e.g., in Orthogonal Frequency Division Multiplexing (OFDM), with the difference being that, in such a case, channel impairments will be represented by vector rather than scalar random variables, which we denote as $\mathbf{h}$ and $\mathbf{n}$. Concerning channel knowledge, or channel state information (CSI), we assume that Alice and Bob can obtain that relative to their channels, even precisely, by exchanging conventional transmissions such as pilot tones. Instead, of Eve's channel we assume that they know only some average quality, which can be derived from general considerations on Eve's possible location.

From the standpoint of confidentiality, the main difference between the wiretap channel setting and that of symmetric cryptography, besides relying on different security notions, is the lack, in PLS, of any form of secret pre-shared between the two legitimate users Alice and Bob. In fact, each transmission channel is unique, random, and reciprocal, and this is what differentiates Bob from Eve in the eyes of Alice under the PLS setting, where the channel features are exploited to achieve confidential transmissions from Alice to Bob. In the PLA setting, instead, messages are no longer confidential, but Bob aims at authenticating Alice by distinguishing her from Eve. In such a setting, therefore, Alice acts a supplicant, while Bob acts as authenticator, and Eve is an opponent who tries to impersonate Alice by forging transmissions. Also, in this case, the differentiation between Alice and Eve is solely based on the differences between their communication channels towards Bob, and does not exploit any pre-shared secrets or credentials.

For such a purpose, in the case of PLA, the communication protocol is divided into the following two stages:

1. *Onboarding*: During this initial stage, it must be possible to assume that Alice is actually transmitting (e.g., by performing this procedure manually), and Alice must transmit a predetermined signal known to everyone, such as a pilot tone. Based on signals received from Alice and knowledge of transmitted signals, Bob can hence collect a set of M time-correlated channel estimates $\hat{\mathbf{h}}_{AB}$, which differ from each other only because of the random channel non-idealities that influence each measure.
2. *Authentication*: In the subsequent authentication step, the assumption that Bob knows for sure who is transmitting is removed. It may in fact be Alice, who correctly transmits the predetermined conventional signal, or Eve, who tries to transmit a forged signal that at the receiver takes a form as similar as possible to Alice's signal. By exploiting his knowledge of the conventional signal, Bob estimates the channel response, $\hat{\mathbf{h}}_{xB}$, and compares it with the reference one obtained during onboarding, for the purpose of deciding whether Alice is actually transmitting or not.

15.3 Confidentiality at the Physical Layer in Practical Settings

Let us consider a wiretap channel model under the simplifying assumption of deterministic encoding, i.e., the assumption that each message $\mathbf{u}$ is deterministically associated to a unique codeword $\mathbf{c}$. The arguments we are going to develop, however, can easily be extended to the case of statistic encoding, in which a component of randomness is combined with the secret message during encoding to confuse the eavesdropper, thus, making the encoding statistical rather than deterministic. This is the case, e.g., of some wiretap coding techniques exploiting coset coding.

When instead encoding is deterministic, such a randomness component is missing, and Eve's total equivocation about the message $\mathbf{u}$ equals Eve's total equivocation about the codeword $\mathbf{c}$, i.e.,

$$s = H(\mathbf{c}|\mathbf{c}_{\mathrm{E}}) = H(\mathbf{c}) - I(\mathbf{c}; \mathbf{c}_{\mathrm{E}}), \tag{15.1}$$

where $I(\mathbf{c}; \mathbf{c}_{\mathrm{E}})$ is the mutual information between $\mathbf{c}$ and $\mathbf{c}_{\mathrm{E}}$, also coinciding with Eve's advantage [5]. Instead, $H(\mathbf{c}) = k' \leq k$ is the source entropy (with $k' = k$ in the special case of a uniformly distributed source). The source entropy rate and Eve's equivocation rate are, hence, defined, respectively, as

$$\begin{cases} R_{\mathrm{s}} = k'/n, \\ R_{\mathrm{e}} = s/n. \end{cases} \tag{15.2}$$

When $R_{\mathrm{e}} = R_{\mathrm{s}}$, we have perfect secrecy, as defined by Wyner [27]. However, in most practical cases, we instead have $0 < \frac{R_{\mathrm{e}}}{R_{\mathrm{s}}} < 1$, which is a condition known as *partial secrecy*. In such a case, despite we do not achieve perfect secrecy, for Eve to correctly decode $\mathbf{c}$ from $\mathbf{c}_{\mathrm{E}}$ she must perform an average number of attempts equal to $2^s = 2^{nR_{\mathrm{e}}}$ [11], with $1 < 2^s < 2^{k'}$. This can be exploited for achieving some level of PLS, as we show next. In fact, a totally blind attacker would have to make k' attempts to guess the secret message, on average, so we can consider $\frac{s}{k'} = \frac{R_{\mathrm{e}}}{R_{\mathrm{s}}}$ as a measure of the achievable PLS level in this setting.

If we look at (15.1), a conservative assumption is to consider that $I(\mathbf{c}; \mathbf{c}_{\mathrm{E}}) \leq \frac{n}{q} C_{\mathrm{E}}$, where C_{E} is the capacity of Eve's channel. Such an upper bound may result rather loose if we consider Shannon's unconstrained capacity for Eve's channel. However, in order to achieve a more realistic bound, we can consider the constraint deriving from the use of some discrete modulation format for Alice's transmissions and use the corresponding input-constrained capacity for Eve's channel. Exploiting the aforementioned bound allows us to eliminate the dependence of Eve's channel mutual information on the specific message distribution, and is equivalent to considering Eve's channel maximum mutual information over all message distributions. We, thus, consider a PLS notion known as mutual information security (MIS), as defined in [5].

One important feature of MIS is that, as proven in [5], achieving it is equivalent to achieving Semantic Security (SS), a notion of security that is commonly used in cryptography, defined as the impossibility for an attacker to compute any function of the secret message with a higher probability than it would have without observing the ciphertext. So, when MIS is achieved, a cryptographically significant security condition is also achieved. In the case of practical transmissions achieving only partial secrecy, however, such a condition is not fully achieved, since some partial information may still leak out. Nevertheless, some cryptographic tools can be exploited along with partial secrecy to re-establish a cryptographically meaningful level of security, as will be shown in Section 15.3.1.

In order to measure the level of PLS under partial secrecy, we can use the aforementioned bound, taking into account that equivocation is a nonnegative function, such that we have

$$\begin{cases} R_\mathrm{e} \geq \max\left\{0, \frac{1}{n}\left[k' - \frac{n}{q}C_\mathrm{E}\right]\right\} = \left[R_\mathrm{s} - \frac{C_\mathrm{E}}{q}\right]^+, \\ s = nR_\mathrm{e} \geq n\left[R_\mathrm{s} - \frac{C_\mathrm{E}}{q}\right]^+ = \tilde{s}. \end{cases} \tag{15.3}$$

Since $R_\mathrm{s} = \frac{k'}{n} \leq \frac{k}{n} = R_\mathrm{c}$, where R_c is the code rate, we have

$$\frac{R_\mathrm{e}}{R_\mathrm{s}} \geq \frac{\left[R_\mathrm{s} - \frac{C_\mathrm{E}}{q}\right]^+}{R_\mathrm{c}} = \widetilde{R_\mathrm{e}}. \tag{15.4}$$

So, $\widetilde{R_\mathrm{e}}$ in (15.4) represents a lower bound on the achievable level of partial secrecy. In other terms, $\widetilde{R_\mathrm{e}}$ provides a lower bound on the ratio of the average number of bits hidden from Eve's eyes to the total number of bits in the secret message.

When we are in the regime of partial secrecy, such a ratio is bounded away from one, and this means that some information about the secret message is leaked to the attacker. On average, in fact, $k' - s$ message bits are known to Eve, which is, therefore, in a less uncertain condition than an attacker who does not have access to the ciphertext. Although such a condition is anyway unfavorable for the opponent, it cannot be considered sufficient to satisfy the confidentiality requirement. Nevertheless, it can still be exploited as a basis for achieving an adequate level of security through the joint use of PLS with cryptography, as we show in Section 15.3.1.

15.3.1 Joining Physical-Layer Security with Cryptography

Let us see how a condition of partial secrecy can be exploited along with appropriate cryptographic tools to achieve adequate levels of confidentiality at the physical layer. The basic idea is to prevent the information leakage due to partial secrecy by applying a transformation to the secret message, or to several consecutive secret messages, before their transmission. Such a transformation must be invertible and not require the use of any pre-shared secret, in accordance with the PLS paradigm.

The most important characteristic that this transformation must possess in order to compensate the information leakage due to partial secrecy is that it can be inverted only upon knowing all the bits of the transformed messages. Otherwise, if even one or a few bits of the transformed messages are missing, the inversion of the transformation must be impossible and require random guessing of the missing bits before performing it.

Transformations of this type have actually existed in cryptography for a long time, and are called all-or-nothing transforms (AONTs) [8, 22]. As its name suggests, an all-or-nothing transform (AONT) is a transformation from a plaintext to a ciphertext and vice versa that can be performed only when the entire plaintext or ciphertext is available, respectively. Focusing on the inverse transformation, which is of greater interest for security purposes, the key features of an AONT are that:

- When the entire ciphertext is known, one can easily apply the inverse transformation and obtain the entire plaintext, without needing to know any pre-shared secrets.
- When the entire ciphertext is not available, because even a few bits of it are missing, the partial application of the inverse transformation is useless, because it does not allow recovering any part of the corresponding plaintext.

As shown in [8, 22], transformations with these properties can be built upon cryptographic primitives, like symmetric ciphers and hash functions. They, therefore, provide a useful tool for avoiding information leakage due to partial secrecy and, at the same time, increase the level of security to one borrowed from cryptography. In fact, good symmetric ciphers and hash functions provide a condition of SS, which is a universally recognized security notion.

In order to show how these tools can be used in the PLS setting, let us consider an AONT $\mathcal{A}(\cdot)$ along with its inverse $\mathcal{A}^{-1}(\cdot)$. For any arbitrary length message $\mathbf{x}$, from the definition of AONT, we have:

- $\mathbf{y} = \mathcal{A}(\mathbf{x}) \Leftrightarrow \mathbf{x} = \mathcal{A}^{-1}(\mathbf{y})$,
- if $\mathbf{y} = \mathcal{A}(\mathbf{x})$ and $\mathbf{y}' \neq \mathbf{y}$, then $\mathbf{x}' = \mathcal{A}^{-1}(\mathbf{y}')$ has no correlation with $\mathbf{x}$.

Let us consider L messages of length k bits each to be transmitted by Alice: $\{\mathbf{u}_1, \mathbf{u}_2, \mathbf{u}_3, \dots, \mathbf{u}_L\}$. They can be concatenated and given as input to the AONT, i.e.,

$$\mathbf{y} = \mathcal{A}(\mathbf{x}), \mathbf{x} = [\mathbf{u}_1 | \mathbf{u}_2 | \mathbf{u}_3 | \dots | \mathbf{u}_L] . \tag{15.5}$$

The output of the AONT has the same length as its input. Therefore, $\mathbf{y}$ from (15.5) can be split back into L messages of length k bits each, that is, $\mathbf{y} = [\mathbf{u}'_1|\mathbf{u}'_2|\mathbf{u}'_3| \dots |\mathbf{u}'_L]$.

The messages $\mathbf{u}'$, obtained by processing the original messages through the AONT, can then be transmitted by Alice over the wiretap channel in the place of the original messages. Due to the properties of the AONT, the original messages can be obtained as $[\mathbf{u}_1|\mathbf{u}_2|\mathbf{u}_3| \dots |\mathbf{u}_L] = \mathcal{A}^{-1}([\mathbf{u}'_1|\mathbf{u}'_2|\mathbf{u}'_3| \dots |\mathbf{u}'_L])$, but only on condition that all the transformed messages $\mathbf{u}'$ are perfectly known. Otherwise, the output of $\mathcal{A}^{-1}$ has no correlation with the original messages. Therefore, using the AONT allows amplifying the differences between messages obtained by Bob and Eve and avoiding any partial information leakage.

15.3.2 Dealing with Variable Channel Quality Through On–Off Transmissions

Another element that differentiates real communication scenarios from ideal ones is the variability of channel quality, which fluctuates based on multiple factors, especially when using low-directivity wireless transmissions.

Let us denote as γ_B and γ_E the signal-to-noise ratio (SNR) of the Alice–Bob and Alice–Eve channels, respectively. Let us consider the simplifying assumption that the SNR remains nearly constant during the transmission of a single message $\mathbf{u}$, i.e., the codeword $\mathbf{c}$ associated with it. In other words, we are assuming that channel variability is not so fast as to be considered significant within the time between the beginning and the end of the transmission of a single message $\mathbf{u}$. Nevertheless, the SNR will change between different transmissions, following some probability distribution.

As explained in Section 15.2, according to the wiretap channel model, the reliability requirement is fulfilled when Bob is able to correctly recover the transmitted message, while the confidentiality requirement is achieved when Eve is far enough away from being able to reconstruct the same message, even in part. Let us denote as $\overline{\gamma}_B$ and $\overline{\gamma}_E$ the minimum (SNR) of Bob's channel and the maximum SNR of Eve's channel that are required to achieve the reliability and confidentiality requirements, respectively. We can then define the minimum ratio between the qualities of the two channels needed to achieve both requirements, or SNR gap, as

$$S_g = \frac{\overline{\gamma}_B}{\overline{\gamma}_E}. \tag{15.6}$$

From (15.6), it follows that a system characterized by very large values of S_g requires Eve to experience a significantly degraded channel quality compared to Bob, which may be unrealistic. When S_g approaches 1, we have that confidentiality and reliability requirements can be achieved even with comparable quality of

Bob's and Eve's channels, which is a desirable property in real settings. When the value of S_g falls below 1, it means that the system is able to achieve the desired confidentiality and reliability requirements even when Eve experiences a better transmission channel than Bob. This clearly is the most desirable condition in practice, although it may be difficult to achieve.

A simple strategy that can be used when channel quality is variable is to let Alice decide whether or not to transmit a message during each time slot, after estimating the quality of her channel to Bob during the same time slot. This allows Alice to avoid transmissions during those time slots in which she experiences a communication channel with Bob having too low quality. Clearly, this lengthens transmission times, but it allows the reliability and security requirements to be met even with a small SNR gap. To put such a strategy into practice requires that Alice and Bob communicate according to an OOT protocol [2].

A protocol of this type is described in Algorithm 15.1, with reference to the system model in Figure 15.1. The main feature of this protocol is to allow Alice, at

---

Algorithm 15.1 OOT protocol over the wiretap channel

Require: $\mathbf{p}$ (public message), $\mathbf{u}$ (private message), $\overline{\gamma}_B$ (lower threshold on Bob's SNR), $\mathbf{s}_1$ (conventional permission signal), $\mathbf{s}_0$ (conventional prohibition signal).

1: **function** $\mathcal{T}$(data to be transmitted $\mathbf{x}$)
2:     return signal to be transmitted $\mathbf{s}$
3: **end function**
4: **function** $\mathcal{E}$(transmitted signal $\mathbf{t}$, received signal $\mathbf{r}$)
5:     return SNR γ
6: **end function**
7: Alice computes and transmits signal $\mathbf{s}_p = \mathcal{T}(\mathbf{p})$
8: Bob receives signal $\widetilde{\mathbf{s}}_p$
9: Bob computes $\gamma_B = \mathcal{E}(\mathcal{T}(\mathbf{p}), \widetilde{\mathbf{s}}_p)$
10: **if** $\gamma_B \geq \overline{\gamma}_B$ **then**
11:     Bob transmits signal $\mathbf{s}_1$
12: **else**
13:     Bob transmits signal $\mathbf{s}_0$
14: **end if**
15: **if** Alice detects $\mathbf{s}_1$ **then**
16:     Alice computes and transmits signal $\mathbf{s}_u = \mathcal{T}(\mathbf{u})$
17: **else**
18:     Alice skips transmission
19:     Alice restarts from line 7 at the next time slot
20: **end if**

---

each transmission slot, to estimate the quality of the channel separating her from Bob, through the transmission of some conventional signal by Alice, and a feedback mechanism from Bob. Clearly, such a decision is made by Alice based on the estimated quality for the channel: if it equals or exceeds a certain threshold $\overline{\gamma}_B$, she transmits the message, otherwise she skips transmission and awaits the next transmission slot.

As shown in [2], when an OOT protocol of this type is applied on series of messages possibly padded with random data and transformed with an AONT, it allows the variability of real channels to be exploited to achieve a predetermined level of security. In fact:

- Processing the messages with the AONT allows getting rid of the information leakage coming from partial secrecy, as shown in Section 15.3.1.
- The OOT protocol allows for coping with channel variability between Alice and Bob, by letting Alice transmit only during the time slots in which the channel has quality above a certain threshold.

A limitation of the above OOT protocol lies in the need to use many time slots and, thus, takes a long time to complete the transmission, when the average SNR gap between Bob and Eve is small. In fact, in such a case, Bob's channel has low quality, thus, it often happens that Alice has to skip the transmission and wait for the next time slot in order for the reliability requirement to be fulfilled. Moreover, Eve's channel has good quality, so for every packet actually transmitted by Alice, Eve's equivocation is low. For this reason, if an acceptable level (e.g., 128 bits) of security is to be achieved, significant random padding of the message is needed in order to spread its contents over many packets processed with the AONT and then transmitted.

A useful idea to improve this aspect is to allow Alice to use the time slots during which Bob's channel has subthreshold quality to transmit random messages aimed at confusing Eve, rather than skipping the transmission. This can be done by introducing *fake packets* that are transmitted by Alice during Bob's channel low-quality periods, as proposed in [23]. The price to pay for this advance is due to the fact that Bob can no longer provide Alice with any feedback on the quality of his channel, otherwise Eve would learn which packets are real and which are fake. Therefore, it is needed that both Alice and Bob independently estimate the quality of the channel. Due to channel reciprocity, their estimates should be identical, but this only happens under ideal conditions. Under real conditions, instead, they need to implement a protocol to reconcile their estimates, which allows them to compensate for the inevitable small differences between them, without letting Eve know about it. However, reconciliation protocols of this type already exist in the literature, e.g., for physical-layer secret key distillation, and can be based on syndrome coding [23]. An on–off transmission with fake packets (OOT-FP) protocol encompassing all these concepts is described in Algorithm 15.2.

Algorithm 15.2 OOT-FP protocol over the wiretap channel

Require: $\mathbf{p}$ (public message), $\mathbf{u}$ (private message), $\overline{\gamma}_{\mathrm{B}}$ (lower threshold on Bob's SNR).

function $\mathcal{T}$(data to be transmitted $\mathbf{x}$)
    return signal to be transmitted $\mathbf{s}$
end function
function $\mathcal{E}$(transmitted signal $\mathbf{t}$, received signal $\mathbf{r}$)
    return SNR γ
end function
function $\mathcal{D}$(bit string $\mathbf{b}$)
    return reconciliation data $\mathbf{l}$
end function
function $\mathcal{R}$(bit string $\mathbf{b}'$, reconciliation data $\mathbf{l}$)
    return reconciled bit string $\mathbf{b}$
end function
function $\mathcal{H}$(bit string $\mathbf{b}$)
    return hash digest $\mathbf{d}$
end function
Alice computes and transmits signal $\mathbf{s}_{\mathrm{p}} = \mathcal{T}(\mathbf{p})$
Bob receives signal $\widetilde{\mathbf{s}}_{\mathrm{p}}$
Bob computes $\gamma_{\mathrm{AB}} = \mathcal{E}(\mathcal{T}(\mathbf{p}), \widetilde{\mathbf{s}}_{\mathrm{p}})$
Bob computes and transmits signal $\mathbf{s}_{\mathrm{p}} = \mathcal{T}(\mathbf{p})$
Alice receives signal $\widehat{\mathbf{s}}_{\mathrm{p}}$
Alice computes $\gamma_{\mathrm{BA}} = \mathcal{E}(\mathcal{T}(\mathbf{p}), \widehat{\mathbf{s}}_{\mathrm{p}})$
Alice sets $\gamma_{\mathrm{B}} = \gamma_{\mathrm{BA}}$ and computes $\mathbf{l}_{\mathrm{B}} = \mathcal{D}(\gamma_{\mathrm{B}})$
Alice computes $\mathbf{d}_{\mathrm{B}} = \mathcal{H}(\mathbf{l}_{\mathrm{B}})$
repeat
    Alice sends $(\mathbf{l}_{\mathrm{B}}, \mathbf{d}_{\mathrm{B}})$ to Bob
    Bob receives $(\widetilde{\mathbf{l}}_{\mathrm{B}}, \widetilde{\mathbf{d}}_{\mathrm{B}})$
    Bob checks if $\widetilde{\mathbf{d}}_{\mathrm{B}} = \mathcal{H}(\widetilde{\mathbf{l}}_{\mathrm{B}})$
until Bob acknowledges
Bob computes $\gamma_{\mathrm{B}} = \mathcal{R}(\gamma_{\mathrm{AB}}, \widetilde{\mathbf{l}}_{\mathrm{B}} = \mathbf{l}_{\mathrm{B}})$
if $\gamma_{\mathrm{B}} \geq \overline{\gamma}_{\mathrm{B}}$ **then**
    Alice computes and transmits signal $\mathbf{s}_{\mathrm{u}} = \mathcal{T}(\mathbf{u})$
    Bob receives it and treats it as a valid packet
else
    Alice generates a random message $\mathbf{f}$
    Alice computes and transmits signal $\mathbf{s}_{f} = \mathcal{T}(\mathbf{f})$
    Bob discards it as a fake packet
end if

In such a protocol, the total equivocation of Eve on each codeword transmitted by Bob is

$$s_{\text{tot}} = s_{\text{e}} + s_{\text{fk}}, \tag{15.7}$$

where s_{e} is the equivocation of Eve following from her channel quality, while s_{fk} is the equivocation of Eve following her uncertainty about the possible fake nature of each packet. Assuming that Bob and Eve experience two channels that are statistically independent, s_{fk} can be easily written in terms of binary entropy, i.e.,

$$s_{\text{fk}} = -p_{\text{fk}} \log_2 p_{\text{fk}} - (1 - p_{\text{fk}})\log_2(1 - p_{\text{fk}}), \tag{15.8}$$

where $p_{\text{fk}} = P\left\{\gamma_{\text{B}} < \overline{\gamma}_{\text{B}}\right\}$ is the probability that Alice transmits a fake packet, based on the channel quality experienced by Bob.

15.4 Authentication at the Physical Layer in Practical Settings

In order to model a PLA setting, we can consider again the system model in Figure 15.1, in which now Eve behaves actively and transmits signals toward Bob. In order to consider wireless transmissions as realistic as possible, we assume that users perform OFDM transmissions over N parallel independent channels. The channel frequency response (CFR) samples of the channels typically consist of three components: a constant part $\mathbf{h}$ that reflects the average channel response over time, a variable component $\boldsymbol{\epsilon}$ with a zero mean, and the Additive White Gaussian Noise (AWGN) noise $\mathbf{w}$ introduced by the receiver. The receiver collects the CFR measured on each of the N subcarriers at time kT in the form of a vector [29] as follows:

$$\hat{\mathbf{h}}[k] = \mathbf{h} + \boldsymbol{\epsilon}[k] + \mathbf{w}[k], \tag{15.9}$$

where the AWGN noise vector $\mathbf{w}[k]$ follows a circularly symmetric complex random Gaussian random distribution $\mathcal{CN}(\mathbf{0}_{N\times 1}, \sigma^2\mathbf{I}_N)$.

The temporal dispersion of a time-varying wireless channel $c(t, \tau)$ is explained using a model that employs a discrete tapped-delay line with L taps [6], which can be written as

$$c(\tau, t) = \sum_{l=1}^{L} A_l(t)\delta(t - \tau_l). \tag{15.10}$$

In this model, t corresponds to the observation time and $A_l(t)$ and τ_l to the complex amplitude and delay of the lth multipath component, respectively. $A_l(t)$ is a time-varying uncorrelated Gaussian random process characterized by zero mean and Doppler power spectrum $S_l(f)$. The Doppler power spectrum considers the fact that the nodes can move within the network.

One of the most popular statistical approaches for PLA is hypothesis testing [4]. By defining as Ψ a general metric (such as a log-likelihood ratio (LLR)) that relates the CFR estimate $\hat{\mathbf{h}}^{(\mathrm{AB})}$ made by Bob during the onboarding phase between Alice and him, and the new estimate of uncertain origin $\hat{\mathbf{h}}^{(\mathrm{xB})}$, hypothesis testing works as follows:

$$\begin{cases} \Psi \leq \theta : & \text{choose } \mathcal{H}_0, \\ \Psi > \theta : & \text{choose } \mathcal{H}_1, \end{cases} \tag{15.11}$$

where θ represents a threshold and hypotheses are defined as:

- $\mathcal{H}_0$: the new message is from Alice;
- $\mathcal{H}_1$: the new message is from Eve.

However, deriving an analytical formulation for PLA in the aforementioned scenario often requires to make some assumptions which can rarely be verified in a real-world scenario. Namely, it is often supposed that the channel is quasi-static, i.e., does not vary during the transmission of a single message and slowly varies over time [29]. Moreover, the nodes mobility is a parameter that is difficult to take into account and solutions involving static nodes are considered [26]. One tool for coping with these non-idealities is machine learning applied to PLA [1, 12], which offers good adaptability to practical contexts [24]. In fact, one notable advantage of these methods is that they are *channel agnostic*, i.e., do not require any prior knowledge of the receiver's CSI. Consequently, they are capable of performing authentication in a non-informed manner.

We can examine two distinct scenarios. In the first, which is more idealistic, the authenticator knows some information about the potential adversary. In this case, onboarding phase coincides with the training stage of a binary classifier [9, 19]. In the second, which is a more realistic scenario, Bob can only gather samples from the legitimate transmitter during Phase I. This scenario necessitates one-class classification (OCC), where the training phase comprises only one class (referred to as the positive or target class), while the other classes (referred to as negative or nontarget classes) are unknown [17]. Among OCC algorithms, we can mention one-class nearest neighbor (OC-NN) and one-class support vector machine (OC-SVM). When applying OCC techniques to authentication, samples from Alice represent the target class, while samples from Eve are considered as outliers. When dealing with a channel that exhibits rapid time-varying characteristics, the utilization of OCC in conjunction with techniques to reduce data dimensionality like principal component analysis (PCA) proves inadequate in capturing the channel's dynamic changes over time. Hence, in such scenarios, one may turn to a specialized type of neural network known as autoencoders [3].

Given the availability of only the target class of data for training, it is possible to employ autoencoders for the purpose of anomaly detection [30]. When it comes to anomaly detection, the main goal is to understand the "normal behavior" of data when only normal data examples are accessible, which in our context refers to Alice's channel variation over time. Subsequently, the aim is to identify and flag samples that deviate from this normal behavior as anomalies, specifically representing samples coming from Eve. Autoencoders possess the ability to extract and learn the underlying characteristics of the normal data samples. Following the training process, they yield a low reconstruction error for normal samples and a high reconstruction error for anomalies. This distinctive property renders them well suited for the task of authentication when working with only one training class.

All the classification methods mentioned here necessitate Bob having precise knowledge of the packet source(s) during the training phase. This prerequisite often entails the implementation of higher-level cryptographic primitive or manual interventions, but this might not always be feasible. To tackle this challenge, one potential approach is to explore the utilization of clustering algorithms [14], like k-means [16]. With such algorithms, Bob can determine the packet origins during the onboarding stage and assign them potentially accurate labels, thereby circumventing other interventions [24].

15.4.1 PLA Metrics

Performance assessment in PLA typically revolves around the measurement of two key factors: the probability of false alarm (FA) and the probability of missed detection (MD). False alarm corresponds to Bob rejecting a message from Alice that is legitimate, whereas MD occurs when Bob accepts a message that has been tampered with by Eve. When measures are taken to decrease the likelihood of false alarms, it often results in an increase in the probability of MDs. As a result, a balance must be struck between these two opposing effects. It is worth noting that time-varying fading has adverse implications for the accurate authentication of valid signals but favors attackers. This is because it directly impacts the likelihood of false alarms, compelling Bob to accept a wider range of inputs, thereby raising the chances for Eve to successfully pass off one of her false messages as authentic.

Concerning statistical approaches, analytical expressions for the aforementioned probabilities are available and rely on the specific test conducted. Conversely, when employing machine learning algorithms, the probabilities of FA and MD can be assessed through the utilization of the confusion matrix [25]. This matrix offers a thorough summary of the outcomes of the classification process. Probability of FA can be derived from the confusion matrix as follows:

$$P_{\mathrm{FA}} = \frac{\mathrm{FN}}{\mathrm{TP} + \mathrm{FN}}, \tag{15.12}$$

where FN and TP represent the number of false negatives and true positives, respectively. Similarly, the probability of MD is computed as

$$P_{\mathrm{MD}} = \frac{\mathrm{FP}}{\mathrm{FP} + \mathrm{TN}}, \tag{15.13}$$

where FP and TN represent the number of false positives and true negatives, respectively.

While supervised classification commonly relies on metrics like accuracy for performance evaluation, these metrics may not be ideal for the OCC scenario. This is primarily due to the highly skewed distribution of negative data in relation to the target data, which can yield misleading results. To address this challenge, we turn to the *Geometric Mean* (g_{mean}) of accuracy [15], calculated separately for each class. It is defined as

$$g_{\mathrm{mean}} = \sqrt{(1 - P_{\mathrm{FA}})(1 - P_{\mathrm{MD}})}. \tag{15.14}$$

One notable advantage of the g_{mean} is its independence from the distribution of positive and negative samples within the test data. This independence enables the assessment of classifier performance based on both classes, rather than being biased toward the predominant class, as is the case with accuracy.

15.5 Numerical Experiments

In this section, we validate the techniques described above for achieving physical-layer confidentiality and authentication in practical settings by providing some examples and numerical simulations that consider constraints typical of real-world systems.

15.5.1 Physical-Layer Confidentiality Examples

Let us consider some real wireless communication scenarios and show how the tools introduced in Section 15.3 can be used to achieve a concrete level of confidentiality at the physical layer. Two fundamental design choices of any wireless communication system concern coding and modulation schemes. In fact, in a real system, deterministic coding schemes based on linear, possibly modern, codes characterized by low-complexity decoding along with discrete modulation formats such as QAM are usually employed. This is opposed to the ideal conditions of random coding and continuous modulation.

Besides considering them in the estimation of Bob's error rate needed to verify the fulfillment of the reliability requirement, these realistic constraints can be taken into account in the estimation of Eve's equivocation on each transmitted packet. For this purpose, we can use (15.4), which takes into account the source entropy rate and the code rate, besides Eve's channel capacity. For the latter, we can use an estimate that takes into account the actual modulation format, such as QAM. This can be done by using an input-constrained capacity estimate for QAM and also the approximations proposed in [2], which provide expressions that are easier to manipulate.

For what concerns wireless channels, a realistic model we can use is that of channels affected by Nakagami-m fading, which also encompasses Rayleigh fading as a special case. Assuming that each transmission slot is short enough compared to the channel variability in time to make the effect of fading negligible during that interval, and using the Nakagami-m fading model, the channel SNR γ during each time slot is characterized by the following probability density function (pdf):

$$p_{\Gamma}(\gamma) = \begin{cases} \frac{1}{\Gamma(m)}\left(\frac{m}{\hat{\gamma}}\right)^m \gamma^{m-1} e^{-\frac{m}{\hat{\gamma}}\gamma}, & \text{for } \gamma \geq 0, \\ 0, & \text{for } \gamma < 0, \end{cases} \tag{15.15}$$

where $\hat{\gamma}$ denotes the average channel SNR and $\Gamma(\cdot)$ is the Gamma function.

By using (15.15), we can easily compute the probability that Bob's channel SNR γ_{B} falls below the threshold $\overline{\gamma}_{\mathrm{B}}$, i.e.,

$$\begin{aligned} \Pr\left\{\gamma_{\mathrm{B}} < \overline{\gamma}_{\mathrm{B}}\right\} &= 1 - \Pr\left\{\gamma_{\mathrm{B}} \geq \overline{\gamma}_{\mathrm{B}}\right\} \\ &= 1 - \frac{1}{\Gamma(m)}\left(\frac{m}{\hat{\gamma}_{\mathrm{B}}}\right)^m \int_{\overline{\gamma}_{\mathrm{B}}}^{\infty} \gamma_{\mathrm{B}}^{m-1} e^{-\frac{m}{\hat{\gamma}_{\mathrm{B}}}\gamma_{\mathrm{B}}} d\gamma_{\mathrm{B}} \\ &= 1 - \left[-\frac{\Gamma\left(m, \frac{m}{\hat{\gamma}_{\mathrm{B}}}\gamma_{\mathrm{B}}\right)}{\Gamma(m)}\right]_{\overline{\gamma}_{\mathrm{B}}}^{\infty} = 1 - \frac{\Gamma\left(m, \frac{m}{\hat{\gamma}_{\mathrm{B}}}\gamma_{\mathrm{B}}\right)}{\Gamma(m)}. \end{aligned} \tag{15.16}$$

Similar arguments can be used to compute the probability that Eve's equivocation falls below some given threshold, taking the modulation format into account [23].

Concerning coding and modulation, the theoretical framework we have introduced allows us to simulate coding and modulation formats that are actually used in wireless communications. As an example, let us consider communications compliant with the IEEE 802.16-2017 standard [13], adopting QAM along with low-density parity-check (LDPC) coding. In order to provide some examples, let us consider transmissions using the standard LDPC codes having length $n = 2304$ and code rate 1/2 and 5/6. This implies that each transmitted packet has length $k = 1152$ and $k = 1920$ bits, respectively. In order to exemplify two transmission schemes characterized by different spectral efficiencies, let us assume that the first code, having lower rate, is used in conjunction with a binary modulation like binary phase shift keying (BPSK), while the second code, having higher rate, is coupled with a denser modulation like 16-QAM. Let us also consider a source entropy rate equal to 0.9, meaning that each packet carries $k' = 0.9 \cdot k$ bits of information.

In Figure 15.2, we show the number of transmission slots required to achieve 128 bits of PLS through AONT and OOT protocols with or without fake packets, obtained with the above formulas, as the SNR gap between Bob and Eve varies. Figure 15.2a considers a channel affected by Rayleigh fading, corresponding to Nakagami-m fading with $m = 1$, while Figure 15.2b considers a less fluctuating channel, affected by Nakagami-m fading with $m = 5$.

Looking at the figures, one aspect that may seem surprising is that, over a wide range of SNR gap values, the least fluctuating channel requires more transmission intervals than the most fluctuating channel to reach the PLS target. Even more, when using the OOT protocol without fake packets, the number of transmission intervals for low SNR gap values tends to diverge faster for the least fluctuating channel (with $m = 5$) than for the most fluctuating channel (with $m = 1$). This is actually not strange, since when the SNR gap is small or even negative, Eve's channel quality is good compared to Bob's channel, and Eve's equivocation is small. In this situation, channel fluctuations become beneficial, because they allow for the occurrence of time intervals in which Eve's channel quality decreases and consequently her equivocation increases. The situation changes for higher SNR gap values, for which, Bob has an advantage over Eve in terms of channel quality, and channel fluctuations become detrimental. In any case, the introduction of fake packets transmitted during the low-quality intervals of Bob's channel produces beneficial effects and succeeds in limiting the number of required transmission slots even in the region of low SNR gap values, where the quality of Eve's channel equals or even exceeds that of Bob's channel. Another observation that might seem counterintuitive concerns spectral efficiency: unlike normal communication applications, in the PLS context, we consider transmission schemes with low spectral efficiency turn out to be advantageous over transmission schemes with higher spectral efficiency.

15.5.2 Physical-Layer Authentication Examples

As a first example of PLA, let us examine a simplified scenario where the communication channel between Alice and Bob does not vary during the onboarding phase and exhibits a slowly time-varying behavior in the subsequent phase. The attacker is supposed to have a highly correlated channel with Alice and Bob's. It is also assumed that all nodes are static.

We assess the performance achieved by binary and one-class versions of SVM and NN algorithms and by a statistical approach, i.e., an LLR test, considering an increasing number of OFDM subcarriers. The SNR during the onboarding phase is equal to 15 dB, while we assume it increases to 20 dB during the authentication phase. All the machine-learning algorithms are trained using a training set of dimension M composed of CFR measured of different subcarriers. As it results

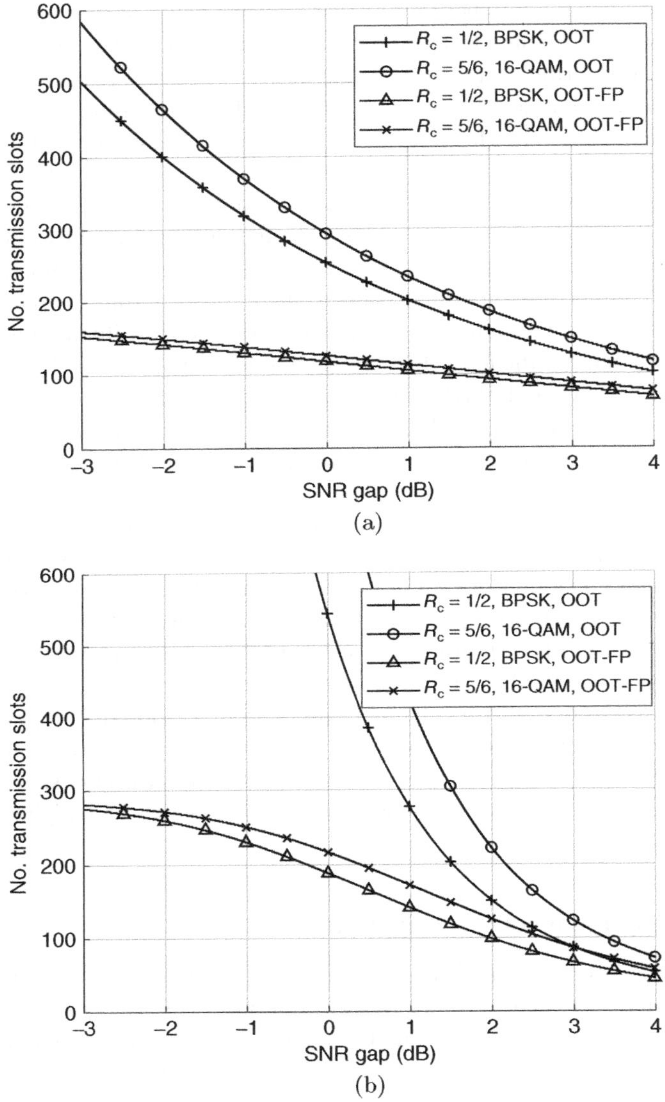

Figure 15.2 Number of transmission slots required to achieve 128 bits of PLS through AONT and OOT protocols with or without fake packet transmissions. (a) $m = 1$ and (b) $m = 5$.

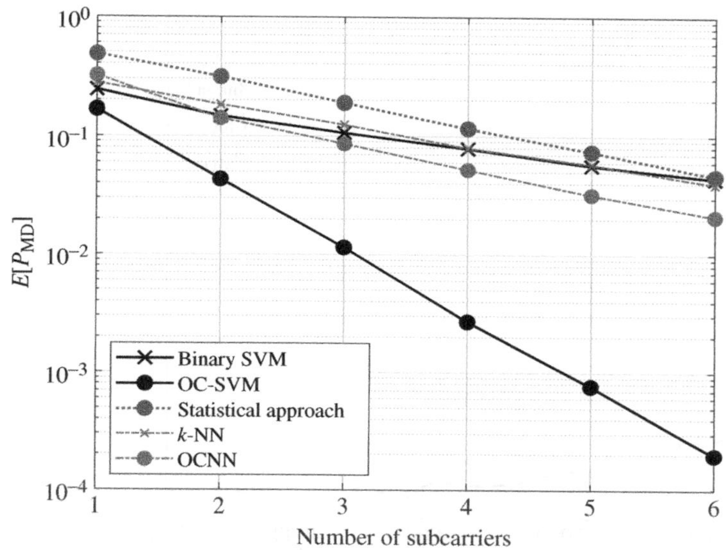

Figure 15.3 Average MD probability with different numbers of subcarriers N, considering $P_{FA}(\text{SVM}) = [1.39 \times 10^{-3}, 2.12 \times 10^{-4}, 6.85 \times 10^{-5}, 1.68 \times 10^{-5}, 5.55 \times 10^{-6}, 1.62 \times 10^{-6}]$, $P_{FA}(\text{1KNN}) = [6.76 \times 10^{-4}, 4.21 \times 10^{-4}, < 10^{-6}, < 10^{-6}, < 10^{-6}, < 10^{-6}]$ and $P_{FA}(\text{stat}) = 10^{-4}$.

from Figure 15.3, given almost the same P_{FA}, machine-learning methods show better performance than their statistic counterpart. Moreover, the one-class version of machine-learning algorithms outperforms the classical binary version, where Eve's samples are available. This is likely a result of the training set for the binary version containing negative samples that are strongly correlated with the positive ones. This correlation leads the algorithm to make mistakes in classifying new samples.

Let us now consider more realistic communication conditions, which include the nodes mobility and more rapidly varying channels. In Figure 15.4, a moving condition for both Alice and Eve is considered, taking into account for simulation the parameters given in Table 15.1 and also supposing to have the same velocity for both. These parameters are derived from the ITU-R P.1411-6 recommendation for outdoor radio communication systems [21]. We suppose that there is no line-of-sight between Eve and Bob and also between Alice and Bob. A subset $\overline{N}$ of 20 subcarriers is chosen randomly out of the total 8192. In this case, we measure the authentication performance achievable by an autoencoder, also evaluating the impact of the training set dimension on the results. Observing that enlarging the training set dimension enhances performance up to 3000 samples, beyond which there is no discernible impact. Indeed, a larger training set facilitates

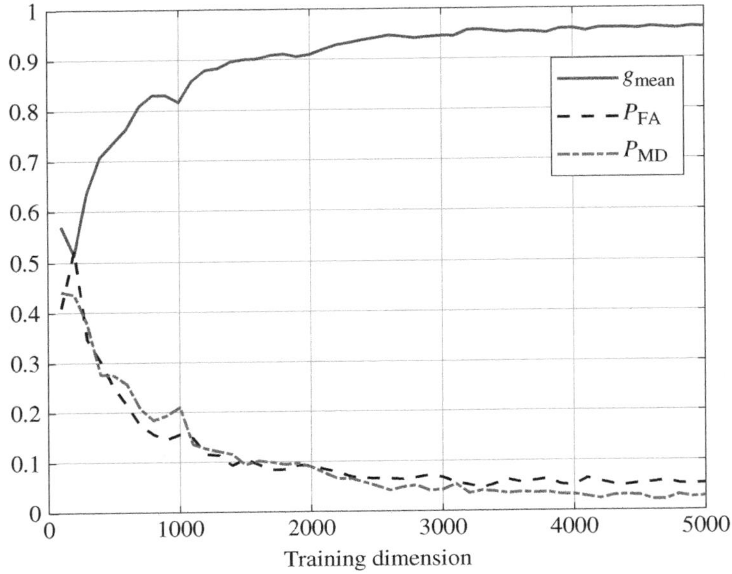

Figure 15.4 Measure of authentication performance with different training set sizes. Case of Alice and Eve both not in line-of-sight with Bob, moving with the same velocity.

Table 15.1 Communication parameters for PLA.

Parameter	Value
Carrier frequency (f_c)	2.4 GHz
Bandwidth	40 MHz
Distance	55 m
N	8192
τ_{Alice} (s)	[0, 7.5e−7, 1.2e−6, 1.7e−6]
Alice paths attenuation (dB)	[0, −25, −28, −30]
τ_{Eve} (s)	[0, 4.5e−7, 1.5e−6, 2e−6]
Eve paths attenuation (dB)	[0, −20, −24, −31]
Doppler shift	8 Hz

the autoencoder in distinguishing between legitimate and malicious signals, as it can better assimilate the channel variation over an extended period of time. As expected, the autoencoder is a more powerful tool than OC-NN and (OC-SVM) but requires to be trained on a larger set of data in order to properly take into account the channel variability and the nodes mobility.

15.6 Conclusion

We studied how to achieve confidentiality and authentication at the physical layer in real systems, where perfect secrecy conditions cannot be achieved, but at most we can only have partial secrecy along with a partial knowledge of transmission channels. Despite this, the use of appropriate tools such as AONT, OOT protocols, and machine learning enables more than acceptable levels of security at the physical layer, demonstrating that these techniques can effectively be used in real-world applications.

References

1 L. Alhoraibi, D. Alghazzawi, R. Alhebshi, and O. B. J. Rabie. Physical layer authentication in wireless networks-based machine learning approaches. *Sensors*, 23(4):1814, 2023.

2 M. Baldi, L. Senigagliesi, and F. Chiaraluce. Achieving semantic security without keys through coding and all-or-nothing transforms over wireless channels. In *Proceedings of the 2016 IEEE Global Conference on Signal and Information Processing (GlobalSIP 2016)*, pages 964–969, Washington, DC, USA, Dec 2016.

3 D. Bank, N. Koenigstein, and R. Giryes. Autoencoders. *arXiv:2003.05991*, Mar 2020.

4 P. Baracca, N. Laurenti, and S. Tomasin. Physical layer authentication over MIMO fading wiretap channels. *IEEE Transactions on Wireless Communications*, 11(7):2564–2573, Jul 2012. ISSN 1536-1276. doi: 10.1109/TWC.2012.051512.111481.

5 M. Bellare, S. Tessaro, and A. Vardy. Semantic security for the wiretap channel. In R. Savafi-Naini and R Canetti, editors, *Advances in Cryptology - CRYPTO 2012*, volume 7417 of *Lecture Notes in Computer Science*, pages 294–311. Springer-Verlag, Berlin Heidelberg, 2012.

6 P. Bello. Characterization of randomly time-variant linear channels. *IEEE Transactions on Communications Systems*, 11(4):360–393, 1963. doi: 10.1109/TCOM.1963.1088793.

7 M. Bloch, M. Hayashi, and A. Thangaraj. Error-control coding for physical-layer secrecy. *Proceedings of the IEEE*, 103(10):1725–1746, Oct 2015.

8 V. Boyko. On the security properties of OAEP as an all-or-nothing transform. In *Advances in Cryptology –CRYPTO' 99*, volume 1666 of *Lecture Notes in Computer Science*, pages 503–518. Springer, 1999.

9 S. Chen, H. Wen, J. Wu, J. Chen, W. Liu, L. Hu, and Y. Chen. Physical-layer channel authentication for 5G via machine learning algorithm. *Wireless Communications and Mobile Computing*, 2018:6039878, 2018.

10 J. Choi. On channel-aware secure HARQ-IR. *IEEE Transactions on Information Forensics and Security*, 12(2):351–362, Feb 2017.

11 P. Cuff. A framework for partial secrecy. In *Proceedings of IEEE Global Telecommunications Conference (GLOBECOM 2010)*, Miami, FL, Dec 2010.

12 H. Fang, X. Wang, and L. Hanzo. Learning-aided physical layer authentication as an intelligent process. *IEEE Transactions on Communications*, 67(3):2260–2273, Mar 2019. ISSN 0090-6778. doi: 10.1109/TCOMM.2018.2881117.

13 IEEE Std 802.16-2017 (revision of IEEE Std 802.16-2012). IEEE Standard for Air Interface for Broadband Wireless Access Systems, Dec 2018.

14 A. K. Jain, M. N. Murty, and P. J. Flynn. Data clustering: A review. *ACM Computing Surveys*, 31:264–323, Sep 1999.

15 M. Kubat and S. Matwin. Addressing the curse of imbalanced training sets: One-sided selection. In *Proceedings of the 14th International Conference on Machine Learning*, pages 179–186. Morgan Kaufmann, 1997.

16 J. B. MacQueen. Some methods for classification and analysis of multivariate observations. In *5th Berkeley Symposium on Mathematical Statistics and Probability*, volume 1, pages 281–297, 1967.

17 M. M. Moya and D. R. Hush. Network constraints and multi-objective optimization for one-class classification. *Neural Networks*, 9(3):463–474, 1996.

18 P. Mu, Z. Li, and B. Wang. Secure on-off transmission in slow fading wiretap channel with imperfect CSI. *IEEE Transactions on Vehicular Technology*, 66(10):9582–9586, Oct 2017.

19 F. Pan, Z. Pang, H. Wen, M. Luvisotto, M. Xiao, R.-F. Liao, and J. Chen. Threshold-free physical layer authentication based on machine learning for industrial wireless CPS. *IEEE Transactions on Industrial Informatics*, 15(12):6481–6491, 2019.

20 H. V. Poor and R. Schaefer. Wireless physical layer security. *Proceedings of the National Academy of Sciences of the United States of America*, 114:19–26, 2017.

21 Recommendation ITU-R P. 1411-7: Propagation data and prediction methods for the planning of short-range outdoor radiocommunication systems and radio local area networks in the frequency range 300 MHz to 100 GHz. *P Series*, 9:71, 2013.

22 R. L. Rivest. All-or-nothing encryption and the package transform. In *Fast Software Encryption*, volume 1267 of *Lecture Notes in Computer Science*, pages 210–218. Springer-Verlag Berlin Heidelberg, 1997.

23 L. Senigagliesi, M. Baldi, and F. Chiaraluce. Semantic security with practical transmission schemes over fading wiretap channels. *Entropy*, 19(9):491, Sep 2017.

24 L. Senigagliesi, M. Baldi, and E. Gambi. Comparison of statistical and machine learning techniques for physical layer authentication. *IEEE Transactions on Information Forensics and Security*, 16:1506–1521, Oct 2021.

25 S. V. Stehman. Selecting and interpreting measures of thematic classification accuracy. *Remote Sensing of Environment*, 62(1):77–89, 1997.

26 J. K. Tugnait. Using artificial noise to improve detection performance for wireless user authentication in time-variant channels. *IEEE Wireless Communications Letters*, 3(4):377–380, Aug 2014. ISSN 2162-2337. doi: 10.1109/LWC.2014.2318731.

27 A. D. Wyner. The wire-tap channel. *The Bell System Technical Journal*, 54(8):1355–1387, Oct 1975.

28 L. Xiao, L. Greenstein, N. Mandayam, and W. Trappe. Fingerprints in the ether: Using the physical layer for wireless authentication. In *Proceedings of IEEE International Conference on Communications (ICC 2007)*, pages 4646–4651, Glasgow, UK, Jun 2007.

29 L. Xiao, L. J. Greenstein, N. B. Mandayam, and W. Trappe. Using the physical layer for wireless authentication in time-variant channels. *IEEE Transactions on Wireless Communications*, 7(7):2571–2579, Jul 2008. ISSN 1536-1276. doi: 10.1109/TWC.2008.070194.

30 B. Zong, Q. Song, M. R. Min, W. Cheng, C. Lumezanu, D. Cho, and H. Chen. Deep autoencoding Gaussian mixture model for unsupervised anomaly detection. In *International Conference on Learning Representations*, 2018.

Index

Physical-Layer Security for 6G, First Edition. Edited by Parthajit Mohapatra, Nikolaos Pappas, Arsenia Chorti, and Stefano Tomasin.

t

u

v

w

z

Printed and bound by CPI Group (UK) Ltd, Croydon, CR0 4YY

17/09/2026

14964308-0004